P9-AGV-114

Clearing the Air

Clearing the Air

The Health and Economic Damages of Air Pollution in China

edited by Mun S. Ho and Chris P. Nielsen

The MIT Press
Cambridge, Massachusetts
London, England

MIT Press books may be purchased at special quantity discounts for business or sales promotional use. For information, please email special_sales@mitpress.mit.edu or write to Special Sales Department, The MIT Press, 55 Hayward Street, Cambridge, MA 02142.

This book was set in Sabon on 3B2 by Asco Typesetters, Hong Kong, and was printed and bound in Hong Kong.

Library of Congress Cataloging-in-Publication Data

Clearing the air : the health and economic damages of air pollution in China / edited by Mun S. Ho and Chris P. Nielsen.
p. cm.
Includes bibliographical references and index.
ISBN-13: 978-0-262-08358-4 (hc : alk. paper)
1. Air—Pollution—Health aspects—China. 2. Air—Pollution—Economic aspects—China. 3. Air—Pollution—Government policy—China. I. Ho, Mun S. II. Nielsen, Chris P., 1960–.
RA576.7.C6C558 2007
363.739′20951—dc22 2006047209

10 9 8 7 6 5 4 3 2 1

11/14

Contents

Preface

The historic rise of the People's Republic of China, driven by an economic transformation that has proceeded almost without pause over the last three decades, is hardly news in 2007. As a result of this rise, and because of its size, it is commonplace now to regard China as a newly arrived global power. Domestically, China's recent trajectory is believed to have lifted hundreds of millions of its citizens from poverty, a transformation of unprecedented scale and pace in the history of nations. While this growth in part reflects an underperforming economy when reforms began, and has brought with it a growing imbalance in the distribution of wealth, in aggregate human terms China's economic progress has been incontrovertibly positive.

It has not had many positive impacts, however, on the natural environment, both within China and across the globe. Fossil energy sources, especially coal, have fueled the economic transformation. As in all nations, burning fossil fuels causes environmental externalities, the term in economics for the costs of individual activities that are inflicted on society as a whole. Among such externalities are the damages of air pollution, foremost its impacts on human health. China has made laudable progress on some fronts of pollution control, for example, forcing the substitution of gas for coal in central cities to reduce large particulate loads in urban air (and limiting, at the same time, carbon dioxide emissions). Addressing many other forms of pollution and their damages, however, is proving more vexing.

Clearing the Air: The Health and Economic Damages of Air Pollution in China presents a new modeling framework for integrating the study of economic growth, energy utilization, and environmental quality in China. Our initial effort in this area, published in 1998, was confined to a projection of Chinese economic growth and carbon emissions, using an aggregate growth model without industry detail. I completed this project in collaboration with Dwight Perkins of the Department of Economics at Harvard and Mun Ho of the Kennedy School of Government. Ho and I then developed a multi-sector model of Chinese economic growth in collaboration with Richard Garbaccio, now of the U.S. Environmental Protection Agency. This model gave special attention to the dual plan-and-market features of the Chinese economy in the 1990s. We developed a version of this model with perfect foresight dynamics in collaboration with Karen Fisher-Vanden, now of Dartmouth College. In constructing these progressively more elaborate models we obtained the invaluable assistance of Li Shantong and Zhai Fan of the Development Research Center of the State Council of China.[1]

Our new economy-energy-environment model of China incorporates population projections including the changing demographic structure, projections of productivity growth, enhancements of labor quality, and changes in household spending and savings behavior. We have analyzed changes in Chinese energy use at the industry level, using data from the 1987 and 1992 input-output tables, and this information is incorporated into projections of energy use per unit of industry output in the model. The initial version of the model incorporated a sub-model of local health impacts of sulfur dioxide (SO_2) and total suspended particulate (TSP) emissions, using information generously provided by Gordon Hughes and Kseniya Lvovsky of the World Bank. This version was first used to examine the reduction in local health damages due to a policy to reduce carbon emissions. Our subsequent analysis focused on the effects of "green tax" policies, designed to reduce local air pollution damages, on economic growth, on reductions in mortality and morbidity, and on carbon emissions.

The model described in the book is the culmination of the improvements to this economic-energy-environment model. We have developed a new environmental submodel that incorporates industry-specific contributions to the TSP and SO_2 concentrations. We have also incorporated the latest economic data for China, including the 1997 input-output table. We are pleased to acknowledge the contributions of Cao Jing, who provided outstanding research assistance to us in completing the new version of the model.

The core of the environmental submodel, initiated in collaboration with Jon Levy of the Harvard School of Public Health, is the assessment of "intake fractions"—a methodological approach to estimating human exposures to pollution in data-constrained contexts such as China's. Intake fractions are derived by modeling pollutant dispersion and human exposures from a sample of real sources. This research was conducted partly in collaboration with Hao Jiming and his colleagues Wang Shuxiao, Liu Bingjiang, Lu Yongqi, and Li Ji from the Department of Environmental Sciences and Engineering of Tsinghua University, Beijing, mainly in residence at Harvard's Division of Engineering and Applies Sciences. Another part of this research was conducted by Zhou Ying, Jon Levy, James Hammitt, and John Evans of the Harvard School of Public Health. Zhou and Hammitt contributed additional research, on the economic value that Chinese citizens place on health, to the environmental submodel.

The ambitiously collaborative nature of the study exemplifies the objectives of the Harvard University Center for Environment (HUCE). This Center grew out of a faculty committee established in 1993 by then-President Neil Rudenstine. The committee's chief mandate was to foster collaboration in environmental research and education across disciplines, and across the ten schools of the university. Under the chairmanship of atmospheric scientist Michael McElroy, the committee evolved into an intellectual and physical center supporting a wide scope of initiatives, bringing students, researchers, and faculty together from across the university. Today the HUCE, under Daniel Schrag, continues to manifest Harvard's vision of how a comprehensive university can best cultivate research and education in a topical area as consummately interdisciplinary as humankind's relationship to the natural environment. Rather than create a self-contained department of environmental studies, the HUCE is designed to draw on the strengths of the entirety of Harvard University, and to evolve as its many initiatives and the interests of participants grow and change.

The China Project has been one of the largest and most sustained single research initiatives of the HUCE, now also supported by Harvard's Division of Engineering and Applied Sciences. Established by McElroy and a team including Xu Xiping of the Harvard School of Public Health, Chris Nielsen and Peter Rogers of the Division of Engineering and Applied Sciences, William Alford and the late Abram Chayes of Harvard Law School, Dwight Perkins and myself of the Department of Economics, Mun Ho of the Kennedy School of Government, and others, the China Project's focus has been the challenge of reconciling China's economic development with protection of the atmospheric environment, and in particular integrating the Chinese domestic priority of air pollution control with the global objective of limiting emissions of greenhouse gases. This topic requires a challenging confluence of knowledge and expertise in economics, law, policy, and natural, applied, and health sciences, among others. It is also a topic that draws together two longstanding academic communities at Harvard: those who study environment, and those who study China.

Crucially, the China Project from the outset has embraced a second collaborative mandate: a full Harvard–China partnership in research with contributions from universities and institutes of the People's Republic of China. In early stages these collaborations were facilitated by designees of Song Jian, State Councilor and

Chairman of the then-State Science and Technology Commission. As the program evolved, the governmental associations also changed, and the Project has consulted over time with senior leaders that have included then-Vice Premier (and now Premier) Wen Jiabao, then-Minister Xie Zhenhua of the State Environmental Protection Agency (SEPA), and others. Qu Geping, the founding Administrator of SEPA and now often called the "father" of environmental protection in China, has long served as a key supporter and senior advisor to the HUCE China Project.

It is important to emphasize that the primary aim of the China Project is to advance independent inquiry, crossing disciplines and nations, and to contribute to basic knowledge judged by traditional standards of peer review and scholarly publication. Independent scholars define their own questions, rather than answer those posed by policy makers. The Project does not provide policy advice nor serve as an advocacy group. It has also sponsored research motivated by competing perspectives, reflecting the conviction that innovative ideas are generated and tested through critical thinking and intellectual challenges. Fostering fruitful debates is essential for mitigating biases due to a specific disciplinary orientation or a particular national perspective. University-to-university collaborations are uniquely suited to pursue this objective. Accordingly, the Project has built relationships over time with a number of schools and departments of Tsinghua University, Beihang University, Hong Kong Polytechnic University, Peking University, and others.

It would be misleading to dismiss the confluence of different scholarly points of view as Ivory Tower irrelevance. What we really know, collectively, about the prospects for reconciling China's economic growth with local and global environmental protection, and the role of wealthier nations in this effort, is still very limited. Studies that merely apply conventional research methods without full understanding of the fundamental underlying assumptions that distinguish Chinese society and the Chinese economy from industrialized economies are unlikely to improve the practice of economic and environmental policy. Independent studies developed with careful attention to these fundamental assumptions and critically scrutinized empirical data have the potential for greater credibility and more original insights. This is the most important explanation of the intense interest in our results by senior leaders in China. Independent research is also the primary means by which the existing centers of scholarly excellence in China can build research capacities meeting international standards and by which non-Chinese scholars can better understand China's unique development.

The first phase of the China Project set the stage for subsequent research by inviting interested scholars to write review studies or initiate preliminary research projects. The resulting papers were presented in a research workshop at Harvard, then revised and published in *Energizing China: Reconciling Environmental Protection and Economic Growth*, edited by Michael McElroy, Chris Nielsen, and Peter Lydon, distributed by Harvard University Press in 1998. A variety of externally funded, multi-year studies in a diversity of fields grew out of this initial effort. These have been reported in scholarly journals, law reviews, book chapters, and published reports, too numerous to list here but given on the HUCE website: www.fas.harvard.edu/~huce/china_project.htm/.

The economic and related work of this book comprise one of three major streams of research under the China Project with roots in the initial phase, each capitalizing on years of development of unique research models, new data resources, and time-tested collaborative relationships. The second is an atmospheric program led by McElroy, who with Wang Yuxuan has developed a high-resolution window over China within the GEOS-Chem global chemical tracer model to analyze poorly understood regional and seasonal dimensions of air quality in China, including complex secondary species such as ozone. The model now makes use of continuous observations of key atmospheric species collected at a measurement station deployed at a site north of Beijing since 2004, in a partnership of the Project with Hao Jiming and colleagues at Tsinghua University that includes William Munger of Harvard.

The third major research component of the China Project is interdisciplinary study of urban transportation, land use planning, air quality, human exposure, and health, now focused on the case city of Chengdu. This venture is partly inspired by and builds on the work of this book, refining some of its exposure and valuation methods, involving several of the same collaborators, and similarly structured as separable but linked modules led by different investigators. Key participants are He Kebin and Wang Shuxiao of Tsinghua University; Zhang Dianye of Southwest Jiaotong University; Shen Mingming of Peking University; and Chris Nielsen, James Hammitt, Guo Xiaoqi, Peter Rogers, and Sumeeta Srinivasan at Harvard.

Clearing the Air is the result of many years of development, data collection, and dedicated research by a large, international team of scholars. It was made possible by numerous funding institutions providing generous support either to individual modules or to the program as a whole. Each chapter contains its own funding acknowledgments, but we will summarize and thank them collectively here.

Three institutions provided major funding. The integrated project was initiated and several elements were funded by a grant from the V. Kann Rasmussen Foundation. Development of the economic model was funded chiefly by the Integrated

Assessment Program of the Office of Biological and Environmental Research, U.S. Department of Energy, under contract DE-FG02-95ER62133. The pollution dispersion and exposure assessment component was supported by a joint grant to the HUCE China Project and Tsinghua University Institute of Environmental Science and Technology from the China Sustainable Energy Program of the Energy Foundation.

Additional funding was provided for individual research components, workshops, or the participation of particular researchers. These include the Task Force on Environmental and Natural Resource Pricing of the China Council for International Cooperation on Environment and Development, the U.S. Environmental Protection Agency, The Henry Luce Foundation, The Bedminster Foundation and the Dunwalke Trust, the Harvard Asia Center, and the Harvard Kernan Brothers Fellowship. We are grateful to all of these institutions for making this program of research possible.

Dale W. Jorgenson
Harvard University

Note

1. A note on the names of Chinese nationals in this book: In this preface they are rendered in correct Chinese name-order, with surnames first. They are reversed in the chapters, however, to prevent erroneous future citations of contributions to this book, a small but serious problem for Chinese in international scholarly literature.

Editors' Acknowledgments

The program of research described in this book grows from a mandate of the Harvard University Center for the Environment (HUCE) to bring scholars together from across disciplines to jointly address environmental research topics. This collaborative effort has involved researchers from different university departments and institutions in China and the United States, and received financial support from a number of organizations, as summarized in the preface by Dale W. Jorgenson.

A central component of the research described here has been led by Jorgenson (Harvard Economics Department) and Hao Jiming (Tsinghua University Department of Environmental Science and Engineering), building on the concept of "intake fractions" introduced to the group by Jonathan I. Levy (Harvard School of Public Health). The intake fraction method for estimating exposures to air pollution is explained in the chapter by Levy and Susan L. Greco, and applied in three chapters involving John S. Evans, James K. Hammitt, Hao, Levy, Li Ji, Liu Bingjiang, Lu Yongqi, Wang Shuxiao, and Zhou Ying (in alphabetical order, using Chinese name-order of surnames first). Hao led research conducted in China that involved the cooperation of various national and local government departments. We are grateful to the Chinese government officials for their contributions.

Hammitt and Zhou additionally conducted research on the valuation of health damages from air pollution in China. Mun S. Ho and Jorgenson used the results of these various components of research to estimate the value of damages due to pollution from various industries, and to examine how environmental policies affect economic performance while reducing local and global pollution.

Building up this collaborative program was no easy task, and was made possible by the willingness of all participants to adjust to, and accommodate, the needs of contributors from other disciplinary fields. We must thank especially the researchers and authors of the chapters for their spirited and patient participation in the venture, from proposal writing, to long coordination meetings, to many rounds of presenting and reworking research, and ultimately to revising reports and journal articles into chapters for this volume. To this we add our gratitude to Dale Jorgenson for his committed oversight of this program of research, and to Michael B. McElroy for sustained support of it as chair of the HUCE and the China Project.

Presenting such a complex economic, energy, and environmental assessment in a single book designed to bridge lay and expert readerships has been a challenging but gratifying undertaking for the editors and authors. We thank Clay Morgan and his colleagues at MIT Press for their constructive advice and commitment to the project. We owe special thanks to three anonymous reviewers for their invaluable, detailed comments, which greatly improved the book.

Progress reports and drafts of what became chapters were presented at numerous meetings including the Ninth Conference of the Parties to the United National Framework Convention on Climate Change in Milan, Tsinghua-Harvard research workshops held in Beijing, workshops in Oslo and Beijing led by the Center for International Climate and Environmental Research-Oslo and the State Environmental Protection Administration, meetings of the U.S. Environmental Protection Agency, and public seminars at Harvard. We are grateful for valuable critiques and comments by participants in all of these events.

Mun Ho would finally like to thank Resources for the Future for hosting him as a Visiting Scholar while much of this work was conducted.

Mun S. Ho
Chris P. Nielsen

Abbreviations and Acronyms

ACS	American Cancer Society
AEEI	autonomous energy-efficiency improvement
AMD	average marginal damage
BR	nominal population-average breathing rate
BRICC	Beijing Research Institute of Coal Chemistry
CB	chronic bronchitis
C_d	pollution concentration
CEPY	China Electrical Power Yearbook
CEY	China Environment Yearbook
CGE	computable general equilibrium economic model
CI	confidence interval
CICERO	Center for International Climate and Environmental Research, Oslo
CMAQ	Community Multiscale Air Quality modeling system
CO	carbon monoxide
CO_2	carbon dioxide
CPC	Climate Prediction Center
CRAES	China Research Academy of Environmental Sciences
CV	contingent valuation
DR	dose response
EC	European Commission
ECON	ECON Centre for Economic Analysis
EIA	environmental impact assessment
EPB(s)	environmental protection bureau(s)

ESP	electrostatic precipitator
FDDA	Four-Dimensional Data Assimilation
FGD	flue gas desulfurization
g	grams
gce	grams of coal equivalent
gce/kWh	grams of coal equivalent per kilowatt-hour
g/m^3	grams per cubic meter
g/s	grams per second
GDP	gross domestic product
GEOS-Chem	a global chemical tracer model
GHG(s)	greenhouse gas(es)
GIS	geographical information system
GW	gigawatt
HAPs	hazardous air pollutants
HEI	Health Effects Institute
HNO_3	nitric oxide
HUCE	Harvard University Center for the Environment
iF	Intake fraction
IIASA	International Institute for Applied Systems Analysis
IPCC	Intergovernmental Panel on Climate Change
ISC	Industrial Source Complex air dispersion model
ISCLT	long-term version of ISC
ISCST	short term version of ISC
IWAQM	Interagency Working Group on Air Quality Modeling
kg	kilograms
kg/s	kilograms per second
kj	kilojoules
kj/kg	kilojoules per kilogram
km	kilometers
km^2	square kilometers
kt, kton	kilotons
ktce	kilotons of coal equivalent
kWh	kilowatt-hours

LHV	latent heat value
m	meters
m/s	meters per second
m^3	cubic meters
m^3/day	cubic meters per day
m^3/kg	cubic meters per kilogram
m^3/s	cubic meters per second
MD	marginal damage
mg	milligrams
mg/m^3	milligrams per cubic meter
mm	millimeters
MM5	fifth-generation Penn State/NCAR mesoscale model of atmospheric circulation
MOP	Ministry of Power
MOST	Ministry of Science and Technology
Mt	million tons
MW	megawatts
μg/day	micrograms per day
μg/m^3	micrograms per cubic meter
μm	micron, also called micrometer
NBS	National Bureau of Statistics
NCDC	National Climatic Data Center
NCEP	National Centers for Environmental Prediction
NEPA	National Environmental Protection Agency (now SEPA)
NMMAPS	National Morbidity and Mortality Air Pollution Study
NO_2	nitrogen dioxide
NO_3	shorthand for nitrate compounds
NO_X	nitrogen oxides
O_3	ozone
OECD	Organization for Economic Co-operation and Development
ORNL	Oak Ridge National Laboratory
Pb	lead
PM	particulate matter

PM_{10}, $PM_{2.5}$, PM_X	particulate matter of less than 10, 2.5, or equal to other specified number of microns in aerodynamic diameter, respectively
POP_1	population within 10 km, or 5 km for mobile sources
POP_2	population within 10–50 km, or 5–50 km for mobile sources
POP_d	at-risk population
ppb	parts per billion
PPP	purchasing power parity
PRCEE	Policy Research Center of Environment and Economy of SEPA
PSD	particle size distribution
R^2	Square of the correlation coefficient
RAINS-Asia	Regional Air Pollution Information and Simulation model for Asia
RFF	Resources for the Future
RHS	right-hand side
SAES	Shanghai Academy of Environmental Sciences
SAM	social accounting matrix
s.d.	standard deviation
SEPA	State Environmental Protection Administration
SH	stack height
SO_2	sulfur dioxide
SO_4	shorthand for sulfate compounds
SPC	State Power Corporation
toe	tons of oil equivalent
TSP	total suspended particulates
TVE(s)	township and village enterprise(s)
TWh	terawatt-hour
UNEP	United Nations Environment Programme
U.S. EIA	U.S. Energy Information Agency
U.S. EPA	U.S. Environmental Protection Administration
VAT	value-added tax
VSL	value of a statistical life
WHO	World Health Organization
WTP	willingness to pay

I

Introduction, Review, and Summary

1

Air Pollution and Health Damages in China: An Introduction and Review

Chris P. Nielsen and Mun S. Ho

1.1 Introduction

By some assessments, one of the leading causes of death in the People's Republic of China is respiratory diseases caused by air pollution. Although some would debate this claim, few would question that air pollution is a very serious problem in China. High-level concentrations of particulates and sulfur dioxide (SO_2) are recorded in many cities, levels that rank a number of them among the worst polluted in the world. This pollution is produced by a variety of sources, including a burgeoning transportation sector, residential heating, and light industry. Much of it is generated by power plants and heavy industry, often located in urban areas by industrial policies predating the post-1978 reform era.

The authorities have responded to this challenge, and there has been clear progress in some areas of pollution control. As a result of these actions, and concurrent changes in economic policies and the structure of the economy, concentrations of SO_2 and total suspended particulates (TSP) declined in many places, especially during the 1990s. For instance, total SO_2 emissions fell 10% between 1995 and 1999, and the averaged ambient concentrations in thirty-two major cities declined from 100 μg/m^3 (micrograms per cubic meter) in 1991 to 62 μg/m^3 in 1998 (World Bank 2001). Another compilation of official data reports that annual average TSP across 140 Chinese cities fell from a mean of 500 μg/m^3 in 1986 to 300 μg/m^3 in 1997 (Florig et al. 2002).

Despite these improvements, the downward trends appear to have largely abated or reversed, and concentrations remain higher than national standards and international guidelines in many localities. The Chinese standard for annual average TSP in residential areas is 200 μg/m^3, for instance, and the World Health Organization's (WHO) guideline was 90 μg/m^3. We should note that the WHO has now stopped setting guidelines for TSP, though it has begun to set guidelines for finer particles, the fractions of TSP that are considered most harmful to health (WHO 2002, p. 186; WHO 2005). Pollutant concentrations are increasing in some major cities. Estimates of the value of total air and water pollution damages nationwide have ranged from 3% to 7.7% of gross domestic product, usually dominated by the air pollution component. Although these estimates are rough and involve subjective judgments, few would dispute that the damages are large compared to China's own past and compared to more developed countries. Furthermore, these damages may well rise with increasing energy use from China's vigorous economic expansion.

We need to think beyond the ambient pollution levels to consider what ultimately concerns us about China's degraded air quality: the damage it causes. These include impacts on human health, on materials such as built structures, on natural resources such as forests, on the productivity of crops, and on ecosystems. All of these, in turn, affect the economy. The assessment of this volume focuses on refining understanding of the damages to human health, which most researchers believe dominate the total impacts. The other, lesser damages are topics for future extensions of this work.

With a focus on health risk, we must keep in mind that what concerns us is not the level of pollution itself, but rather the level that reaches human lungs—a distinction that fundamentally motivates and shapes this assessment. We must consider such factors as the rapid urbanization in China and how expanding cities and rural-to-urban migration locate an increasing proportion of the Chinese citizenry nearer to pollution sources. In terms of total population risk, this could even outweigh whatever gains in pollution control have occurred. The crucial factor of human *exposure* is not often considered carefully in pollution damage assessments in China.

Whereas it is generally accepted that Chinese health and economic performance have been harmed by its degraded air quality, the power of many existing damage estimates to influence policy has been undermined by their uncertain and aggregate nature. More reliable and comprehensive estimates based on detailed energy use and emissions could help guide environmental policy and law and orient public and official opinion to expanded preventive action. Disaggregating the estimates to specific industries, or geographical areas, can help target policies.

To make such detailed estimates, the prevailing approach in more developed countries has been (1) to use air-dispersion models to characterize the link between emissions of all sources and atmospheric concentrations of pollutants in a target region; (2) to estimate human health impacts on the basis of exposure-response functions; and (3) to monetize these damages by using some valuation method. A large body of such research has been developed over the course of many years and at considerable expense in many nations.

Researchers in this field in China, however, face major limitations compared to those in the West. A key challenge is that basic pollution data are sparse, because the national emission inventories and monitoring infrastructure are still developing. Local institutions have only recently begun to use advanced air-dispersion models, and have applied them comprehensively—that is, to estimate all sources—for a few localities only. For example, the State Environmental Protection Administration (SEPA) and the U.S. Environmental Protection Agency (U.S. EPA) have in recent years commissioned studies using such models and exposure assessments in major cities such as Shanghai (Shanghai Academy of Environmental Sciences [SAES] et al. 2002) and Beijing (Tsinghua University et al. 2005). For health valuation, such studies have had to try to translate Western results to Chinese conditions or use results from one of just a few small unpublished studies completed in China. As we will describe in section 1.6, a small number of additional international collaborations with Chinese institutes have used these approaches to investigate the impact of air pollution in other cities and in provinces. Additional studies have investigated specific economic sectors, such as electric power generation.

All of these assessments provide valuable information and are useful to informing and improving control strategies. Few of them, however, identify *national* priorities, evaluating the best use of the pollution control yuan across the entire economy. To set national pollution-control priorities, the Chinese government has had to rely on simplified analyses and rules of thumb rather than comprehensive assessments that permit systematic benefit-cost analysis.

As we will summarize later, a World Bank report from 1997, *Clear Water, Blue Skies*, was a first international collaboration to estimate aggregate damages from air and water pollution, as high as 7.7% of Chinese gross domestic product (GDP). This provided decision makers with a useful estimated scale of damages, but its methods were debated in China and beyond and, in any case, were understood to provide rough approximations. Furthermore, it was limited for national priority setting because it was not designed to identify sectors that contribute most to this pollution or to specify impacts of alternative control policies systematically on both the public health and the economy.

The effort of this volume was conceived to build, and improve, on this important first effort by the World Bank with a more extensive and deeper air pollution damage assessment. We believe it also complements several ongoing collaborations involving SEPA and its affiliated bureaus and research entities, Tsinghua and other universities, and a number of international partners. The latter include the Center for International Climate and Environmental Research—Oslo (CICERO), the Organization for Economic Cooperation and Development (OECD), the World Bank, and the U.S. EPA. It contributes to the body of policy-targeted air pollution research in a number of ways that are also intended to advance and promote development of the scholarly literature in this area:

- by developing more emission data and air-dispersion estimates on the basis of local meteorological conditions and actual source characteristics;
- by applying a new method to make a best-available assessment, given data limits, of national human exposure to key ambient air pollutants from major sectors;
- by conducting one of China's first systematic contingent valuation studies to help build a Chinese health valuation literature;
- by providing a framework for integrated analyses of economy-wide costs and benefits of air pollution control policies; and
- by conducting such analyses of two control policies: damage-weighted taxes on fuels and sector outputs.

We do not seek the greatest possible precision in the damage assessment, which would take far more effort and resources than we could provide. We aim, however, to make reasonable estimates of the main sources of pollution, to permit interindustry comparisons (and not just estimate aggregate damages, as in the abovementioned World Bank study), and to create a basis for more detailed evaluation as data and research capacities improve. Our health damage estimates are directly linked to economic activity and energy use on a sector basis, allowing us to identify the sources of damage and allocate responsibility. This can help to prioritize pollution sources for emphasis in national energy and emission control policies, and to understand their effects throughout the economy.

Our assessment focuses on particulate matter (PM) and SO_2 in determining health effects in China, for two reasons. First, we believe that they likely dominate other pollutants as the source of air pollution health damages in China. Second, all such studies are limited by research practicalities, and the data for estimating the effects of PM and SO_2 are more readily available in China—though not necessarily simple to obtain—than for other pollutants. It is nevertheless important to note other pollution types as additional concerns for control and future research. In particular, as the transportation sector continues its explosive growth, we expect the health risks associated with nitrogen oxides and ozone to grow swiftly. China has already made impressive progress against a third mobile-source pollutant, lead, with a national ban on production and sale of leaded gasoline that became effective in 2000.

We further note the wide awareness that emissions of local air pollution and global greenhouse gases (GHGs) are closely related and ideally should be studied and addressed jointly. Our module for benefit-cost analysis explicitly evaluates both local pollutants such as PM and SO_2, and the primary anthropogenic GHG, carbon dioxide. This allows us to assess a benefit to the global environment of reducing China's local air pollution.

1.2 How to Read This Book

This book is explicitly targeted at a wide range of audiences, from curious lay readers, to informed nonspecialists and policy makers, and ultimately to researchers and scholars with either interdisciplinary or more narrowly specialized interests. The structure of the volume reflects this effort to appeal to many communities. The authors themselves are environmental health scientists, engineers, and economists. The research detail and methodological presentation deepen as the book progresses, from the introductory (chapters 1 and 2) to the specialized (chapters 5–10), with two transitional chapters in between (chapters 3 and 4). The scope of the book also evolves, because it addresses a consummately interdisciplinary topic by building up from a set of more focused studies. It begins with a broad, summarizing perspective (chapters 1–3), proceeds through more narrow ones (chapters 4–8), and then returns to the wider, integrated scope at the end (chapters 9–10).

Readers should approach the book accordingly. Nonspecialists should focus first on part I, including the introduction and background presented here in chapter 1 and a summary of the research project and its conclusions for policy purposes in chapter 2. If their interest is piqued, they can gain a much deeper understanding of the assessment by at least beginning part II, which presents the research itself. Chapter 3 is specifically written for nonexpert readers as well as expert ones, and as a transition between parts I and II. It repeats the scope of chapter 2, summarizing the project and its conclusions, but much expanded to its full methodological and research context. (In fact, chapter 2, as a summary for policy, is mainly just an abbreviation of chapter 3.) We hope that many lay readers will want to read this far, to learn about the nature of these types of studies and how they produce their conclusions. (Environmental assessments, after all, are often used in public and political debates about environment and can become the basis for lasting government policies. They are sometimes reported in popular media—imagine a headline: "Study Says Air Pollution Kills 1,500 Per Year; Costs Estimated in Billions"—and it is useful to know how to think critically about them.) At the conclusion of chapter 3, readers will have a sense of the remaining chapters that might most interest them.

Specialist readers and scholars may be more interested in the academic studies at the core of the assessment and may quickly choose to focus their attention on part II. They can treat chapter 3 as their introduction to the integrated aims of the project, chapter 4 as an introduction to central human exposure and epidemiological issues in the book, and then proceed through the original research of chapters 5–10 as their interests dictate. These chapters are written in detail so that our methods and results, including their limitations, are clear and can inform future research. They are still meant, however, for engaged readers of diverse expertise, with limited disciplinary jargon and introductory sections. The most specific technical and methodological details are placed in appendices. This is not to suggest that part I and this chapter offer specialists nothing, especially if they are unfamiliar with China's energy and air-quality conditions or with the existing literature on the topic. Researchers may want to skip chapter 2 altogether, however, because they will find chapter 3 a much more informative version of the same material.

The remainder of this chapter provides background on the topic, introduces our analytical approach, and summarizes related analyses by others to date. Specifically, sections 1.3 and 1.4 describe recent energy use and air pollution conditions for those who may be unfamiliar with China. In section 1.5, we outline how our research investigates health damages of air pollution and analyzes policies to reduce them. In section 1.6 we briefly mention other studies that have been conducted in this field, from which we have gained important insights for our own program.

1.3 Energy Use in China

Two prominent features of China's energy structure warrant a brief overview. One is China's very high, and continuing, dependence on coal. The second is a swift decline in energy intensity since economic reforms began in 1978.[1]

Economic growth has far exceeded growth of energy consumption in the reform period since the late-1970s, as shown in figure 1.1. In the ten years prior to 1997, the primary energy consumption per yuan of GDP fell by 37%. If one accepts the somewhat questionable official figures for the late 1990s and early 2000s, showing a sudden decline in coal use in that period (discussed below), it fell another 26% by 2000, before leveling and then rising (Sinton et al. 2004, table 4B.2). (Many statistics reported in this section are from this source, which compiles energy-related data from official sources and reports them in English. Table and figure numbers are

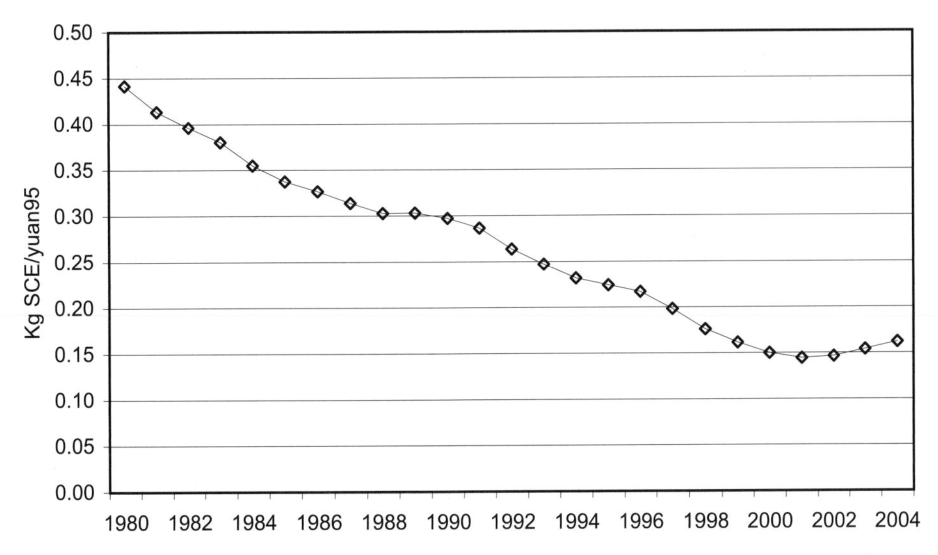

Figure 1.1
Energy consumption per unit GDP. Total primary energy is the sum of coal, oil, gas, and hydroelectric power, all converted to kilograms of standard coal equivalent (SCE) units. Yuan95 is yuan in constant 1995 value. Source: Sinton et al. 2004, table 4B.2.

included so it can be consulted for citations of the original data sources of the Chinese government.)

Despite this remarkable overall transformation, China remains one of the most highly energy-intensive major economies in the world, defined by the amount of energy needed per dollar of output. That noted, China's energy use on a per capita basis remains exceedingly low compared to more developed countries. In 2000 energy use per capita in China was about one-twelfth that of the United States, and about one-sixth that of Japan (U.S. EIA 2003).

Total energy consumption by source is represented over time in figure 1.2a. China is the world's largest producer of coal, generally the worst fossil fuel for both local and global pollutant emissions. Coal supplied 70–76% of its commercial energy consumption from 1980 until 1997, after which the disputed data suggest a share decline to 61–64%. The main users of coal are power generation and industry (mostly manufacturing), accounting for about 48% and 35%, respectively, in 2002[2] (Sinton et al. 2004, tables 4A.1.4 and 4A.17.2). According to official data, annual consumption of coal exceeded 1300 million tons during 1995–1997, fell sharply to 982 million tons in 2000, and rose again to 1579 million tons by 2003 (NBS 2005).

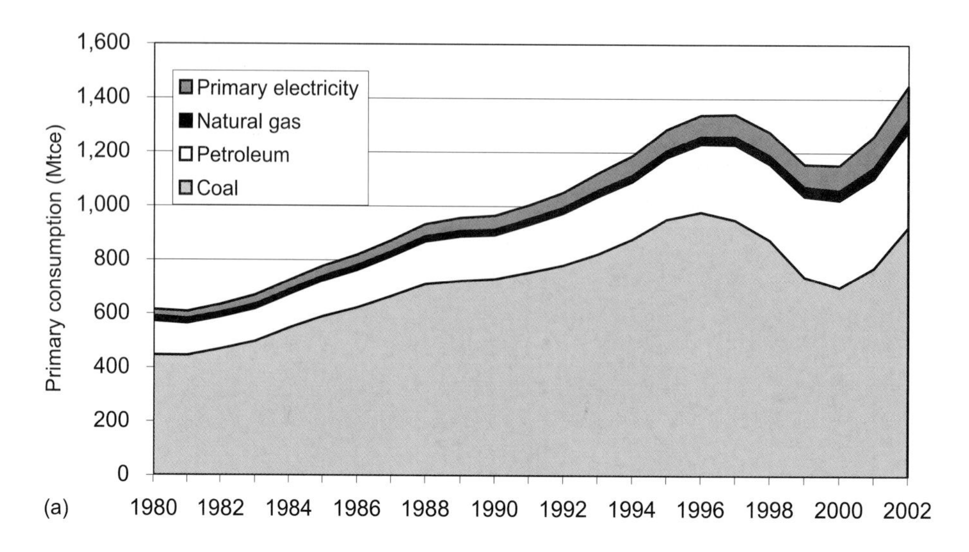

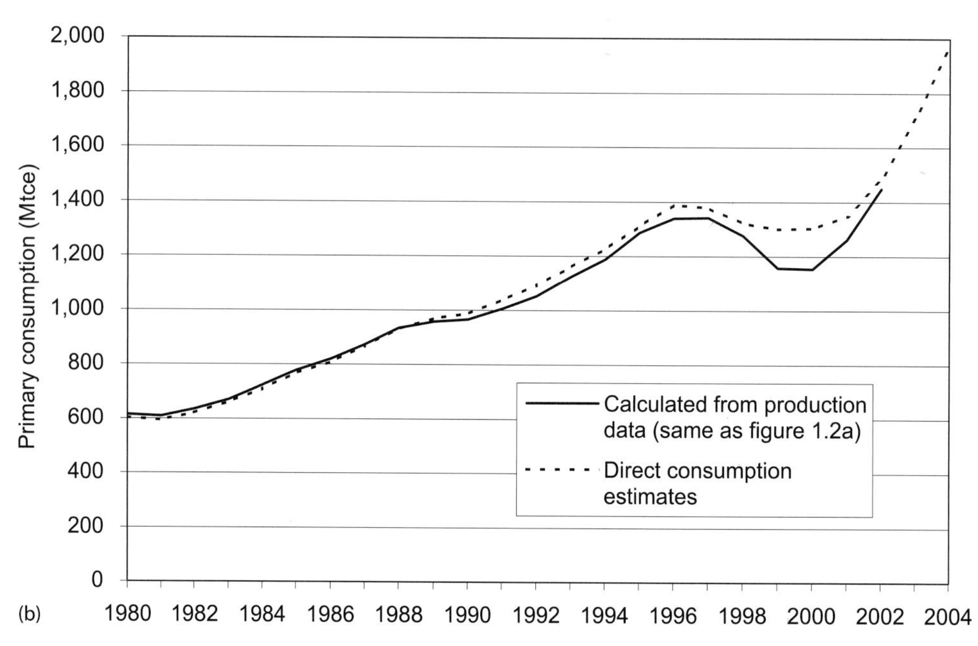

Figure 1.2
Primary energy consumption in China. Consumed energy is converted to million tons of standard coal equivalent (Mtce) units. Primary electricity is power generated from sources other than combustion of fossil fuels. It is chiefly hydroelectricity and a small amount of nuclear power. (a) Consumption calculated from production data. (Source: Sinton et al. 2004, table 4A.1.2.) (b) Comparison to direct consumption estimates. (Source: NBS 2006, table 7-2.)

Concerning the data in the late-1990s into 2001, the dramatic dip in coal consumption shown in figure 1.2a came at the same time that official statistics show the economy growing at more than 7% per year. If this is accurate, it is an historic event in decoupling economic growth from energy demands.

Sinton and Fridley (2000) reviewed a number of reasons that might explain such a decline, including closure of small factories and changes in economic structure, gains in end-use efficiency, fuel-switching, a rise in coal quality, and a policy of the central government to close small mines. In light of a high-profile, international spotlight on the inferred environmental benefits of this sudden downward dip in coal use, including for greenhouse gases—e.g., articles in *Science* (Streets et al. 2001) and the *New York Times* (Eckholm 2001)—we note that a wide debate about the accuracy of China's official data for the late-1990s promptly followed. A variety of doubts about the energy and economic statistics have been described (Rawski 2001; Sinton 2001; U.S. Embassy 2001). Others have continued to propose economic and other explanations for this remarkable trajectory of energy consumption, such as transformations occurring in anticipation of China's accession to the World Trade Organization (Fridley et al. 2003; Fisher-Vanden et al. 2004).

We will not belabor this debate here but will note that the National Bureau of Statistics (NBS) includes a balance term in its official coal tables that represents the difference between direct estimates of consumption and those based on production data. In principle, these will be equal; the latter are more often reported because they are generally considered more accurate. The size of this residual term grows somewhat in 1996–1998 compared to earlier years in the decade and then rises dramatically for the years 1999–2001 (three-fold and more, to *hundreds of millions* of tons), falling back to the previous scale in 2002.[3] Thus the direct consumption estimates (NBS 2006) show a shallower dip in 1999–2001 compared to the production-based data of figure 1.2a, as shown in figure 1.2b. Nevertheless, even this alternative measure shows an unusual decline in coal consumption at a time of strong overall growth in the economy.

The reporting of the ballooning balance term indicates that the NBS is not unaware of growing inconsistencies in its coal data for those years but, rather, acknowledges and indeed quantifies them. The cause of these data problems is certainly of pertinent interest to the government; a contributor may be disruptions from reform of systems to compile and report statistics under the broader government reorganization in 1998 and as the objectives of data provision continue a shift from support for a planned economy to informing a market one. It would be unsurprising if this process encountered difficulties in implementation in view of the size and inertia of this system, and we assume that efforts to improve the statistical infrastructure continue today. Regardless of possible explanations, for purposes of research we take the NBS at its word about the existence of large statistical inconsistencies for this period. For this reason we are cautious about using national data from the years in which this residual term is anomalously large—1999–2001—to indicate trends in coal use and related emissions over time.

Another important trend in China's energy structure is the growth in petroleum consumption and imports, driven by a rapidly motorizing transport sector. From 1990 to 2002, the national stock of passenger vehicles increased more than 18% annually, to 12 million units, and motorcycles increased an average of 21% annually, to 43 million units. This contributed to a growth in petroleum consumption of 6.7% annually over these years and an increase in petroleum share of total primary energy from 16.9% to 24.6%. China is a significant petroleum producer but became a net importer in 1996, and by 2003 net imports of crude oil had reached 83 million tons (Sinton et al. 2004, tables 5B.3, 4A.1.4, and 7A.1.1).

China is usually described as geologically limited in natural gas resources, though this energy resource was also historically underexplored and underdeveloped. Driven in part by air-quality concerns, the government in recent years has promoted the extraction and distribution of this cleanest of fossil fuels, particularly for domestic heating in major cities. A major pipeline from gas fields in western China to central and eastern demand centers is partially completed. Even with the recent government efforts, however, gas comprised only 3% of total primary energy consumption in 2002 (Sinton et al. 2004, table 4A.1.4).

Other sources of commercial energy in China include substantial hydropower, which provided 17.6% of gross electricity generation in 2002, and nuclear, which provided 2.3%. These two constituted 9.1% of total primary energy production in 2002. Biomass—chiefly crop wastes and firewood—is a major fuel in rural China and a source of indoor air pollution, but it is not generally a commercial fuel and is left out of most energy tables. When counting all sources of energy, not just commercial forms, biomass constitutes 15% of total energy, compared to 56% by coal and 20% by petroleum in 2002 (Sinton et al. 2004, tables 2A.4.2, 2A.1.3, and 4B.1).

Although they are somewhat imposing, we offer two more figures on energy, because they allow the reader to appreciate the relative scales of both the sources of energy and its end uses. Figure 1.3 represents the entire energy system of China, with flows translated into equivalent units across energy types and scaled accordingly. We also include a similar diagram for the United States (figure 1.4) to contrast

China's energy structure with that of the largest developed country. The end-use categories are not precisely the same, reflecting different conventions in economic and energy data in the two countries. We also caution that construction of such diagrams requires many assumptions, especially in estimates of the efficiency of conversion into "lost" versus "useful" energy. One should view these diagrams as representations of the big picture of energy flows in the two countries rather than sources of detailed information.

The tremendous coal dependence in China compared to the United States, and the minimal role of natural gas, is clear in these figures. Petroleum has grown to a sizable share of total energy use in China, though still lower than richer countries such as the United States. Notable for China on the consumption side is the huge share of energy consumed by industry, via direct combustion of petroleum and especially coal, along with electric power. China's transportation sector consumes a relatively small share of total energy, despite its enormous recent growth, whereas it is the largest end-use in the United States. Hydroelectricity provides a larger share of energy consumed in China than in the United States, whereas the reverse is true for nuclear power. We must spotlight the role of biomass in the form of crop wastes and fuel wood as a sizable energy source in rural China, repeating the caveat that this flow is noncommercial and left out of most energy tables for China, so one should not assume that these data are comparably collected.

Finally, we should note that, if the scaling were consistent across the two diagrams, the U.S. flows would collectively be roughly one and a half times the size of China's, to fuel an economy that was eight times the size (unadjusted for purchasing power) and to support a population less than one-fourth as large.

1.4 Air Pollution Emissions and Ambient Concentrations

As mentioned above, China's air quality is generally poor, with two-thirds of 338 monitored cities out of compliance with at least one of the nation's air-quality standards for residential areas in 1999 (World Bank 2001). We limit this section to a brief summary of key characteristics and trends in the emissions and ambient conditions for the pollutants that are the subject of the current study.[4] A more detailed discussion of China's air pollution is PRCEE et al. 2001, a background study for World Bank 2001.

We note an important distinction about these data: the emission data are for the entire country, whereas the ambient concentration data are only for major cities. In any case, the link between emissions and concentrations are more complex than

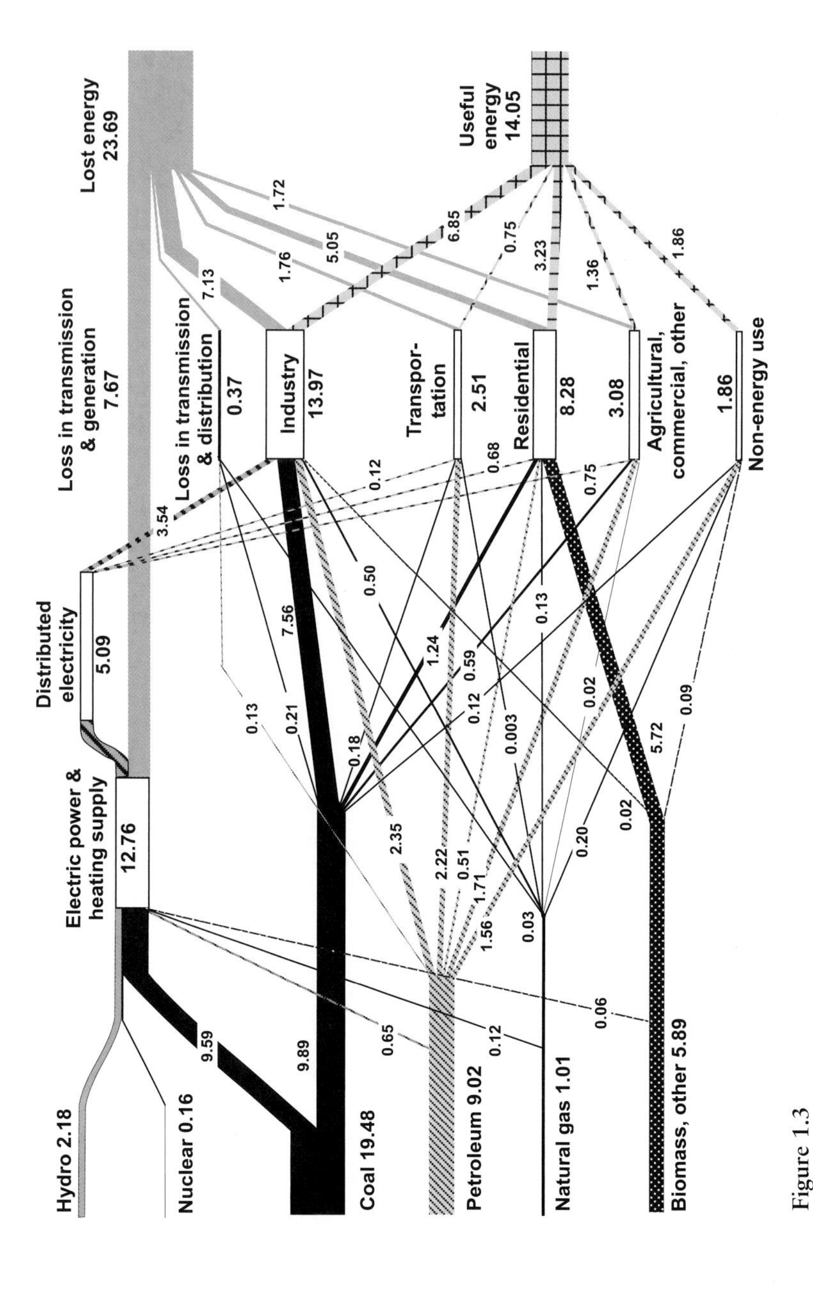

Figure 1.3
People's Republic of China energy flows for 2000: Total primary energy consumption of approximately 37.75 quadrillion British thermal units. Source: NBS 2004. These data were compiled and the figure was first drafted for the authors by Qiaomei Liang, Yiming Wei, and Ying Fan of the Institute of Policy and Management, Chinese Academy of Sciences.

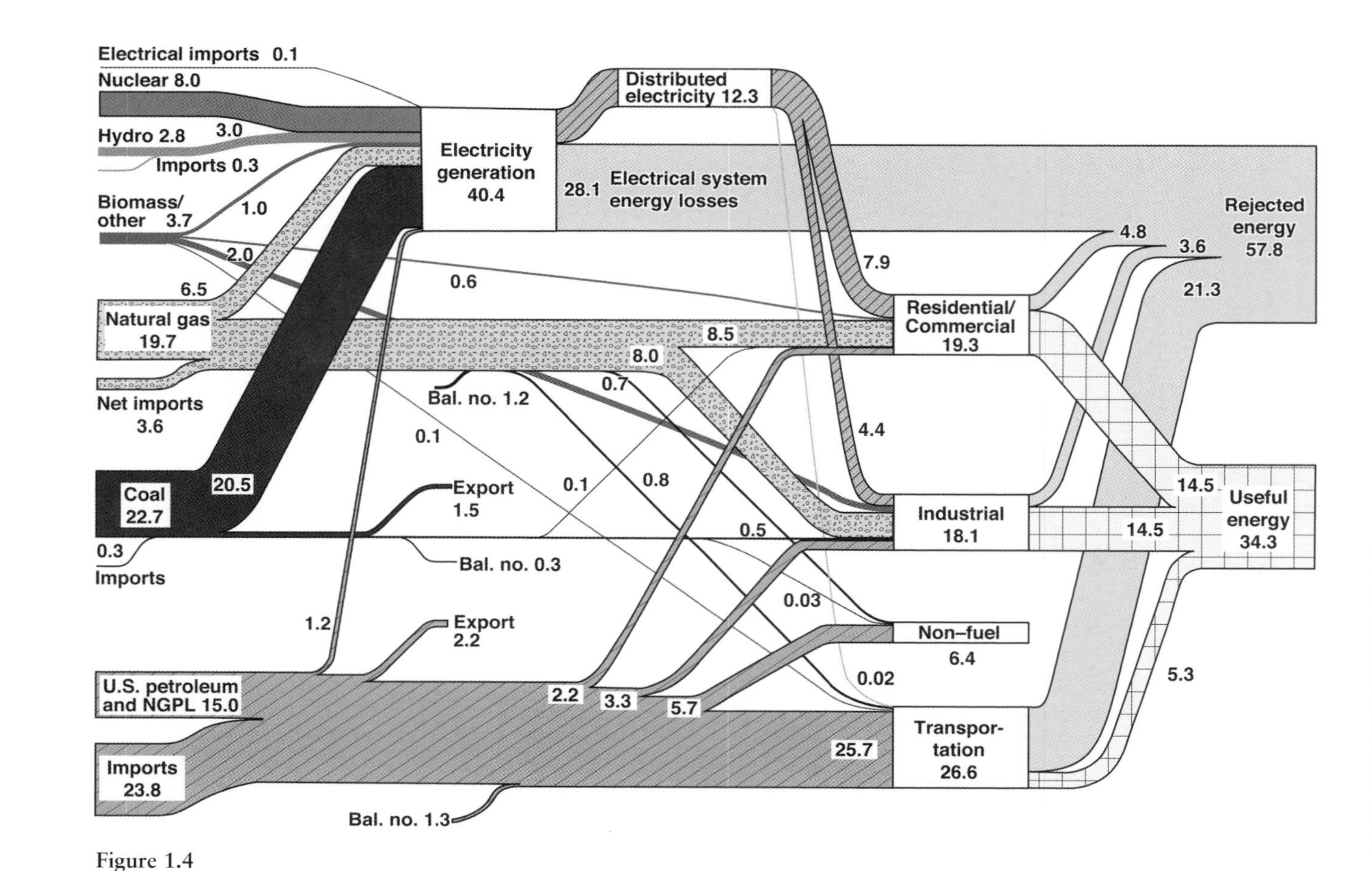

Figure 1.4
U.S. energy flows for 2000: Net primary energy resource consumption of approximately 98.5 quadrillion British thermal units. Source: EED 2001.

many realize. Investigating this is one of the primary motivations of the research of this volume, as there are many factors that can influence atmospheric dispersion, removal, and transformation of pollutants after they are emitted.

1.4.1 Particulates

The SEPA data on particulate matter have, until recently, only systematically covered TSP. There has been less information on finer particulates such as PM_{10} or $PM_{2.5}$ (particles less than 10 or 2.5 microns in diameter, respectively), though reporting of PM_{10} ambient concentrations has increased in recent years. These are the forms most closely associated with adverse health effects and that most current particulate epidemiology investigates. For health studies in China the long reliance on the TSP measure has necessitated an estimation of the fraction that is PM_{10} and/or $PM_{2.5}$.[5] TSP is classified as combustion emissions ("soot" in official publications) or process emissions ("dust").[6] The coverage of this TSP data is being improved, with an effect that historical data are not comparable over long periods. Earlier data included just medium and large enterprise sources, and only in the last decade were the environmentally significant township and village enterprises (TVEs) gradually added.

National particle emissions from combustion sources are given in figure 1.5 (SEPA 1992–2004), where the break in 1995 reflects the inclusion of data from the

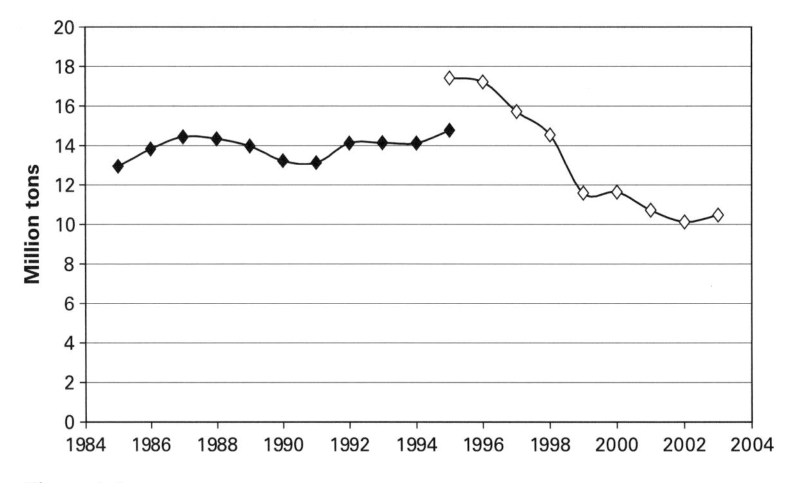

Figure 1.5
Combustion emissions of total suspended particulates in China. Data coverage extended in 1995 to include township and village enterprises. Source: SEPA 1992–2004.

TVEs. The data show stability in the early 1990s and a considerable reduction since 1995. When noncombustion emissions are included, the initial level is doubled, but the downward trend is sharper because of the more rapid reduction in process emissions. We are not aware of a systematic study of these trends, but it appears clear that improved pollution control played a role in the decline in emissions over the 1990s, on top of the effects of the slowing of total energy consumption shown in figure 1.2.

Particulate emissions by sector are reported later in chapter 9, table 9.1. The combustion emissions are mostly from electricity generation (32% of the total in 1997) and cement and related products (21%). The primary fuel source is coal. Process emissions are largely from cement production and iron smelting, at roughly 70% and 15% of the total, respectively.

Urban ambient concentrations of TSP result from the above emissions and secondary transformation of other pollutants, such as sulfur and nitrogen oxides (NO_X) described below. SEPA compiles data for cities, contrasting northern and southern ones, and they are reproduced in figure 1.6. On the basis of these cities, the national average TSP concentrations for urban areas declined about 30% from 1990 to 1999.

Although this trend was encouraging, it has leveled since 1999, with the average concentration in 2003 reaching 256 $\mu g/m^3$. As important, the national TSP standard for residential areas (termed "class II") is 200 $\mu g/m^3$, and the WHO guideline was 90 $\mu g/m^3$ until it was suspended for this particulate form. About 60% of the surveyed

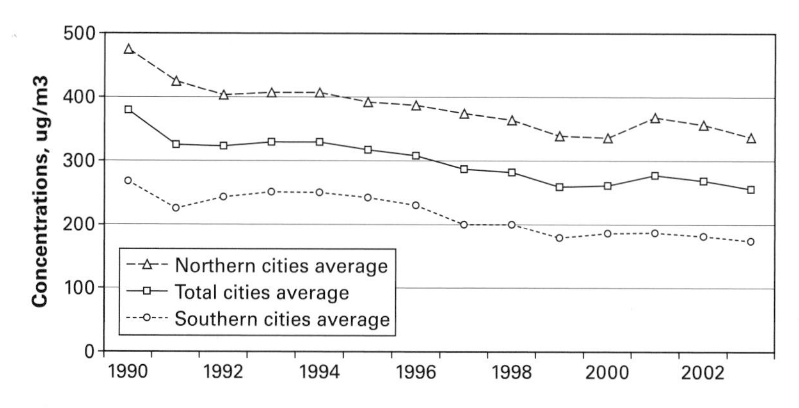

Figure 1.6
Average ambient concentrations of total suspended particulates in cities of China. Source: Sinton et al. 2004, SEPA 2004.

Table 1.1
Annual average measured PM_{10} concentrations in major cities for 2003

	Number of Cities	Annual Average Concentrations (μg/m^3)
Northern cities	52	140
Southern cities	54	102
Total cities	106	121

Source: SEPA 2004.

cities in 1999 exceeded the Chinese class II standard. Indeed, 36% exceeded the class III standard, designed for industrial areas (PRCEE et al. 2001). The figure also illustrates a marked difference in ambient TSP conditions between southern and northern cities, the southern ones, on average, meeting the class II standard in recent years with northern ones still far from compliance. It is difficult to explain the higher TSP levels in the north. It may be obvious that the colder north requires heating, but one also has to consider the effect of higher background concentrations and wind-blown dust in addition to human emissions.

As noted above, there was for a long time little information available on the smaller particulate matter that is known to be most harmful to health. Since 1999, ambient PM_{10} has been subject to an official class II (residential) air-quality standard of 100 μg/m^3, and hourly monitoring of it is now carried out in many cities, with annual averages now reported in official sources in addition to TSP. The recently announced WHO annual average guideline is 20 μg/m^3 (WHO 2005). In table 1.1 we summarize annual average PM_{10} concentrations for major cities in 2003, again classified as either northern or southern (SEPA 2004). Seventy-eight percent of northern cities and 52% of southern ones showed annual averages above the class II standard.

1.4.2 Sulfur Dioxide

The officially reported national SO_2 emissions are plotted in figure 1.7, again a break in 1995 reflecting the addition of TVEs to the estimates. Emissions rose from 13.2 million tons (Mt) in 1985 to a peak of 19.5 Mt in 1995, equivalent to 23.7 Mt under the new definitions. It then fell and leveled off around 2000, before shooting up again to 21.59 Mt in 2003 (SEPA 1986–2004). This trend follows that of coal consumption in figure 1.2 reasonably closely—more than 90% of SO_2 emissions are from coal combustion—with only a small reduction in emissions per ton of

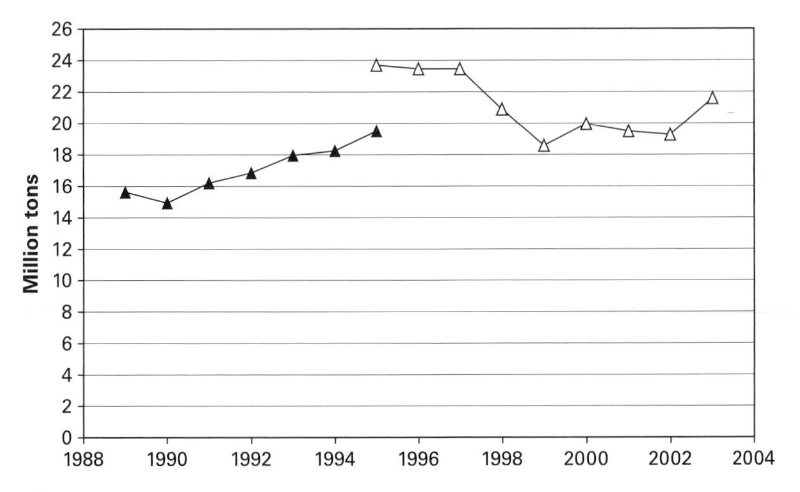

Figure 1.7
Emissions of sulfur dioxide in China. Data coverage extended in 1995 to include township and village enterprises. Source: SEPA 1986–2004.

coal. Official emission estimates are based on energy data, and such "bottom-up" inventories may thus be subject to some of the same uncertainties and errors discussed in section 1.3 on energy use.

Streets et al. 2000 conducted an independent assessment of SO_2 emissions through 1997 that was based on extrapolation of a detailed emission inventory developed for the RAINS-Asia project (Regional Air Pollution INformation and Simulation-Asia), which used energy-use data from the International Energy Agency. A follow-up inventory, adjusting official Chinese figures with omitted emission sources, estimated a total for 2000 (Streets et al. 2003). These figures are consistently higher than official data but show a comparable trend: from 17.90 Mt in 1985 to a peak of 26.21 Mt in 1996 (Streets et al. 2000), followed by a decline to 20.38 Mt in 2000 (Streets et al. 2003). An improved and updated inventory covering the more recent increases in SO_2 emissions may be forthcoming from this research group.

Nevertheless, conducting inventories of pollutant emissions in China is challenging work and uncertainties, while narrowing, remain large. Zhang (2005) included a review of the Chinese and international literature and plotted widely varying estimates of total annual SO_2 (and NO_X) emissions in China, from 1984–2004. It is noteworthy that for each year, the official estimates of SO_2 emissions are lower than all estimates by independent research teams. The emission uncertainties may

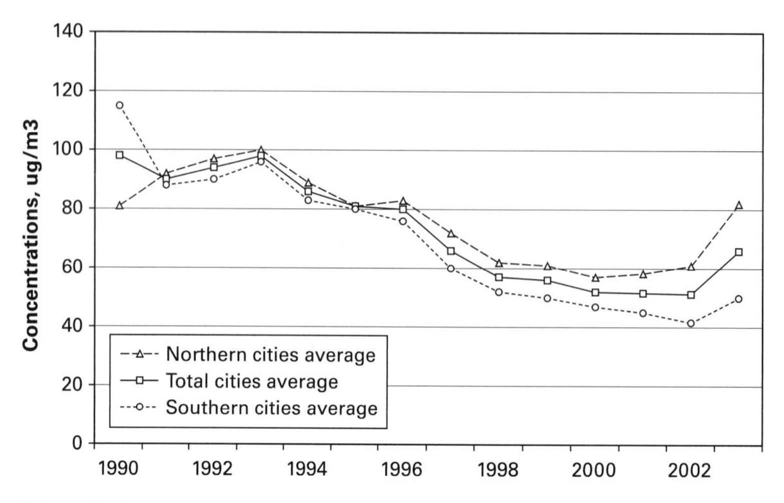

Figure 1.8
Average ambient sulfur dioxide concentration in cities of China. Source: Sinton et al. 2004, SEPA 2004.

soon be reduced further by new lines of evidence derived from satellite-based observations of SO_2 levels over China (Richter, Wittrock, and Burrows 2006).

Figure 1.8 gives average annual concentrations of ambient SO_2 reported for Chinese cities, similarly differentiated by SEPA into southern and northern cities. The concentrations steadily decreased from 1992 to 2000, leveling at a national average of 51 $\mu g/m^3$ in 2002 before shooting up to 66 $\mu g/m^3$ in 2003. Through much of the 1990s there was less geographical variance compared to TSP concentrations, with a steady pace of decline that is similar across regions. In recent years, however, SO_2 concentrations rose, an increase that preceded the rise in TSP concentrations, and the rise was greater in the north than the south. More than TSP, furthermore, the earlier progress in SO_2 concentrations in many major Chinese cities has now clearly and sharply reversed. Forty-two percent of cities did not attain the class II standard of 60 $\mu g/m^3$ in 2003, compared to 29% in 1999 (SEPA 2004; PRCEE 2001). The annual average guideline of WHO is 50 $\mu g/m^3$ (WHO 2002, p. 196).

The decline in SO_2 concentrations through the 1990s is attributed chiefly to the changing fuel structure of household energy use, in particular switching from coal combustion to natural gas, washed coal, liquid petroleum gas, and electricity (PRCEE et al. 2001). This took place despite the rise in aggregate emissions of the early 1990s shown in figure 1.7. These sources tend to be at ground level, directly

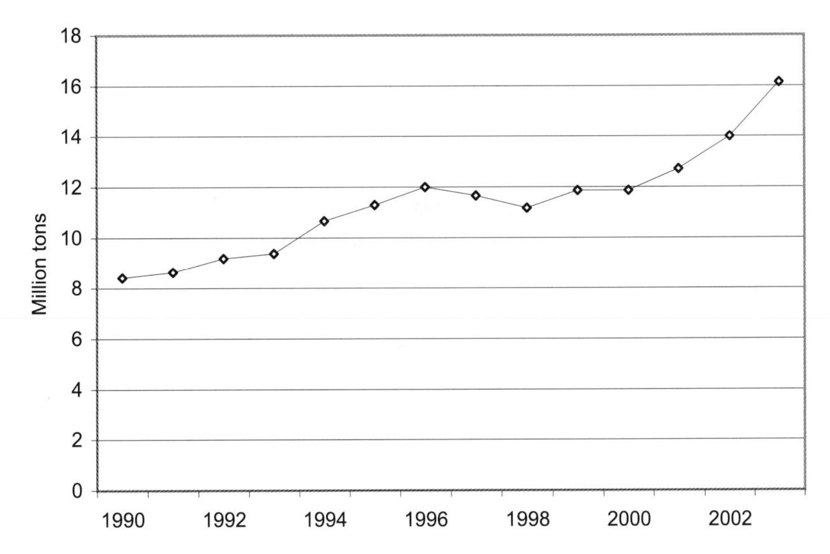

Figure 1.9
Emissions of nitrogen oxides in China. Source: Hao et al. 2002 for data through 1998; 1999–2003 data from personal communication with Hao research group at Tsinghua University.

affecting concentration measurements. Successful pollution control in urban areas may have been offset by growing emissions in nonurban sources and by higher stacks for urban sources allowing greater dispersion away from the cities. More recently, however, the trends of emissions and urban concentrations appear to have recoupled, the rise in SO_2 concentrations following rising emissions from higher coal use.

1.4.3 Nitrogen Oxides

NO_X emissions are associated with health damages, but they likely pose more health risk in their contribution to secondary chemical formation of ozone and fine particles. This volume considers NO_X only briefly, in chapter 5. There is no systematic accounting reported by the government of NO_X emissions, and we do not describe it in much detail here. Hao et al. (2002) estimated total anthropogenic emissions over two decades until 1998 on the basis of commercial energy consumption, NO_X emission factors, and fuel types, and the Hao research group has continued to calculate emissions in subsequent years. They estimate that total emissions increased from 4.76 Mt in 1980 to a peak of 12.03 in 1995, fell to 11.18 Mt in 1998, and rose again to 16.14 Mt by 2003, which reflects the coal and total energy trends described above (and which are the basis for the inventory). These trends are depicted in figure 1.9.

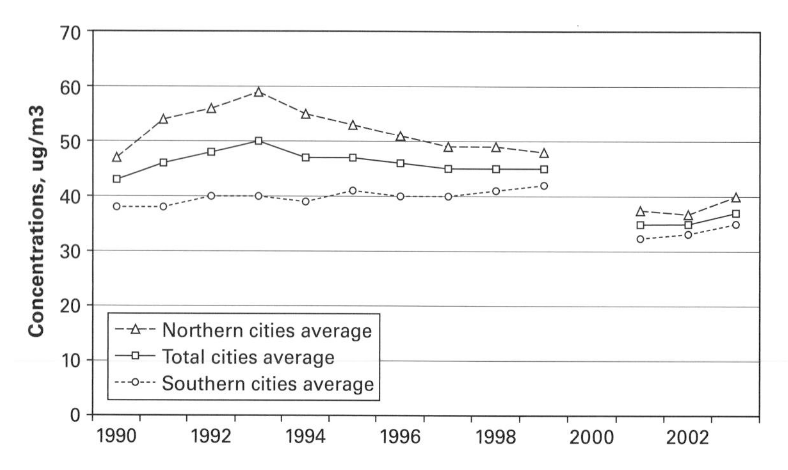

Figure 1.10
Average ambient concentrations of nitrogen oxides in cities of China. The measurements were switched from all nitrogen oxides (NO_X) to just nitrogen dioxide (NO_2) in 2001. Source: Sinton et al. 2004; SEPA 2004.

Hao et al. also analyzed source contributions of NO_X, in 1998, with industry estimated to have contributed 41% (4.59 Mt) of emissions, electric power 38% (4.23 Mt), and transportation 13% (1.45 Mt). Streets et al. 2003 provided an estimate for 2000 of 11.34 Mt, including emissions from biofuel and biomass combustion not included in the Hao et al. 2002 inventory.

Measured ambient concentrations reported by the government suggest that, when taken alone as a primary pollutant, NO_X has not been a serious risk in most Chinese cities. The average annual urban concentrations compiled by SEPA for all major cities did not exceed 50 μg/m^3 in any year from 1990 to 1999, as shown in figure 1.10. The break in the graph in 2001 was when measurements were switched from NO_X to its major constituent species, NO_2. These have been within the WHO guidelines for annual average NO_2, of 40 μg/m^3 (WHO 2002, p. 179). Until 2002 the official data showed offsetting trends of NO_X or NO_2 nationally, a steady downward path of concentrations in northern cities and an upward one in southern cities. PRCEE et al. 2001 noted that the trend also was better for smaller than larger cities and attributed this to rising emissions from the rapidly expanding vehicle populations in larger, southern cities with the swiftest economic growth.

This suggests that viewing the NO_X trends from a national perspective masks what is likely a swiftly worsening problem in a number of major urban areas with growing vehicle stocks and congested road networks. Recent official data shown in figure 1.10 indicate that the long decline of NO_X in northern cities has reversed, whereas gradual growth of concentrations in southern ones continues. A new line of independent evidence gained a high profile in 2005 when satellite measurements by the European Space Agency reported in *Nature* indicated that, by 2004, an urban and industrial region of China stretching from Beijing to Shanghai had the worst tropospheric NO_2 levels in the world (Richter et al. 2005).[7] Annual average levels in this region had risen around 50% since 1996, driven by especially large and accelerating increases in winter months. The Richter et al. estimates are of NO_2 in the lower atmosphere, not just at the surface, but they reflect the same emission sources and provide independent, scientific evidence of worrisome trends that official emission inventories and government-sited and -operated monitoring stations may not be picking up.

In summary, the emissions and urban concentrations of our target pollutants show, on balance, considerable progress in control through the 1990s but leveling or negative trends more recently. TSP emissions and ambient urban concentrations declined steadily and substantially before stabilizing in the current decade. SO_2 emissions rose and fell through the 1990s but have been rising again, whereas ambient urban concentrations steadily declined, leveled, and then by 2003 were sharply rising again. In many cities concentrations of these two pollutants exceed air-quality standards, especially TSP. NO_X emissions are not officially reported but, after some decline in the late-1990s, are likely rising again, and, after a long period of stability, ambient urban concentrations are also increasing, possibly dramatically.

1.5 Health Damages of Air Pollution in China and Research Framework

We are concerned about China's degraded air quality because it causes damages. The research of this volume focuses on health effects, the type of damage that most researchers believe dominate total pollution impacts, and the implications for the economy of policies to protect health. (Other studies of China have incorporated some non-health damages of pollution, including Hirschberg et al. 2003, O'Connor et al. 2003, and Wang and Smith 1999. We aim to consider nonhealth damages in future research.)

The burning of fossil fuels that cause much of this local air pollution also emits greenhouse gases, and these are widely expected to produce varied and pervasive global damages (IPCC 2001). Quantifying these effects is very difficult at this stage, and we do not attempt such estimates in this book. We do, however, discuss the effect of pollution control policies on the quantity of greenhouse gas emissions.

We now introduce our assessment framework, the details of each part given in separate chapters. Our research strategy follows what is commonly referred to as a pollution causal chain:

Component	Description	Chapter
1	From economic activity and energy use to emissions	1, 9
2	From emissions to concentrations (atmospheric transport)	4–7
3	From concentrations to human exposure	4–7
4	From exposures to health impact (dose-response)	4
5	Economic valuation of health impacts	8
6	National damage assessment by sector	9
7	Benefit-cost analysis of policies to reduce emissions	10

Our research (represented graphically in figure 1.11), makes primary contributions to components 1–3 and 5–7. Although we could not contribute here to component 4—air pollution epidemiology, which requires costly and time-consuming field

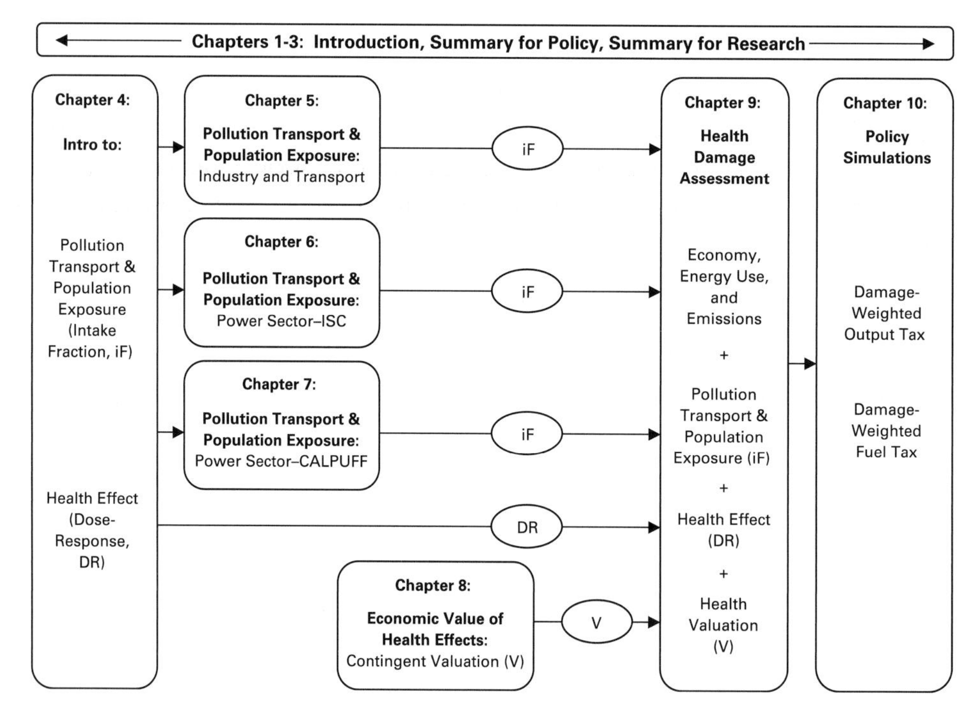

Figure 1.11
Research framework.

survey work—we do introduce the issues of defining and estimating functions that describe the human health response to pollution exposures. These are drawn from studies that have already been published. Components 1, 6, and 7 are implemented by using an economic model of China that identifies thirty-three distinct production sectors and one nonproduction sector (households), and traces the evolution of the economy over time. Earlier versions of this model have been used in previous studies of the effects of controlling carbon emissions and of the local health effects of controlling global pollutants (Garbaccio, Ho, and Jorgenson 1999a, 2000). This model generates the output of each sector for each period and with it the use of fossil fuels. The combustion of fuels and the production of goods generate emissions of PM, SO_2, and CO_2, and these are estimated using time-varying emission factors. This is a fairly well known procedure, as we describe in detail as part of the national analyses of chapter 9.

Components 2 and 3 are the focus of chapters 4–7. The requirement here is to translate emissions from a range of economic sectors into human exposures. Past assessments of this sort have typically adopted a damage function framework. In this approach, the emissions of the relevant pollutants from *all* sources are first estimated, along with characteristics of the source (e.g., stack height and diameter, pollutant exit temperature, or velocity). Researchers use atmospheric models to estimate how these emissions influence ambient concentrations or pollutant deposition at a number of locations. These typically incorporate detailed meteorological characteristics, sometimes allowing for chemical interactions among pollutants. Population distributions and locations are then used to determine at-risk populations. Finally, in a step summarized by component 4, concentration-response functions from epidemiological studies are combined with the exposure assessment to estimate population health risk.

Although this methodology is well established (ORNL and RFF 1994; EC 1995; Rowe et al. 1995; U.S. EPA 1999; Levy et al. 1999), it requires a large amount of very detailed information that is currently unavailable for a national-scale analysis in China.

Our aim therefore is to develop a framework to estimate health impacts on the basis of fundamental physicochemical properties but without all of the detailed input data that are now lacking. We wish to make a reasonable approximation of nationwide health impacts with a substantial, but not overwhelming, data-collection effort. By adopting a modular approach, we can identify methodological improvements to be tackled in turn. Additional refinements can be added as more data become available and more advanced atmospheric models are developed in China.

To make this approximation, we apply the concept of "intake fraction," as developed by researchers at the Harvard School of Public Health and a number of colleagues at other universities and organizations. Intake fraction will be formally defined and explained in the chapter 4, but its basic meaning is the amount of material released from a source that is eventually ingested or inhaled. The concept has been developed for precisely the problem at hand: to conduct reasonable, if not precise, risk assessments with limited input information, as is often the only option in developing countries. Once the intake fraction of a source type has been estimated, population health risk can be simply quantified as the product of intake fraction, emission rate, and the unit health risk of the pollutant. The key research challenge is to determine reasonable intake fractions for different pollutants and source types.

Chapter 5 estimates intake fractions for four highly polluting sectors: iron and steel, chemicals, cement, and transportation. National inventories were not available, so our team collected detailed data from five cities across China. These data are supplemented by sample information for sources in the entire country. In assessing exposure for these sectors, we employed an air-dispersion model of relatively simple (Gaussian) form, ISCLT (Industrial Source Complex Long Term), which could be applied to the many sources of our database without overwhelming time and effort.

Chapter 6 estimates intake fractions from the electric power industry, which is by far the biggest user of coal. This uses a very large national emission inventory compiled by our team, chiefly from environmental impact assessment reports. This chapter also made use of the air-dispersion model ISCLT.

Chapter 7, like chapter 6, considers intake fraction and health risk in the power sector. This research was initiated to test the feasibility of applying a more advanced air-dispersion model, CALPUFF, that could cover a larger spatial domain and include secondary chemistry. The time and computational requirements to apply this model to our entire power plant database proved imposing, so we relied chiefly on the ISCLT results for the national power sector exposure assessment. The CALPUFF analysis, however, provides other information for the purposes of this volume (and stands also as a valuable independent study). It allows us to incorporate approximation of the health risks due to long-range air transport and secondary particles that the model of chapters 5 and 6 does not estimate. It also allows comparison to similarly conducted analyses in the United States, showing how higher population densities and more urban plant sites in China raise health risks for the same quantity of emissions.

With estimates of human exposure to air pollution for the five sectors, the next step is component 4, to apply dose-response functions from epidemiological studies to estimate health effects. As mentioned above, we do not offer new air pollution epidemiology, which requires expansive and costly field research. It instead relies on results in the published literature. A number of such studies were conducted by colleagues of our team with support of earlier stages of our program (Xu 1998; Wang et al. 1999; Xu et al. 2000; Venners et al. 2001, 2003).

Attributing the incremental morbidities and mortalities (e.g., the cases of actual respiratory disease and related death) to pollution emissions from particular industries provide a powerful quantified case for pollution control. Many policy makers, including those in China, also like to see such health impacts translated into economic terms, to help compare costs and benefits of different policy interventions.

The monetization of health damages—component 5—is acknowledged even by its advocates as an imprecise science. Recognizing the uncertainties, we nevertheless devote considerable effort to it here because it offers useful information to help communicate and prioritize environmental risk in real-world policy processes.

A number of estimation methods, with associated terminologies, have been developed to value damages to health. These include assessments of "willingness to pay" (WTP), determined by "revealed preference" observed from actual market transactions, or by interview-based surveys ("contingent valuation," or CV). They also include a more conservative approach favored by many Chinese researchers, the "human capital" method that values health outcomes chiefly by their effect on expected earnings. The assessment includes a new contribution to this field in chapter 8, a CV study, and a brief overview of valuation methods and underlying theory is included in that chapter for interested readers.

To bring all of the foregoing together and assess national health and economic damages on a sector basis—component 6—chapter 9 begins with estimates of national emissions of TSP and SO_2 for each sector. This is done in our environmental-health submodel, as described in that chapter. It then brings the intake fractions of chapters 5–7 and the dose-response coefficients recommended in chapter 4 together to estimate the health effects. Our methodology draws on World Bank (1997) and examines the impact of pollution on eleven health effects, the two most important being premature mortality and chronic bronchitis.

Because of uncertainties in methodological approaches to valuation, chapter 9 applies a range of estimates from the literature, including those of chapter 8 and values derived in studies from other countries and scaled to Chinese incomes. The result is valuation of the health damage from air pollution in each sector. This allows us to derive the marginal damage per unit output for each sector as well as the marginal damage per unit fuel combusted. The estimated total damages for each sector and their distinct marginal damages help us rank sector priorities in pollution reduction.

The estimated marginal damages indicate the relative importance of each unit of output and fuel use to the total air pollution problem. We use these measures to analyze policies that tax output and fuels in proportion to the damage caused (a "green tax," or, in economic jargon, a Pigovian tax). This policy analysis is reported in chapter 10 and is accomplished by use of the economic model combined with the environment-health submodel. With the integrated model, we are able to analyze both the benefits and costs of control policies of component 7. The benefits have the form of reduced health damage, and the costs are in terms of lower consumption and GDP.

We emphasize again that this research is an incremental advance, not a definitive one, in the development of a framework to assess health damages of air pollution and associated policies in China. Its design is such that future refinements in any component may be immediately used in subsequent applications to improve the policy analysis. As such, we note now that it suffers a number of gaps. The first caveat is that our analysis has a limited treatment of secondary pollutants, using a simple approximation method based on results of chapter 7 to estimate exposures to secondary particulates, and leaving ozone to future research.

Second, we carefully investigate exposures in five major sector sources of ambient air pollution but use simpler approximations for the others. Because of several factors, including the scale of energy consumption in these other sectors, they are likely to have substantially lesser effects on air pollution than the ones modeled, save one: the residential sector. Unlike industrial and transportation sources, households have complex emission and exposure characteristics and thus pose an analytical challenge meriting a separate assessment that could easily be as large and challenging as the one reported here. It is essential to spotlight indoor air pollution, particularly in the rural residential sector, which could be one of largest sources of health damages of air pollution in China.[8] Very different sorts of policies, however, are generally needed to target pollution control in the residential versus industrial and transport sectors.

This is one of several reasons we do not believe it is accurate or even especially useful in a policy sense to reduce estimates of health damage from air pollution to a single number, the percentage loss to GDP, as is often reported and emphasized in the literature. Rather than estimating aggregate national cost, we are more interested in—and such models are better suited for—differentiating the relative health and economic impacts of emissions from the many sectors and examining the impacts of prospective policy interventions on emissions, public health, and growth of the economy at large.

1.6 Review of Other Assessments of Health Damages of Air Pollution in China

A substantial literature estimating the costs of environmental degradation in China has developed in the past fifteen years, varying widely in ambition. In light of the focus of the current study on advancing methods of estimating the health damages of air pollution—which typically dominate costs of environmental degradation more generally—we spotlight studies of these damages, especially international collaborations with Chinese research institutes employing methods from the worldwide literature. The brief descriptions of the different contributions here are intended as a reference for interested readers and permit us to credit prior efforts and contrast several with our assessment. We do not provide a comprehensive discussion or critique of the methods and results of each.

Table 1.2 summarizes key features of these studies. We list a number of representative characteristics across the different analyses but caution against direct comparisons, because each study differs in objectives, methods, scope, base year, and assumptions. In addition, many key elements are not included in the table (e.g., health damages from SO_2 and morbidity). These are each complex research efforts, and the reader is directed to the original sources for a fair understanding.

We emphasize a key modeling limitation inherent in the following summary. To some degree, researchers in this field, as in many others, must trade off two interests: high resolution in analysis of the energy-to-exposure pathway versus broader approaches that include regional air dispersion and take into account economic feedbacks, so-called general equilibrium effects.

The high resolution of a "bottom-up" assessment allows a detailed analysis, for example, of specific power-generation technologies or of all emission sources in a city. Environmental health scientists increasingly emphasize that the health impacts of pollution exposure depend critically on not just small spatial scales, but also small temporal ones, such as the timing of daily activities. Such assessments require data resolved at fine grids, such as emissions at the plant level and population by small districts.

To date, a number of researchers have conducted bottom-up assessments to assess exposure and health in China on subnational geographical scales. The studies listed below include several conducted for municipalities or provinces from the bottom up. Although there are also large approximations in these models, and no research team

Table 1.2
International collaborations on health damage of air pollution in China

Name (Year)	Geographical Domain	Sectors	Air Pollutants	Types of Air Pollution Damages	Mortality PM Dose-Response Coefficient	Method(s) and Central Value of Statistical Life or Year of Life Lost	Central Estimate of Aggregate Effects
National studies							
World Bank (1997)	China	All, based on measured ambient concentrations	PM_{10}, SO_2, Pb	Mortality and morbidity (urban ambient and rural indoor air); also acid rain on crops, materials, ecosystems	6 deaths per 1 million people per μg/m^3 of PM_{10}	U.S. WTP scaled to China by GDP ratio: $31,762/life (rural), 60,000/life (urban); human capital method: $4,748/life (rural), 9,970/life (urban)	178,000 annual urban premature deaths from exceeding ambient standards, 111,000 from rural indoor air levels; equivalent to 7.1% of GDP in early 1990s using WTP method
ECON (2000)	China	Not applicable; study develops a model but does not apply it to data	PM_{10} or SO_2, O_3, Pb	Mortality and morbidity (ambient and indoor air); also acid rain and ozone on crops, materials	0.4 (0.0, 0.6) percent increase mortality per μg/m^3 of PM_{10}; equivalent to 24 deaths per 1 million	100× per capita GDP (based on Western ratios): $77,000/life	Not applicable; study develops a model but does not apply it to data

O'Connor et al. (2003)	China, based on Guangdong	All, 62-sector CGE economic model	PM, SO_2, NO_X, O_3, VOCs, CO_2	Mortality and morbidity (ambient air). Also O_3 on crops	2.2 (0, 4.1) deaths per 1 million adults per μg/m^3 of PM_{10}; 0.7 for infants	Taiwan WTP scaled to China by GDP ratio: $43,275/life (for all China), $68,973/life (for Guangdong only).	Net benefits of 0.05% of output due to a 10% carbon reduction
Hirschberg et al. (2003)	China, based on Shandong	All, based on electric power	PM, SO_2, NO_X, NH_4	Mortality and morbidity (ambient air). Also acid rain on crops, climate	1.57E-4 YOLL per year-person-μg/m^3 of PM_{10}	Western WTP scaled to China by PPP GNP: $430,000/life, $15,710/YOLL	1 million premature deaths annually from all air pollution; 9.1 million YOLL per year: equivalent to 6–7% GNP
Sector studies							
Wang and Smith (1999)	China	Electric power, households	PM_{10} (including SO_4^{-2}), CO^2	Mortality and morbidity (ambient and indoor air)	0.1 (0.04, 0.3) percent increase mortality per μg/m^3 of PM_{10}	24,000× average daily wage (based on U.S. ratio), adjusted by PPP: $123,000/life	75,400–122,500 annual premature deaths avoided from BAU in 2010, depending on scenario

Table 1.2
(continued)

Name (Year)	Geo-graphical Domain	Sectors	Air Pollutants	Types of Air Pollution Damages	Mortality PM Dose-Response Coefficient	Method(s) and Central Value of Statistical Life or Year of Life Lost	Central Estimate of Aggregate Effects
Feng (1999)	China, Korea, Japan	Electric power	TSP, $PM_{2.5}$, SO_2, SO_4^{-2}, NO^2, NO_3^-, CO, CO_2, CH_4, N_2O	Mortality and morbidity (ambient air); also multiple pollutants on crops forests, ecosystems, materials, visibility, climate	Multiple coefficients, for both chronic and acute mortality, and for different ages	U.S. WTP scaled to China by PPP GNP: $339,500/life	Not applicable; study estimates damages from a hypothetical power plant
Provincial study							
Aunan et al. (2004)	Shanxi	Industry, electric power, coking, households	PM_{10}, SO_2, CO_2	Mortality and morbidity (ambient air)	2.2 (0, 4.1) deaths per 1 million adults per μg/m^3 of PM_{10}; 1.2 (0.8, 1.7) for infants	100× per capita local GDP (based on Western ratios): $63,000/life	Not applicable, study estimates avoided deaths and savings of six technology options

City studies							
ECON et al. (2000)	Guangzhou	All sources in urban domain	PM_{10}, SO_2, NO_X, CO_2	Mortality and morbidity (ambient air); also acid rain on materials	2.2 (0, 4.1) deaths per 1 million adults per μg/m^3 of PM_{10}; 0.7 (0.4, 0.9) for infants	Taiwan WTP scaled to China by GDP ratio: 620,288 yuan/life	1,420 premature adult deaths annually from all air pollution; 110 premature infant deaths annually
Lvovsky et al. (2000)	Shanghai	Large industry, small industry, electric power, households, urban transportation	PM_{10}, SO_2, NO_X, CO_2	Mortality and morbidity (ambient air); also reduced visibility, soiling, materials, and climate	0.194 percent increase mortality per μg/m^3 of PM_{10}	U.S. WTP scaled to China by income ratio: $72,859/life	3,979 premature deaths annually from air pollution; equivalent to 5.5% local income
SAES et al. (2002)	Shanghai	All sources in urban domain	PM_{10}, SO_2, NO_X, CO_2	Mortality and morbidity (ambient air)	0.43 (0.26, 0.61) percent increase mortality per μg/m^3 of PM_{10} (long-term); 0.028 (0.01, 0.046) percent (short-term)	WTP by CV for Chongqing, income adjusted to Shanghai: US $108,500/life	647–5,472 avoided deaths in 2010 depending on energy scenario; 1,265–11,130 in 2020
Kan and Chen (2004)	Shanghai	All sources in urban domain; based on measured ambient concentrations	PM_{10}	Mortality and morbidity (ambient air)	0.43 (0.26, 0.61) percent increase mortality per μg/m^3 of PM_{10}	WTP by CV for Chongqing, income adjusted to Shanghai: US $108,500/life	4,780 premature deaths annually from air pollution; equivalent to 1.0% local GDP

Table 1.2
(continued)

Name (Year)	Geo-graphical Domain	Sectors	Air Pollutants	Types of Air Pollution Damages	Mortality PM Dose-Response Coefficient	Method(s) and Central Value of Statistical Life or Year of Life Lost	Central Estimate of Aggregate Effects
Wang and Mauzerall (2006)	Zaozhuang, and region	All sources in urban domain (biogenic in addition to anthro-pogenic)	$PM_{2.5}$, PM_{10} (including SO_4^{-2}, NO_3^-), SO_2, NO_X, CO, NH_3, VOCs	Mortality and morbidity (ambient air)	0.58 (0.2, 1.0) percent increase mortality per µg/m^3 of $PM_{2.5}$; 4.7E-4 YOLL per year-person-µg/m^3 of $PM_{2.5}$	WTP by CV for Chongqing, inflation adjusted: US $34,235/life	6,000 prema-ture deaths annually from air pollution; equivalent to 10% of local GDP
Peng et al. (2002)	Shijiazhuang	All sources in urban domain	PM_{10} (including SO_4^{-2})	Mortality and morbidity (ambient air)	6 deaths per 1 million people per µg/m^3 of PM_{10}; 0.08 percent increase mortality per µg/m^3 of PM_{10}	WTP by CV: $160,000/life	251 annual premature deaths from exceeding ambient stan-dards; equiva-lent to 4.3% of local GDP

Notes: Base years: We are unfortunately unable to list all of the base years of these assessments, required for careful comparisons across studies. The base years often differ by study component (e.g., emissions vs. valuation) and in many cases are unclear because they are not described clearly or at all in published reports. We reemphasize that this table is merely introductory and researchers should consult the original literature. General abbreviations: CGE, computable general equilibrium (economic model); GDP, gross domestic product; GNP, gross national product; PPP, purchasing power parity; µg/m^3, micrograms per cubic meter; YOLL, year of life lost; WTB, willingness-to-pay; CV, contingent valuation; and BAU, business as usual. Pollutant abbreviations: PM, particulate matter; and PM_{10} and $PM_{2.5}$, categories of PM under aerodynamic diameters of 10 and 2.5 microns; TSP, total suspended particulate matter; SO_2, sulfur dioxide; SO_4^{-2}, sulfate particles (of various chemical forms); NO_X, nitrogen oxides; NO_3^-, nitrate particles (of various forms); O_3, ozone; CO, carbon monoxide; Pb, lead; VOCs, volatile organic compounds; CO_2, carbon dioxide; CH_4, methane; and N_2O, nitrous oxide.

has modeled literally every single source of pollution in a jurisdiction, such studies can provide valuable guidance for local decision making.

There is also a pressing need, however, to understand effects at the regional or national level, especially for a full economic analysis. Some pollutants such as SO_2 travel hundreds of kilometers, and a comprehensive accounting of health damages requires a regional scope. More important, a consideration of only the direct local or industry-specific effects of damages and policies may miss out a larger total effect on the rest of the economy. For example, a policy that affects the electric power industry will change the price of electricity. This may affect the cost of making steel and cement needed to construct power plants, which in turn may lead to further changes in the demand for electric power. In general such policies would have ramifications across the entire economy today and, by influencing the rate of investment, would affect the economy in the future. A comprehensive benefit-cost assessment of national pollution control policy needs to consider not just the impact of policies on directly affected enterprises and sectors, but also the potentially large indirect effects such interventions may have throughout the economy both now and in the future.

We must recognize that localities in some circumstances can set their own pollution-control policies, but the Chinese government must also make environmental policy choices on a national basis. Without a truly massive data-gathering and research effort, it is impossible to conduct a bottom-up modeling assessment for the entirety of China, even if one were to accept the assumptions made by the teams below in studies of more limited domains. In practice, the setting of national environmental policies can be analyzed only using more abstract "top-down" models that take the national damages and economic interactions into account. It is to advance the very limited body of this type of analysis in China that this volume, structured around a national economic model with general equilibrium features, was conceived and conducted.

1.6.1 National Studies

China Research Academy of Environmental Sciences (CRAES 1999) The estimation of environmental losses in China can be traced to the end of the 1980s. Most of the first studies were modest in scale and scope, usually conducted for a geographical region or city. A report by CRAES (1999) lists a large number of these studies, some quite informal and all in Chinese. For instance, Guo and Zhang (1990) conducted one of the first national assessments of the losses of environmental and ecological damages. By use of human capital valuation methods and other approaches, they estimated an average annual loss of 38 billion yuan from 1981 to 1985, about 6.75% of GDP in 1983 (CRAES 1999). Air pollution was estimated at an average of 12.4 billion yuan, or 2.19% of GDP.

World Bank (1997) The first large-scale collaboration to apply methods common in international research to assess health damages of air (and water) pollution was led by the World Bank (1997). *Clear Water, Blue Skies* was a valuable initial effort to estimate impacts on a nationwide basis. Faced with data constraints but unable to conduct major field efforts, the team employed approximation methods that included borrowing key relationships in the emission-to-exposure pathway from analogous studies in Eastern Europe. It is impossible to quickly summarize the study's results, but for example it estimated 178,000 premature deaths per year from exceeding China's urban air-quality standards.

The World Bank (1997) employed two approaches to valuation of health effects. The first was based on "revealed preference" studies in the United States, transferred to China by linearly scaling to the ratio of GDP per capita. This yielded an estimated unit value of mortality in urban China of US$60,000, one factor in a subsequent estimate that air pollution costs China nearly US$50 billion per year, or 7.1% of GDP.[9] This result was controversial in China, where, among other reservations, experts generally preferred a second, more conservative, "human capital" form of health valuation. This alternative approach reduced the loss estimate to $20 billion per year, or 2.9% of GDP.

The World Bank study was an important first assessment and was updated in previous work of members of the current project (Ho, Jorgenson, and Di 2002, see below). As mentioned earlier, the effort of this volume was initiated in part to improve more extensively on World Bank 1997, with a more detailed analytical approach that is also rooted in more data gathered from fieldwork in China. To enhance the applicability of results, the assessment here is also designed to attribute damages differentially to emissions and concentrations modeled from key economic sectors. This allows a national-level damage attribution and sector prioritization that the World Bank approach cannot provide. We based our decision to focus initially on health impacts of PM and SO_2 in part on the overwhelming dominance of these damages over those of other environmental pathways (e.g., agricultural impacts and acid rain) in World Bank 1997.

ECON Centre for Economic Analysis (ECON 2000) The World Bank and SEPA subsequently initiated a program to test and revise the methods and estimates of

World Bank 1997, and eventually develop a common framework for annual damage assessment in China, on national and even provincial scales. Titled the "China Environmental Cost Model," it intends to assess a wide range of both air- and water-pollution costs. This encompassing initiative was nearing completion at the time of writing, and it cannot be cited or detailed here. One of the initial undertakings, however, was to commission consultants to develop a modeling framework that includes impacts on human health, materials, and agricultural production. This is described in the ECON 2000 report.

The ECON 2000 model framework focuses on PM_{10} and SO_2, taking into account their interrelationships, and also adds independent impacts of ozone and lead. For valuation, it reviews the worldwide literature and suggests a convenient formula that values a statistical life at 100 times the per capita annual GDP. The purpose of the study was to develop the model, not to apply it, and thus it did not yet attempt to generate national damage estimates in its report. Its comparison to World Bank 1997 is limited to constituent elements of the assessments, such as the number of morbidity cases per 1 million people per 1 $\mu g/m^3$ increase in ambient PM_{10}.

By applying directly the ratio of GDP per capita between China and the United States, the ECON model converted the unit economic loss from respiratory illness in the United States to the unit economic loss from the same illness in China. However, the calculated annual medical cost is three times higher than the actual costs in Beijing. Thus, a study of Tsinghua University et al. (2002) modified the model by taking one third of the unit economic loss from morbidity or death in the model as the unit economic loss from respiratory illness in Beijing.

A primary strength of the pending China Environmental Cost Model, based on ECON 2000, should be its breadth, assessing water quality in addition to air quality, and including lesser damages from additional pollutants and nonhealth pathways. It is also designed for relatively easy and routine application. The assessment of this volume differs in its underlying economic structure, distinguishing health damages by sector and integrating them directly into a model with general equilibrium features. This allows it to consider full impacts throughout the economy in assessing the costs and benefits of given policies. An additional difference is its more extensive consideration of population exposures.

Ho, Jorgenson, and Di 2002 This earlier analysis of health damages from local air pollution used information and estimates in World Bank 1997 and calculated national health damages caused by each industry. It examined the costs and benefits of national green tax policies, the effects on both emissions and the economy. This study noted the very high damages estimated for the transportation sector, something that had not been greatly emphasized previously.

O'Connor et al. 2003 This is a co-benefits study using a two-region economic model (in the jargon, a computable general equilibrium model, or CGE; "co-benefits" refers to the association of both global and local environmental benefits of emission control). It considers air-dispersion characteristics in Guangdong province in relative detail and then extrapolates them to the rest of China. O'Connor et al. employed two dispersion models, both a simple "stack-height-differentiated" one as in Ho, Jorgenson, and Di 2002 for primary pollutants and health impacts and another for estimating formation of ozone, a secondary pollutant that damages crop productivity. It asked how much the welfare costs of an energy tax for 5–30% carbon reductions in 2010 from a baseline might be offset by benefits to health and agriculture.

One result of this study that is especially noteworthy for its divergence from conventional wisdom is its conclusion that pollution damages to agriculture (from ozone in particular) may outweigh pollution effects on health (from particulates and SO_2) in economic terms. It thus asserts that damage avoidance in agriculture provides a larger proportion of the ancillary benefits of emission reductions.

O'Connor et al.'s incorporation of the damage estimations in a CGE model to consider the economywide effects of emission-control policies makes it closest of all the studies to the integrated analysis in our program reported in chapter 10. It has an advantage in its consideration of agricultural impacts, whereas chapter 10 uses our team's more elaborate estimate of pollution dispersion and exposure.

Hirschberg et al. 2003 A study by a Swiss-Chinese-American team conducted a detailed cost assessment of ambient air pollution from the power sector in Shandong province, extended in unspecified manner to the economy of China. Capturing local features that the high resolution of its provincial scope would allow, this "full scope, bottom-up" effort applied the EcoSense model to Shandong power sector sources and scenarios into the future. This model packages atmospheric transport and chemical conversion, population-based exposures, dose-response relationships, and valuation models, the latter two adapted from industrialized countries.

Among Hirshchberg et al.'s conclusions are that, extrapolated to all of China, roughly one million Chinese die prematurely each year of ambient air pollution from all sources, with costs corresponding to 6–7% of GDP. Health costs, particularly mortality, dominate other damages, such as to crops. A unique contribution of this study is noting a strong influence of agriculturally produced ammonia on health damages, because of its central role in chemical transformation of SO_2 into sulfate particles. They conclude that the major part of damage to health is caused by these secondary pollutants. The ultimate focus of this study was estimating external costs of electric power for use in analyzing power-sector technology scenarios in the future.

1.6.2 Sector Studies

Wang and Smith 1999 These researchers conducted a national health damage analysis for two sectors, electric power and households. This was a co-benefits study, the aim of which was analysis of four energy scenarios for China to achieve 10% and 15% reductions of greenhouse gases by 2010 and 2020, respectively. They compared technology pathways to reach least-cost reductions in both global warming potential and unit dosages of SO_2 and particulates, which were then translated into changes in health risk. Finally, the team estimated marginal net economic costs of achieving GHG targets by alternative coal use in these sectors, calculated from valuation of health benefits and incremental costs of energy changes. Health valuation was adapted from U.S. figures, which were adjusted by relative wage levels and purchasing power in China.

The results spotlighted enormous averted health damages as a co-benefit of achieving a GHG target, particularly through fuel substitution in the household sector. For instance, central estimates of avoided premature deaths in 2020 of the three scenarios over business as usual ranged from 154,400 to 185,200, against a projected total mortality in that year of 14 million. The study was designed to compare health co-benefit levels of different ways to achieve GHG reduction, not to calculate aggregated health damages of air pollution.

Feng 1999 This study, a published Ph.D. dissertation, assessed a hypothetical, characteristic power plant, at two possible locations in China. It merits inclusion here for its ambitiously comprehensive scope. The study modeled emissions and concentrations of a wide range of primary and secondary pollutants, both chronic and acute mortality and morbidity effects, and additional impacts on crops, forests, ecosystems, materials and visibility and included some damages in neighboring countries. It also considered emission of several GHGs, and attributed costs of climate change based on estimates from global studies. For valuing health effects, it translated results of WTP studies from other countries and adjusted them for population age groups. Feng calculated the total environmental damages of the plant without pollution control and then evaluated the cost-effectiveness of current and prospective policies.

1.6.3 Provincial Study

Aunan et al. 2004 This project had a similar co-benefits objective and approach as Wang and Smith 1999 but applied to the single province of Shanxi and not to China as a whole. They posed a number of options to abate emissions of SO_2, particulates, and CO_2 in industry, power generation, and rural households. These included precombustion options of coal washing and coal briquetting, and combustion options of improved management, boiler replacements, cogeneration, and modified boiler designs.

Aunan et al. 2004 made conservative estimates of health impacts of these prospective interventions by focusing on eighteen cities in Shanxi for which data were available. Valuation of premature deaths avoided by adoption of each alternative used a rule of thumb of 100 times per capita local annual GDP, drawn from the analogous ratios resulting in Western studies. The study concluded that all six options are profitable in a socioeconomic sense. The least-cost avoidance of health damages, however, happened to be the most costly form of carbon abatement, which indicates a divergence of priorities depending on local or global environmental objectives.

1.6.4 City Studies

ECON et al. 2000 The aforementioned research by ECON 2000 was incorporated into "Guangzhou Air Quality Action Plan 2001," which was conducted by Norwegian and Guangzhou researchers to advise a package of least-cost pollution control options that would meet SO_2, NO_X, and TSP targets already set by municipal officials. Assessment of a number of technology and policy options yielded cost curves for SO_2 and NO_X control that demonstrated the relative ease of meeting the former (along with TSP, which is assessed somewhat differently) versus the latter.

The study developed its own comprehensive emission inventory on the basis of a large survey and estimated gridded pollutant concentrations with a dispersion modeling package. It found encouraging correspondence with measured levels. Combined with population data, the resolution of the model allowed the team to distinguish exposure characteristics of different pollutants and sources. Modeling health effects of given control options was then methodologically straightforward in such a comprehensive, "bottom-up" model. The study employed WTP valuation from research in Taiwan, scaled to China by per capita GDP.

Lvovsky et al. 2000 In this study, a World Bank–based team developed a "rapid assessment model" to estimate environmental damages and attribute them to various sources. It investigated six cities around the world, including Shanghai. The damages it assessed included health, reduction in visibility, soiling, material damages, and climate change.

The spreadsheet-based methodology of this study was purposely simple. Using a dispersion model that aggregated large sources in the center of a city and dispersed pollutants equally in all directions, for instance, the model was designed for quick use to generate a rough characterization of damages in data-limited contexts. Valuation of a statistical life was based on U.S. WTP scaled to China by income ratio. By use of these methods, health damages from air pollution in Shanghai totaled $730 million in 1993 dollars, 5.5% of Shanghai's income. Chiefly because of exposure characteristics, health damages were deemed greatest from small stoves and boilers in households and industry in Shanghai.

Shanghai Academy of Environmental Science (SAES) et al. 2002 Another study of Shanghai was supported by the China Council for International Cooperation on Environment and Development, SEPA, the Shanghai Environmental Protection Bureau, U.S. EPA, and others. It estimated health impacts of air pollution in 2010 and 2020, under a number of energy scenarios diverging from a base case. The study employed an atmospheric model originally designed to investigate acid rain that was adjusted to an urban scale. It estimated health effects of PM_{10} exposures, including acute and chronic mortality, chronic bronchitis, and several others, by use of a mix of Chinese and Western epidemiology. SAES et al. monetized premature mortality using a simple CV study by Wang et al. (2001, from an early study of Chongqing by our Harvard University China Project, with multiple collaborators). For morbidity endpoints, the study applied income-adjusted values from the West or "cost-of-illness" estimates. The project then reported the total economic damage of its scenarios, but not as percentage of GDP. An associated study of Beijing has now taken place, but a full report was unavailable at the time of writing (Tsinghua University et al. 2005).

Kan and Chen 2004 Health scientists from Fudan University who participated in SAES et al. 2002 above followed with another assessment of Shanghai. Rather than model emissions and air dispersion, they evaluated the effect of measured annual average daily PM_{10} for 2001 from concentrations observed at a background station. Applying the same concentration-response coefficients and valuations as the prior study, they attributed nearly 11% of annual adult deaths to the incremental PM_{10}. Premature mortality comprised nearly 83% of the estimated aggregate economic damage, 1% of Shanghai local GDP. This brief study concluded with a valuable discussion of uncertainties for consideration in future research.

Wang and Mauzerall 2006 This study examined impacts on health of particulate matter from sources in the city of Zaozhuang, Shandong. One of its primary strengths was its comparatively sophisticated treatment of atmospheric transport, chemical transformation, and deposition. After constructing an inventory of both anthropogenic and biogenic emissions, the study used the Community Multiscale Air-Quality Modeling System (CMAQ) to simulate ambient concentrations of primary and secondary particulates ($PM_{2.5}$) across a multiprovince domain. It investigated the regional health damages from Zaozhuang emissions both in 2000 and in 2020 under business-as-usual and two alternative technology scenarios. Unlike many studies that focus on acute mortality, this one included chronic effects. It calculated premature mortality by using two methods and then applied the Wang et al. 2001 CV results to value the damages. The result was an estimate of damages equivalent to 10% of Zaozhuang's GDP in 2000, a proportion growing to 16% by 2020 absent interventions like the modeled technology options.

Peng et al. 2002 This study examined health damages of sulfate particulates in the city of Shijiazhuang, by using an urban-scale, high-resolution, atmospheric puff model that featured secondary chemistry and took account of background concentrations. For valuation, it employed results of an unpublished CV study by Li, Schwartz, and Xu (1998). Peng et al. concluded that mortality and morbidity costs of sulfate amount to more than 4% of urban GDP in the year 2000, and their assessment of control options spotlighted the cost-effectiveness of adopting low-sulfur coal.

1.7 Conclusion

Two observations are in order to conclude this introduction and review. It is clear that China's growing economy and dependence on fossil fuels, especially coal, have engendered a number of serious air-quality problems and that these have substantial impacts on human health. China has made considerable strides in addressing some of the most critical types of air pollution, as evidenced by declining measured concentrations of SO_2 and particulates in urban areas through the 1990s. Few living in China's cities through this decade would dispute this. This advance has undoubtedly benefited public health. It is partly testimony, however, to how severe air quality was to begin with, and conditions remain worse than official standards in many places. Moreover, despite the progress, the positive trends have recently leveled off (TSP) or reversed (SO_2 and NO_X), possibly sharply, as the economy continues to boom. Newer forms of pollution, such as photochemical smog, are also now becoming serious problems in a number of cities and their downwind regions.

What is also clear is that there is no shortage of interest in China's air pollution. Aside from public and official concern, a body of research is developing to better understand both the nature of the hazard itself and the damages that it can cause, to human health, to the local economy, and to the global environment. The research reported in this volume is one extensive effort to assess the costs of pollution and to incorporate the results in tools to inform policy, but it is one of many studies to date on this set of issues. The next chapter provides a brief summary of the results of our research collaborations for such policy inferences and sets the stage for the full reporting of our assessment in the second part of the book.

Acknowledgments

The research of this chapter was generously supported by the Harvard Kernan Brothers Fellowship; a grant from the China Sustainable Energy Program of the Energy Foundation to the Harvard University China Project and Tsinghua University Institute of Environmental Science and Engineering; and a grant from the V. Kann Rasmussen Foundation to the Harvard China Project. These funding sources are gratefully acknowledged. We also thank three anonymous referees for their review comments and Shuxiao Wang for excellent suggestions on final revisions.

Notes

1. Intensity measures include both narrow concepts like tons of coal per ton of steel and broader ones like total primary energy per yuan of GDP.

2. Coal transformed to coke is counted as industrial consumption.

3. Specifically, the balance terms—"coal available for consumption" minus "coal consumption"—have magnitudes of tens of millions of tons over 1990–1998 (33–42 Mt in 1990–1995, 61–75 Mt in 1996–1998; NBS 2001, table 4.5) and rise suddenly to hundreds of

millions of tons for 1999–2001 before dropping back to the previous scale in 2002–2003 (228 Mt, 264 Mt, 278 Mt, 70 Mt, and 58 Mt, respectively; NBS 2005, table 4.5). We note another indication of data problems for this period that our team happened upon. Adding up 1999 coal consumption listed *by province* in table 5.16 of NBS 2001 yields a total that exceeds by 165 million tons the oft-cited *national* coal consumption listed in table 4.5 of the same source. Such inconsistencies also occur in prior years, but on a much smaller scale.

4. As we will explain later, we do not consider severe indoor air pollution problems in rural China, which pose a major additional threat to public health. Considerable air pollution information is now conveniently available in the online version of the *State of the Environment in China*, published by SEPA, at www.zhb.cn.

5. We do this in chapter 4.

6. The term used for combustion emissions in the official books, *yanchen*, literally means "smoke dust," whereas *fenchen* is the dust from production processes.

7. The *Nature* article was overinterpreted in some international media coverage, for instance suggesting it indicated that Beijing had become the "air pollution capital of the world" (Watts 2005). These largely ignored that the data covered the full lower atmosphere over a region containing many large cities, including Shanghai, Tianjin, and Jinan and that they concerned only NO_2 and not other air pollutant forms believed to play a larger role in total population health risk.

8. Our program has conducted some research on indoor air pollution in earlier phases, both in rural and urban settings. These included a multidisciplinary social survey on the human dimensions of air quality and environmental policy implementation in rural Anhui (Alford et al. 2001), coupled with support for epidemiological research on indoor air pollution in the same area (Venners et al. 2001).

9. This does not include the valuation of damages from water pollution, which together with the air pollution categories add up to the 7.7% of GDP noted in the report (World Bank 1997).

References

Alford, W. P., and Y. Y. Shen. 1998. Limits of the law in addressing China's environmental dilemma. *Stanford Environmental Law Journal* 16 (1):125–148.

Alford, W. P., R. P. Weller, L. Hall, K. R. Polenske, Y. Y. Shen, and D. Zweig. 2001. The human dimensions of environmental policy implementation: Air quality in rural China. *Journal of Contemporary China* 11 (3):495–513.

Aunan, K., J. Fang, G. Li, H. Vennemo, K. Oye, and H. M. Seip. 2004. Co-benefits of climate policy: Lessons learned from a study in Shanxi, China. *Energy Policy* 32:567–581.

Bennett, D. H., T. E. McKone, J. S. Evans, W. M. Nazaroff, M. D. Margni, O. Jolliet, and K. R. Smith. 2002. Defining intake fraction. *Environmental Science and Technology* 36:206–211.

Chinese Research Academy of Environmental Science (CRAES). 1999. *Handbook of Sustainable Development: Index System of Urban Environment*. Beijing: Chinese Environmental Press. In Chinese.

Eckholm, Erik. 2001. China said to sharply reduce emissions of carbon dioxide. *New York Times*, June 15, late edition-final, A1.

ECON Centre for Economic Analysis. 2000. An environmental cost model. ECON report no. 16/2000. May. Oslo, Norway.

ECON Centre for Economic Analysis, Norwegian Institute for Air Research, Institute for Energy Technology, Centre for International Climate and Energy Research, Guangzhou Research Institute of Environmental Protection, Guangzhou Environment Monitoring Center, Guangzhou Environmental Protection Bureau, Guangzhou Environmental Supervision Institute. 2000. Guangzhou air quality action plan 2001. ECON report no. 9/2000. February. Oslo, Norway.

Energy and Environment Directorate (EED), Lawrence Livermore National Laboratory. 2001. U.S. Energy Flow Trends. December. Available at http://eed.llnl.gov/flow/00flow.php/.

European Commission (EC). 1995. *ExternE: External Costs of Energy*, vol. 4: *Oil and Gas*. Brussels: EC Directorate-General XII, Science, Research, and Development.

Feng, Therese. 1999. *Controlling Air Pollution in China: Risk Valuation and the Definition of Environmental Policy*. Cheltenham: Edward Elgar.

Fisher-Vanden, Karen, Gary H. Jefferson, Hongmei Liu, and Quan Tao. 2004. What is driving China's decline in energy intensity? *Resource and Energy Economics* 26:77–97.

Florig, H. K., G. D. Sun, and G. J. Song. 2002. Evolution of particulate regulation in China: Prospects and challenges of exposure-based control. *Chemosphere* 49 (9):1163–1174.

Fridley, D. G., J. Sinton, F. Q. Zhou, B. Lehman, J. Li, J. Lewis, and J. M. Lin, eds. 2001. *China Energy Databook 5.0*. Lawrence Berkeley National Laboratory, China Energy Group, LBNL-47832. Berkeley, Calif.: LBNL.

Fridley, D., J. Sinton, and J. Lewis. 2003. Working out the kinks: Understanding the fall and rise of energy use in China. Lawrence Berkeley National Laboratory. 5 March. Unpublished document.

Garbaccio, R. F., M. S. Ho, and D. W. Jorgenson. 1999a. Controlling carbon emissions in China. *Environment and Development Economics* 4:493–518.

Garbaccio, R. F., M. S. Ho, and D. W. Jorgenson. 1999b. Why has the energy output ratio fallen in China? *Energy Journal* 20 (3):63–91.

Garbaccio, R. F., M. S. Ho, and D. W. Jorgenson. 2000. The health benefits of carbon control in China. In *Proceedings of Workshop on Assessing the Ancillary Benefits and Costs of Greenhouse Gas Mitigation Strategies*. Organization for Economic Co-operation and Development, 27–29 March, Washington, D.C.

Guo, X. M., and H. Q. Zhang. 1990. *Study on Prediction and Strategy of China's Environment for the Year 2000*. Beijing: Tsinghua University Press. In Chinese.

Hao, J. M., H. Z. Tian, and Y. Q. Lu. 2002. Emission inventories of NO_X from commercial energy consumption in China, 1995–1998. *Environmental Science and Technology* 36 (4):552–560.

Harrison, K., D. Hattis, and K. Abbat. 1986. Implications of chemical use for exposure assessment: Development of an exposure-estimation methodology for application in a use-clustered priority setting system. Report Number CTPID 86-2, for the U.S. EPA. Cambridge, Mass.: MIT Center for Technology Policy and Industrial Development.

Hirschberg, S., T. Heck, U. Gantner, Y. Q. Lu, J. V. Spodaro, W. Krewitt, A. Trukenmuller, and Y. H. Zhao. 2003. Environmental impact and external cost assessment. In *Integrated Assessment of Sustainable Energy Systems in China*, ed. B. Eliasson and Y. Y. Lee. Dordrecht: Kluwer.

Ho, M. S., D. W. Jorgenson, and W. H. Di. 2002. Pollution taxes and public health. In *Economics of the Environment in China*, ed. Jeremy J. Warford and Li Yining. Boyds, Maryland: Aileen International Press.

Intergovernmental Panel on Climate Change (IPCC). 2001. *Third Assessment Report*. Cambridge: Cambridge University Press.

Kan, Haidong, and Bingcheng Chen. 2004. Particulate air pollution in urban areas of Shanghai, China: Health-based economic assessment. *Science of the Total Environment* 322:71–79.

Levy, J. I., J. K. Hammitt, Y. Yanagisawa, and J. D. Spengler. 1999. Development of a new damage function model for power plants: Methodology and applications. *Environmental Science and Technology* 33:4364–4372.

Li, J., J. Schwartz, and X. Xu. 1998. Health benefits of air pollution control in Shenyang, China. Research report, Harvard School of Public Health, Boston.

Lvovsky, K., G. Hughes, D. Maddison, B. Ostro, and D. Pearce. 2000. Environmental costs of fossil fuels: A rapid assessment method with application to six cities. World Bank Environment Department Paper no. 78. October. Washington, D.C.

Morgenstern, R., R. Anderson, R. Greenspan Bell, A. Krupnick, and X. Zhang. 2002. Demonstrating emissions trading in Taiyuan, China. *Resources* 148:7–11.

National Bureau of Statistics (NBS). 2001. *China Energy Statistical Yearbook*, 1997–1999. Beijing: China Statistics Press. In Chinese.

National Bureau of Statistics (NBS). 2004. *China Energy Statistical Yearbook*, 2000–2002. Beijing: China Statistics Press. In Chinese.

National Bureau of Statistics (NBS). 2005. *China Energy Statistical Yearbook*, 2004. Beijing: China Statistics Press. In Chinese.

National Bureau of Statistics (NBS). 2006. *China Statistical Yearbook*, 2005. Beijing: China Statistics Press. In Chinese.

Peng, C. Y., X. D. Wu, G. Liu, T. Johnson, J. Shah, and S. Guttikunda. 2002. Urban air quality and health in China. *Urban Studies* 39 (12):2283–2299.

Policy Research Center of Environment and Economy (PRCEE) of SEPA, China National Environmental Monitoring Center (CNEMC), and Chinese Academy of Environmental Sciences (CRAES). 2001. New countermeasures for air pollution control in China. Background report for World Bank 2001.

Oak Ridge National Laboratory and Resources for the Future (ORNL and RFF). 1994. *Estimating Fuel Cycle Externalities: Analytical Methods and Issues*. Washington, D.C.: McGraw-Hill/Utility Data Institute.

O'Connor, D., F. Zhai, K. Aunan, T. Berntsen, and H. Vennemo. 2003. Agricultural and human health impacts of climate policy in China: A general equilibrium analysis with special reference to Guangdong. Technical Paper Series No 206. OECD Development Centre. Paris, France. March.

Rawski, T. G. 2001. What is happening to China's GDP statistics? *China Economic Review* 12 (4):347–354.

Richter, A., J. P. Burrows, H. Nuss, C. Granier, and U. Niemeier. 2005. Increase in tropospheric nitrogen dioxide over China observed from space. *Nature* 437 (Sep. 1):129–132.

Richter, A., F. Wittrock, and J. P. Burrows. 2006. SO_2 retrieval from GOME and SCIAMACHY nadir measurements. Presented at Atmospheric Science Conference of the European Space Agency, May 10, Frascati, Italy.

Rowe, R. D., C. M. Lang, L. G. Chestnut, D. A. Latimer, D. A. Rae, S. M. Bernow, and D. E. White. 1995. *The New York Electricity Externality Study*, vol. I: *Introduction and Methods*. New York: Empire State Electric Energy Research Corporation.

Schmidt, C. W. 2002. Economy and environment: China seeks a balance. *Environmental Health Perspectives* 110 (9):A517–522.

Shanghai Academy of Environmental Sciences (SAES), Fudan University School of Public Health, National Renewable Energy Laboratory, and World Resources Institute. 2002. The integrated assessment of energy options and health benefit. Report of the Integrated Environmental Strategies Program of the U.S. EPA, China Council for International Cooperation in Environment and Development, and World Resources Institute. December. Available at http://www.epa.gov/ies/shanghai.htm/.

Sinton, J. E. 2001. Accuracy and reliability of China's energy statistics. *China Economic Review* 12 (4):347–354.

Sinton, J. E., and D. G. Fridley. 2000. What goes up: Recent trends in China's energy consumption. *Energy Policy* 28:671–687.

Sinton, J., D. Fridley, J. M. Lin, J. Lewis, N. Zhou, and Y. X. Chen, eds. 2004. *China Energy Databook 6.0*. Lawrence Berkeley National Laboratory, China Energy Group, LBNL-55349. Berkeley, Calif.: LBNL.

Smith, K. R. 1993. Fuel combustion, air pollution, and health: The situation in developing countries. *Annual Review of Energy and the Environment* 18:529–566.

State Environmental Protection Administration (SEPA). 1986–2004. *China Environment Yearbook*. Beijing: Chinese Environmental Yearbook, Inc. In Chinese.

State Statistical Bureau (SSB). 2000. *China Statistical Yearbook 1997–1999*. Beijing: China Statistical Press. In Chinese.

Streets, D. G., N. Y. Tai, H. Akimoto, and K. Oka. 2000. Sulfur dioxide emissions in Asia in the period of 1985–1997. *Atmospheric Environment* 34 (26):4413–4424.

Streets, D. G., K. J. Jiang, X. L. Hu, J. E. Sinton, X. Q. Zhang, D. Y. Xu, M. Z. Jacobson, and J. E. Hansen. 2001. Recent reductions in China's greenhouse gas emissions. *Science* 294 (November 30):1835–1837.

Streets, D. G., T. C. Bond, G. R. Carmichael, S. D. Fernandes, Q. Fu, D. He, Z. Klimont, S. M. Nelson, N. Y. Tsai, M. Q. Wang, J.-H. Woo, and K. F. Yarber. 2003. An inventory of gaseous and primary aerosol emissions in Asia in the year 2000. *Journal of Geophysical Research* 108 (D21):8809.

Tsinghua University, Beijing Urban Planning and Design Research Institute, Beijing Academy of Environmental Protection, Beijing Institute of Labour Protection Science, China Research

Academy of Environmental Science. 2002. Study on the strategy and countermeasures of air pollution control in Beijing. Technical report of project funded by the Ministry of Science and Technology, People's Republic of China. Beijing. In Chinese.

Tsinghua University Department of Environmental Science and Engineering, Peking University School of Public Health, Yale University School of Public Health, and National Renewable Energy Laboratory. 2005. Energy options and health benefit: Beijing case study: Executive summary. Report of the Integrated Environmental Strategies Program of the U.S. EPA and the State Environmental Protection Agency of China. November. Available at http://www.epa.gov/ies/beijing.htm/.

U.S. Embassy. 2001. The controversy over China's reported falling energy use. Report of the Environment, Science, and Technology section. August. See also http://www.usembassy-china.org.cn/sandt/energy_stats_web.htm/.

U.S. Environmental Protection Agency (U.S. EPA). 1999. *The Benefits and Costs of the Clean Air Act: 1990 to 2010*. Washington, D.C.: U.S. EPA Office of Air and Radiation and Office of Policy.

U.S. Energy Information Agency (U.S. EIA). 2003. *International Energy Annual 2001*. Washington, D.C.: U.S. Government Printing Office. Available at http://www.eia.doe.gov/iea/.

Venners, Scott A., Binyan Wang, Jiatong Ni, Yongtang Jin, Jianhua Yang, Zhian Fang, and Xiping Xu. 2001. Indoor air pollution and respiratory health in urban and rural China. *International Journal of Occupational and Environmental Health* 7 (3):173–181.

Venners, S., B. Wang, Z. Peng, Y. Xu, L. Wang, and X. Xu. 2003. Particulate matter, sulfur dioxide and daily mortality in Chongqing, China. *Environmental Health Perspectives* 111 (4):562–567.

Wang, B. Y., Z. G. Peng, X. B. Zhang, Y. Xu, H. J. Wang, G. Allen, L. H. Wang, and X. P. Xu. 1999. Particulate matter, sulfur dioxide, and pulmonary function in never-smoking adults in Chongqing, China. *International Journal of Occupational and Environmental Health* 5 (1):14–19.

Wang, H., J. Mullaly, D. Chen, L. Wang, and R. Peng. 2001. Willingness to pay for reducing the risk of death by improving air quality: A contingent valuation study in Chongqing, China. Conference paper presented at The Third International Health Economic Association Conference, July 22–25, York, U.K.

Wang, X. D., and K. R. Smith. 1999. Near-term health benefits of greenhouse gas reductions: A proposed assessment method and application in two energy sectors of China. World Health Organization report no. WHO/SDE/PHE/99.01. March. Geneva.

Wang, Xiaoping, and Denise L. Mauzerall. 2006. Evaluating impacts of air pollution in China on public health: Implications for future air pollution and energy policies. *Atmospheric Environment* 40 (9):1706–1721.

Watts, Jonathan. 2005. Satellite data reveal Beijing as air pollution capital of the world. *Guardian*, October 31. Available at http://www.guardian.co.uk/international/story/0,3604,1605041,00.html/.

World Bank. 1997. *Clear Water, Blue Skies: China's Environment in the New Century*. Washington, D.C.: World Bank.

World Bank. 2001. China. *Air, Land, and Water: Environmental Priorities for a New Millennium.* Washington, D.C.: World Bank.

World Health Organization (WHO). 2002. Air quality guidelines for Europe, 2nd ed. WHO regional publications, European series, no. 91. Copenhagen. Available at http://www.euro.who.int/document/e7192.pdf/.

World Health Organization (WHO). 2005. WHO air quality guidelines global update 2005. Report on a working group meeting, October 18–20, Bonn, Germany.

Xu X. P. 1998. Air pollution and its health effects in urban China. In *Energizing China: Reconciling environmental protection and economic growth*, ed. M. B. McElroy, C. P. Nielsen, and P. Lydon. Cambridge, Mass.: Harvard University Committee on Environment/Harvard University Press.

Xu, Z. Y., D. Q. Yu, L. B. Jing, and X. P. Xu. 2000. Air pollution and daily mortality in Shenyang, China. *Archives of Environmental Health* 55 (2):115–120.

Zhang, Qiang. 2006. Study on Regional Fine PM Emissions and Modeling in China. Ph.D. dissertation, Tsinghua University. In Chinese.

2

Summary for Policy

Chris P. Nielsen and Mun S. Ho

2.1 Introduction

Every society faces constraints on the physical and human resources it can commit to public purposes. Like it or not, there will be some limit to the yuan available in the collective social budget for protection of China's environment. Opinions will differ on what budget level is justified and on the true costs of different environmental policy options. Nevertheless, all can presumably agree that any yuan committed to environmental protection should be deployed as efficiently as possible, to maximize its effect.

In considering air-quality policy, we might therefore pause first to clarify the objective. What concerns us most is not, ultimately, pollution itself, but rather the damage it causes. This is a subtle but important distinction for policy, because every microgram of particulate matter in the air, or of any other pollutant, is not in fact equal in its likelihood to bring about harm. If we want to use the pollution-control budget wisely and efficiently—whatever the total level that has been decided—we might try to design policies accordingly, to differentiate and control emissions that cause the most damage, taking into account the relative costs of control.

As mentioned in the first chapter, although the damages from air pollution to agriculture, forests, and buildings can be large, the damage to human health usually dominates and should arguably be the foremost concern. The study of health damages and mitigation options can be usefully done at a number of scales: at the plant level, at the level of a city, or at the national level. In this volume, we are interested in the health damages of air pollution on a national scale in China. Specifically, we seek to identify the sources of damage by sector: for instance how the damages of emissions from electric power generation compare to those from the iron and steel industry, the chemical industry, and so forth. This is designed to inform pollution-control strategies at the national scale, by indicating which sectors pose the greatest total air pollution risk to public health. To date, there is little information beyond conventional wisdom attributing national health risk to sector sources. The reader will see that this assessment of China yields some unexpected results, challenging some of the usual assumptions about the worst pollution sources.

Because there is a strong policy interest in considering such damages in economic terms, our project also translates the estimates of sector health damages into monetized values, for instance valuing additional cases of chronic bronchitis in yuan. We then incorporate this information into our national economic model of China. This provides a tool for a number of unique analyses. Our team can consider not only how policy interventions may limit growth in emissions, reduce health damages, and affect target sectors, but also consider the important question of how those policies will reverberate throughout China's national economy. This permits a fuller understanding of the costs and benefits of such policies than most analyses to date have done. More specifically, our integrated analysis allows us to estimate the health benefits of environmental protection policies and, at the same time, the costs in terms of lower output and consumption.

We will examine two market-based tax policies to reduce air pollution and estimate their effects on economic performance, over both the short and long runs. Moreover, our results provide important insights into not only the ramifications of environmental policies, but also other pressing fiscal policy questions facing the government, including tax reform. For instance, pollution-tax revenue could be used to reduce highly distortionary existing taxes or diverted to investment, which would lead to higher future economic growth.

We remind the reader that this book is structured to appeal to several readerships, as described in "How to Read This Book" in section 1.2 of chapter 1 (which some may want to reread). This chapter—a brief summary of the results and conclusions of the integrated assessment reported in the book—is written specifically for non-expert and lay readers most interested in the policy implications of the results of our research, and who may not have the time to consider carefully the methodological issues in generating these results.

This summary for policy is largely an abbreviated and simplified version of the summary for research of chapter 3, which is also written for the general reader but presents results and policy conclusions in the full context of the research methods that produced them. After reading the current chapter, those focused on policy whose interest is piqued will find chapter 3 accessible and valuable. It can help them appreciate the assumptions and uncertainties behind such analyses, their strengths and weaknesses for advising policy, and how support for data and research might itself rank as a policy priority.

For more expert readers interested in the detailed studies of our program reported in chapters 4–10, we advise them to skip the simplified summary presented here and instead begin with the much fuller version, in chapter 3 "Summary for Research."

2.2 Results

The research components of our assessment are described in chapter 1, section 1.5. Following what is called the "pollution-causal chain," we summarize these as the following steps in sequence:

- From economic activity to energy use to pollutant emissions to concentrations in the air to human exposures (section 2.2.1);
- From exposures to health impact and their economic value (section 2.2.2);
- National damage assessment (section 2.2.3); and
- Benefit-cost analyses of policies to reduce emissions (section 2.2.4).

The bulk of this summary will concern the results and conclusions of the integrated national damage assessment and our benefit-cost analyses of two emission-reduction policies. We must first lay a little groundwork from the earlier links in this chain to facilitate understanding of these policy implications, with a few intermediate policy insights along the way.

2.2.1 From Economic Activity to Human Exposure to Pollutants

The assessment is structured around an economic model of China featuring thirty-four distinct sectors that traces the evolution of the economy over time. With this representation of the economy, our team can generate the output of each sector for each period and with it the use of fossil fuels. The combustion of fuels used in the production of goods generates emissions of pollutants that we want to investigate. We focus on the pollutants that are likely to have the greatest impact and for which data are available: particulate matter (PM), sulfur dioxide (SO_2), and carbon dioxide (CO_2).

With emissions of key air pollutants determined by the model for each sector, we must evaluate how they are transported and transformed in the air and then reach human populations. (This does not include the greenhouse gas [GHG] CO_2, of which we will estimate only emissions under different policies, not damages.) Doing this for all sources in China is, of course, impossible, so we must find ways to approximate these processes reasonably, focusing on the dominant sector contributors of these pollutants: electric power, chemicals, iron and steel, cement, and transportation. For the other sectors that are smaller contributors to ambient pollution levels, we use a simpler procedure that is derived from our detailed work on these major industries.

Although it is clear that different sectors produce different levels of emissions, it may be less obvious that, for instance, each ton of particulates emitted from different sectors may cause different amounts of health damage. There are a variety of possible factors influencing this, including the proximity of sources to population centers, high smokestacks helping to diffuse pollution into the air, and different distribution of particle sizes affecting their settling rates, among others. This differentiation of the exposure characteristics of sectors is a central aim in our assessment and a dimension by which our project differs from other major pollution studies in China to date.

To evaluate the emission-to-exposure pathway, we use a research concept developed by health scientists called "intake fraction." As introduced in chapter 4 by Jonathan I. Levy and Susan L. Greco of the Harvard School of Public Health, one may think of the intake fraction from a given source as the amount of a pollutant emitted that is eventually inhaled by people before it dissipates from the atmosphere. An intake fraction value can be interpreted simply as: for every metric ton (10^6 grams) of a pollutant emitted, x grams will be inhaled by the population exposed to the source. The intake fraction concept has been developed for precisely the problem at hand: to conduct informative, if approximate, health-risk assessments given limited input data and research resources, as is the case in China and most developing countries. The scarcity of data, we should emphasize, is a primary reason that researchers must resort to such approximating steps in China.

How does one use the intake fraction for, say, SO_2? For a particular factory, multiplying it by the annual emissions of SO_2 will give the annual dosage of the pollutant. Knowing the dosage, we can estimate the health effects using the steps described in the next section. If one knows the average intake fraction for the entire cement industry, multiplying it by the total annual emissions of SO_2 from that industry gives an estimated total dosage, and from this one can estimate annual health damages due to cement production.

To determine these intake fractions, we first had to collect extensive field data (emission characteristics, meteorology, and population) from sources in our five sectors and then model the dispersion of the pollutants in the atmosphere in relation to the location of human populations. To obtain good estimates of the averages, we did this for a sample of sources in each of our five sectors. This step required the development of a national power-sector database, field studies in five cities of China to gain information for the other four sectors, and the development of detailed population-density maps. Combining this information with air-dispersion models required several man-years of computer modeling, covering more than 600 sources.

This was the largest of the various research components of this assessment and was undertaken through coordinated research by Chinese and U.S. participants. The essential field data were collected by a Tsinghua University team led by Jiming Hao, with Ji Li, Bingjiang Liu, Yongqi Lu, and Shuxiao Wang. Three chapters report modeling studies conducted with these and other data: chapters 5 and 6, led by Shuxiao Wang and Bingjiang Liu, respectively, in residence at Harvard and using the inputs from the Tsinghua fieldwork; and chapter 7, by Ying Zhou, Jonathan I. Levy, James K. Hammitt, and John S. Evans of the Harvard School of Public Health, using these and additional data. We emphasize again that this ambitious effort was required specifically to take into account the important exposure dimension of proximity of pollution to population. We used data on actual emission sources to try to differentiate pollution likely to be inhaled from that unlikely to be inhaled.

Once we characterized the intake fraction for our sampled plants, we used statistical techniques to generalize the exposure relationships for each industry and different pollutants. We did this for fine PM and SO_2 produced directly from combustion (primary pollution) and fine particulates that form chemically in the air from SO_2 (secondary particulates). The estimates for secondary particulates are based on rougher approximations but are important to our final results. Finally, we used this information to develop similar human-exposure characteristics for the other, less-damaging, sectors that we could not model directly.

These steps produce only intermediate results in our assessment, but they have some initial relevance to policy. Our intake fractions suggest that there are indeed significant differences in the pollution exposure (and therefore health damage) characteristics of sectors. In particular, a given microgram of primary PM emitted from the power sector appears to have considerably less chance of reaching human lungs than one from our manufacturing sectors, and this difference is more pronounced for SO_2. This is a notable result because much prior pollution research has focused on the power sector, the largest user of coal in China. This is not to suggest that electricity generation has little health effect, but rather that prioritizing based on only the scale of emissions—the convenient assumption in lieu of further information—may overestimate this sector's contribution to total damages.

Moreover, the intake fractions for transportation sources are roughly an order of magnitude higher than the manufacturing sectors. This may be explained by closer proximity to people, because of both the ground-level height of vehicle exhaust pipes—contrasted with elevated stacks in the other sectors—and the high population densities around urban streets. These results highlight the value of considering other sectors that might have comparatively high human exposure characteristics and thus health effects.

2.2.2 From Exposure to Health Impacts and Their Economic Value

To put all of this information together with the sector emissions from our economic model, we need two additional ingredients. First, we need to know the effects of changes in exposure on health outcomes of concern, eleven types of health effects in total, chiefly premature deaths and respiratory disease. This is done with environmental epidemiology, the observational studies of human populations that identify relationships between exposure to measured pollutants and health effects, for example, the percent increase in daily visits to the emergency room per unit increase in PM concentration.

Though relying on existing literature rather than new research, this is a tricky step, with great uncertainties and use of subjective judgment. This is particularly true because most of the world research in this area has been conducted in industrialized countries, and the dose-response relationships to relate health to pollution exposure do not transfer easily to populations with very different environmental characteristics. In the end, we had little choice but to consider chapter 4's review by Levy and Greco of the scientific evidence of risks of mortality and morbidity from the limited Chinese studies that exist and from Western studies and pick the best available coefficients. We have chosen coefficients that are somewhat different from those used in the well-known 1997 World Bank study of the topic, *Clear Water, Blue Skies*, giving us lower damage estimates.

Attributing additional cases of death and disease to emissions from particular industries is a vital quantitative basis for evaluating pollution control policies, and we report these in our assessment. Decision makers also like to know what changes in health risks might mean in economic terms, to compare costs and benefits of possible policy interventions. For this we need estimates of the value of health risk reduction, a sometimes controversial but necessary step if policies are to be evaluated in a benefit-cost framework.

This too is a very underdeveloped research field in China, and so Ying Zhou and James K. Hammitt conducted a survey-based investigation into the "willingness to pay" to reduce health risks in China (chapter 8). From this they derived average values that citizens place on avoiding such health endpoints as colds, chronic bronchitis, and premature mortality. These values are also somewhat lower than those assumed in *Clear Water, Blue Skies*.

2.2.3 National Damage Assessment

We now bring all of the foregoing results together to estimate the health damage due to air pollution from each sector of the economy, returning to the economic framework and research of Mun S. Ho and Dale W. Jorgenson of Harvard and reported in chapter 9. These estimates are needed to identify the major sources of damage, including the sector share of total damages and the damages per unit output. This information is directly related to our thirty-four-sector environment-economic model of China, and we are thus able to estimate the change in health damages over time. This will also enable us to examine the effects of pollution control policies in our last component, as described in the final section below and in chapter 10. We present only key results here; the last two chapters present a full discussion of the health and economic results of the integrated assessment, and the policy implications.

National damage by sector We can combine the estimates of emissions of total suspended particles (TSP) and SO_2 for each of the thirty-four sectors of our economic model with the previously described intake fraction results and dose-response relationships to estimate various health outcomes for each sector (e.g., 2500 excess deaths per year due to primary particulates from electric power). By including the valuations of each health effect, we then obtain the total value of health damages from air pollution for each sector by adding over all eleven health effects. This permits us to derive the marginal damage per unit output for each sector, in terms of yuan of damages per yuan of output. For example, we estimate that there are 0.02 yuan of health damages per yuan of cement output and 0.09 yuan per yuan of electricity. Most of the damage is from TSP (primary and secondary) and less from primary SO_2.

Summarizing the key sector comparisons, electric power alone is estimated to cause more than 26% of all health damages from ambient air pollution from combustion. Perhaps unexpectedly, the nonmetal mineral products (cement) sector, which is responsible for more than 12% of damages, ranks second. Transportation ranks third, at nearly 10%, and this was for a base year that is well before personal vehicle ownership in China skyrocketed. Other major emitting sectors—chemical production and iron and steel—contribute less to population health risk, at less than 7% and 5%, respectively, though still more than most remaining sectors.

We can now draw some stronger conclusions. First, our consideration of population exposure in assessing health damage affects recommended sector priorities. Electric-power generation, which consumes the most coal, is likely the largest source of health damages from ambient air pollution, as is often assumed. Its tall stacks and other characteristics reduce the local health impact of its primary pollution emissions by limiting proximity to populations, but this is offset by its high levels of SO_2 emissions. These generate enormous secondary particle formation, which spread over great distances and raise regional damages. The 26% share of national damages from electric power, nevertheless, is considerably less than its share of total emissions (35% of combustion TSP and 45% of combustion SO_2), as hinted earlier by the more abstract intake fraction results.

Second, cement production may be underemphasized as a source of health risk from fossil fuel combustion. It appears to far outrank all industrial sectors except power generation. Third, although our estimates for the transportation sector are weak due to a limited database, damages from this sector appear very large even before the vehicle population of China began to explode. Furthermore, this does not incorporate impacts of the more complex secondary mobile-source pollutant, ozone. This suggests that pollution control efforts give increased priority to transportation and that researchers try to improve data and estimates of mobile-source damages (as we have begun to do in our own program).

We must emphasize one critical qualification about the sector results: these estimates are for pollutants generated by combustion only and do not consider emissions from chemical and mechanical processes of production. There are reasons to question whether the larger-sized particles of process emissions have the same health impact as those of combustion emissions, as discussed in chapter 9. If future research indicates the effects are similar, the enormous process emissions of the cement industry would vault it far above other sectors, including electric power, as a contributor to total damages. In any case, cement manufacturing appears to be a leading pollution control priority to reduce health damages.

National damage per unit fossil fuel We can also estimate the damage due to primary fuels, to provide guidance for possible fuel tax policies. Although different sectors produce different emissions per ton of fuel burned and we could produce damage estimates accordingly, we do not differentiate damages by both fuel and sector. Taxing fuels has proven politically feasible in some countries, but setting different fuel taxes for different sectors would be impractical. Instead, we estimate the average national damage per ton of coal, ton of oil, or cubic meter of gas.

The estimated damage from coal is very high, 94 yuan per ton in 1997 and 0.58 yuan of health damage per yuan of coal. This means that the damages approach the market cost. The damage from oil is much less, only 0.025 yuan per yuan of oil, whereas the damage from gas is negligible. We will use these results in one of our policy simulations.

Aggregate national damages Policy actors often prefer to have an aggregate national damage value, typically expressed as a percentage of gross domestic product (GDP). This is simple for us to estimate, in light of the research described so far. Readers may be surprised that we report it here to warn against regarding this value as a particularly useful result of any such assessment. Although we have conducted one of the most detailed analyses to date, aggregating all of the foregoing research into a single number also aggregates the many uncertainties. In view of the limits of data in China and the assumptions implicit in the many modeling steps, such percentage-of-GDP figures should be regarded very cautiously. We discuss sensitivity analyses and various uncertainties and assumptions in chapter 3 and give additional detail in each chapter. After reading these chapters, a reader will rightly gather that we can legitimately report only a broad range of plausible values. There is a chance that the true value may be off by 100 percent or more from our central estimate, which is also true for other similar assessments, and not just of China. In light of the magnitude of damages, however, even a large underestimate will advise a proactive pollution control policy. More data and information would sharpen the total damage estimates, though the degree of uncertainty may remain unavoidably large.

Readers disappointed by this warning should understand that such a figure was never the objective of this project in the first place. There is little that policy makers can do with such a value, as top environmental officials in China have themselves noted, except to advertise the seriousness of the air pollution problem as a whole. Even if such a figure were strictly accurate, by itself it would recommend little about the targeting of pollution control.

An analysis like ours is more useful for its indications of relative damages within the Chinese economy. The absolute numbers are only approximations, but, because our methods are consistent across the sectors, we are on firmer ground drawing conclusions about how they relate to each other. Indeed this is far more useful in a policy sense. Decision makers need to know not just how much air pollution impairs public health or the economy, but also how to focus policies on those parts of the economy that are most responsible. As we will see in the analyses of policies that follow, differentiating more harmful industries (or fuels) from less harmful ones enables us to investigate pollution control policies that are better targeted.

With this strong caveat, our central estimate of the value of national health damages due to air pollution is 137 billion yuan for 1997, or 1.8% of GDP. Of this figure, primary particulates account for only 56 billion yuan, with the remainder due to secondary particulates and primary SO_2. To give a sense of the range of uncertainty, when the low-end parameter estimates are used, national health damages come to only 0.65% of GDP. Upper-end parameters—which incorporate effects of chronic pollution exposure estimated from epidemiology in other countries—raise the national damage estimate to 4.7% of GDP.

2.2.4 Benefit-Cost Analyses of Policies to Reduce Emissions

In light of the severe damages caused by air pollution in China, there have been a large number of studies examining options to control or reduce emissions. In contrast to many of these studies, we make an integrated estimate of the health benefits and costs of pollution-control policies throughout the entire economy. In chapter 10, by Ho and Jorgenson, we focus on economywide tax policies and examine how they affect fuel use and, hence, emissions and health damages. At the same time we estimate how these taxes affect output, allocation of resources, existing taxes, and, over time, economic growth. We also consider the effects on GHG emissions.

Why analyze market-based policy instruments when more traditional pollution-control approaches, such as the technology mandates familiar to most officials and researchers, are more the norm in China? Market-based instruments such as the taxes we analyze are increasingly preferred for pollution control in Western countries, in those circumstances where the nature of the hazard suits the approach and where institutional and legal capacities are in place to support them. Such conditions are often lacking in developing countries, notably for emission trading. Chinese officials have strong interest in environmental policy innovations of the West, however, and a track record of trying them in carefully limited real-world policy experiments. In some cases, market-based approaches are increasingly workable in China now. In others, they suggest possible future policies, as market, institutional, and legal reforms proceed.

The full effects of such policies extend beyond the environment. Ideally, for instance, taxes can rationalize energy inputs throughout the economy, including in sectors with few emissions at all, and can establish a revenue source that itself may benefit the economy if it displaces more distortionary taxes. Most pollution-control assessments—including nearly all conducted to date in China, many described in chapter 1—are not designed to evaluate such secondary economic impacts of a policy intervention. These aggregate impacts, however, can be enormous and tend to accumulate over time. An appropriate way to estimate the full effects of such policy alternatives is with a dynamic national economic model, such as the one employed in our policy simulations.

Analyzing pollution taxes We analyze policies by beginning with a base case that assumes no new policy interventions and uses the estimates of damages due to the current patterns of output and fuel use in each industry. The economic model is dynamic and permits us to estimate the value of damages each year as the projected economy expands and changes in structure. In the base case, or "business-as-usual" scenario, the model projects GDP to grow at 5.1% per year during the next thirty years while energy use only grows at a 3.3% rate. The growth of coal use is projected at 2.3%, whereas oil use is at 5.2%, which is consistent with the expectations of energy-efficiency improvements and rapid growth of motor vehicle use.

To evaluate our alternative policies of the green (or Pigovian) taxes, we estimate their effects and compare them to the base case. In light of the size of the estimated health damages, these pollution taxes can be large and generate large revenues. To maintain revenue neutrality, we reduce existing taxes. The results of these final policy analyses are presented in table 2.1 and discussed next—illustrating first the modest effects of damage-weighted taxes on outputs and then the stronger potential benefits of such taxes on fuels.

Damage-based output taxes The first policy imposes green taxes on industry output equal to the estimated marginal damages. This is a high tax on electricity, a modest one on cement, and a very low one on most other outputs. The initial effect of the taxes is to raise the price of electricity by almost 6% and the prices of the other dirty commodities by about 1.5–2.5%. This change in prices leads to a very small reduction in aggregate GDP but a shift in consumption to investment. The economywide effects of this policy are given in columns 2 and 3 of table 2.1 for the first and twentieth year of the simulation.

In the first year, the output of electricity falls by 6.7% while the other dirty sectors—cement, transportation, and health-education-services (which has relatively high damages due to heating with boilers at its many facilities)—fall by 1–3%, thus

Table 2.1
Effects of damage-based taxes on outputs or fuels

Variable	Output Tax		Fuel Tax	
	Effect in 1st Year	Effect in 20th Year	Effect in 1st Year	Effect in 20th Year
GDP	−0.04	+0.18	−0.04	+0.05
Consumption	−0.06	−0.01	−0.06	−0.02
Investment	+0.06	+0.55	+0.07	+0.20
Coal use	−3.95	−3.36	−13.2	−18.2
CO_2 emissions	−3.41	−2.41	−10.7	−12.5
Primary particulate emissions	−3.08	−2.30	−6.0	−9.0
Combustion TSP	−4.25	−4.00	−11.8	−16.3
From electricity	−9.21	−7.42	−12.4	−16.3
From low-height sources	−1.84	−0.92	−11.0	−15.3
SO_2 emissions	−4.56	−4.00	−10.2	−14.0
Premature deaths	−3.21	−2.88	−10.7	−14.3
Value of health damages	−3.53	−3.01	−10.7	−14.0
Change in other tax rates	−8.5	−6.1	−2.8	−2.1
Reduction in damages/GDP	0.07	0.09	0.20	0.43
Pollution tax/total tax revenue	9.1	6.0	2.72	1.99

Note: The entries are percentage changes between the counterfactual tax case and the base case, except for the last two rows, which are changes in percentage shares.

leading to a reduction in the demand for electricity and coal. Inputs released from these shrinking sectors go to cleaner industries, which leads to a slight expansion in finance, food products, and apparel.

These changes lead to a reduction in total primary particulate emissions of 3.1%, with the power sector having the greatest reduction, 9.2%, whereas emissions for sectors with low stack heights only fall 1.2% in aggregate. For SO_2, total emissions fall by a significant 4.6%.

The effect of these pollution reductions is to lower the value of health damages by 3.5%. The number of cases of premature mortality falls by 3.2%, and those of chronic bronchitis fall 3.7%. The value of this reduction in damages in the first year comes to about 0.07% of GDP of that year. This is a very modest reduction compared to the base-case estimated total health damages of 1.8% of GDP.

In terms of the effects on government finances, the revenue raised from this broad-based tax is substantial, 9.1% of total government revenues in the first year. With our requirement that government revenues and spending be kept fixed at base-case levels, this allows a reduction in value-added and capital income taxes of 8.5%. This tax cut eventually leads to higher retained earnings and, hence, investment. The higher rate of investment leads to a higher stock of capital and a greater productive capacity for the entire economy. As shown in table 2.1, GDP is 0.18% higher than the base case by the twentieth year, separate from the health benefits.

The emissions of the GHG CO_2 are related to the local pollutants. In the first year, CO_2 emissions from fossil fuels fall by 3.4%, a little less than the decline in coal use, because there is some switching to oil and gas via reallocation of industry activity. This is a relatively inefficient policy to reduce emissions of both PM and CO_2, so we turn next to our second policy, which will prove to be more efficient.

Damage-based fuel taxes Emissions are a function of output levels, fuel choice, energy efficiency, and control strategies. We now consider a more sharply targeted air pollution–control policy, a tax on fossil fuels in proportion to the health damages that result from current combustion technologies.

The heavy tax on coal initially reduces its use by 13%, whereas the modest tax on oil reduces refining output by 1.3%. Heavy users of these fuels must raise their output prices to compensate, which causes a reduction in demand for these goods. The electricity sector is the greatest user of coal. Its price rises by 2.4% and output falls by 2.7%. Iron and steel and cement are the next-highest consumers of coal, and their outputs fall by 0.8 and 0.6%, respectively.

The total additional tax burden is relatively small. Pollution-tax revenue comes to only 2.7% of total revenue, and the offsetting cuts are correspondingly small. The small additional retained earnings of enterprises cause aggregate investment to rise by 0.07% in the first year. This is accompanied by a decline in real consumption of 0.06% as households face mostly higher prices of goods. The changes in composition of aggregate output lead GDP to fall by a modest 0.04%.

These changes in industry structure and fuel switching reduce total primary PM emissions by 6.0%, with those from electricity falling the most (12%) and those from manufacturing (in aggregate) the least (3.9%). Large reductions in SO_2 and primary particle emissions reduce total PM, which in turn generates an 11% reduction in health damages. The value of reduced damage comes to a substantial 0.20% of GDP, recalling that our base-case total damage is about 1.8% of GDP.

Because only the three fossil fuels are taxed, the revenue is much smaller than the 13% collected in the output tax policy, which taxes all commodities. The output tax case, however, generates only a 3.2% reduction in health damages, compared to the

11% here. Thus, a narrowly based but well-targeted tax that raises only a modest amount of revenue can lead to a sizable reduction of pollution and related health damages.

The patterns over time are also very different from the output tax case. Here the reduction in emissions and damages rise over time, comparing the first- and twentieth-year columns in table 2.1. By the twentieth year, even though GDP is 0.05% higher than the base case, the reduction in primary PM emissions is 9.0%, greater than the 6.0% reduction achieved in the first year, when GDP fell. Similarly, the reduction in SO_2 emissions over the base case grows over time. Health damages, driven by these emissions and the growing urban population, fall even more. By the twentieth year, the total health damages are down 13% from the base path—reemphasizing that this occurs under a policy that simultaneously increases GDP.

Considering the effects on CO_2, we see that the reduction in emissions is smaller than the fall in coal use due to the switch to other fossil fuels. In the first year, when coal falls by 13%, CO_2 falls by 11%. Over time, as the tax rises, the CO_2 share from oil rises even more. The reduction of CO_2 in the twentieth year is only 12%, whereas coal use falls 18%. Although this may seem an inefficient instrument to reduce CO_2, it is actually an effective "second-best" instrument, even when the very large additional health benefits are ignored. In our view, one should see substantial reductions in CO_2 emissions as a critical side benefit of dealing with the urgent issues of local air quality and public health, and vice versa.

2.3 Conclusion

Our assessment consists of an integrated series of studies, exploring and linking many aspects of the health and economic damages of ambient air pollution in China. As the reader of this book will see, understanding the causes, effects, and policy options in air quality is a complex analytical challenge that should not be underestimated. This is partly because the problem is inherently multifaceted, encompassing economics, engineering, atmospheric science, health science, and public policy. It is certainly more difficult (and the results less precise) if underlying information and data are limited. For a fuller understanding of these and other challenges to estimating damages of air pollution and devising informed air-quality policies, we encourage the reader to continue on to the more detailed integrated summary of research presented in the next chapter and to the original studies that follow in the rest of the book.

We summarize the assessment by spotlighting a number of key conclusions:

- The damage to health of air pollution in China is considerable, even under conservative choices in the ranges of our assumptions. Our central estimate of damages comes to around 1.8% of GDP, though we caution against overconfidence in the strict accuracy of any such estimates. Using low-end or high-end parameter inputs produces health damage estimates of 0.65% and 4.7% of GDP, respectively.
- It is feasible and valuable to focus on damages—and not on energy use or emissions per se—to understand properly the adverse effects of air pollution. Industries that generate the most emissions are not necessarily those that cause the most damages. Our assessment finds the electric power sector to be the largest source of damages, followed by the cement industry and the transportation sector.
- The intake fraction methodology, and our specific intake fraction results, can produce a simple, quick, but reasonable approximation of exposures to TSP and SO_2 from emission sources, if data, time, or other resources to run air-dispersion models are lacking.
- The regional, as opposed to the local, dimensions of air quality are critical and should not be overlooked, because damages from primary and secondary pollutants far from the source of emissions are considerable. They are of similar magnitude as the local damages that are most often (and most easily) calculated.
- The benefits of pollution control in China likely far exceed the costs, if market instruments are used. This high benefit-cost ratio is robust to the many uncertainties. Short-run adjustment costs can be mitigated by gradual implementation.
- In light of the high level of damages, the whole range of pollution-reduction policies, beyond those considered here, should be analyzed. Wealthier countries generally used end-of-pipe regulations to reduce emissions of particulates and SO_2 before developing and beginning to adopt market-based policy instruments more recently. The former include mandated control technologies, emission standards, and other regulatory policies. Analysis of such policies would give authorities a wider range of options and should be high on the priority list of action. A careful assessment of these regulations would require data on costs of operating various scrubbers, costs of washing coal, costs of low-sulfur coal, and so forth.
- A green tax policy could generate substantial new revenues for the government that could be used for tax reform (i.e., the reduction of other, distortionary taxes). This tax rationalization over the long run could produce higher retained earnings and investment, leading to higher capital stock and greater productive capacity for the entire economy. Such policies in principle could not only reduce health damages of air pollution, but also potentially improve economic performance over time.
- We strongly support efforts to analyze global and local pollution in concert, which give a more comprehensive accounting of total costs and benefits. The pollution-control policies that we examined produced substantial concurrent reductions in

emissions of CO_2, the most important anthropogenic GHG. In our fuel tax simulation, this reduction reached 12% annually over business as usual by the twentieth year. The magnitude of the gain to global welfare should spur serious consideration by wealthier nations concerned about climate change to aid China in these control efforts.

Those who read the chapters of original research in this book will take note of the wide range of uncertainty described at many steps in our assessment. In response, we emphasize two points. First, assessing environmental damage anywhere in the world is by nature an uncertain science, and the difficulties grow as the scope of an assessment grows. Analyses at the city or province level can be more precise, but other systematic studies of damages at the national scale in China include approximating steps (and thus uncertainties) like our own. It was the challenge of improving on the best-known of these, the World Bank's *Clear Water, Blue Skies* report, and extending the aims to evaluate the effects of particular policies that inspired the more scholarly research venture reported here. The following chapters describe the research in complete, transparent, and replicable detail so that others can critique, apply, adapt, or improve any and all elements of our method.

Second, at every step of this and similar assessments, uncertainty results from limited available data and underlying research. This is understandable and hardly unique for a developing country such as China. Nevertheless, we emphasize that for any country to confront its environmental hazards efficiently and successfully, an essential first step is to build systematic basic information on the nature of the physical, economic, and institutional challenges. This in itself is an important policy suggestion to safeguard China's environmental future.

II

Studies of the Assessment

3

Summary for Research

Chris P. Nielsen and Mun S. Ho

3.1 Introduction

When assessing the impacts of air pollution in China, one needs to think beyond ambient concentrations in the air to what ultimately concerns us about degraded air quality: the damages it causes. These include the effects on human health, crop productivity, forests and water bodies, ecosystems, and physical infrastructure. All of these factors are parts of the economy, and changes in any part will have ramifications throughout the economic system.

The assessment of this volume focuses on refining understanding of the damages to human health, which most experts believe dominate the total impact. (Other, lesser damages are topics for future extensions of the research.) Our team is interested in the health damages of pollution on a national scale in China and specifically seeks to identify the sources of damage by sector. We want to know, for instance, how the damages of emissions from the electric-power sector compare to those from the iron and steel industry, the chemicals industry, and so forth. This is intended to inform pollution-control strategies at the national scale, by indicating which sectors pose the greatest total air pollution risk to public health. To date, beyond conventional wisdom there is little information attributing national health risk to sector sources. Our assessment of China yields some unexpected results, in part because of this unique consideration of how pollution exposure may differ by sector, which challenges some of the usual assumptions.

Because there is a strong policy interest in considering such damages in economic terms, our assessment also translates the estimates of sector health damages into monetized values, for instance valuing the estimated cases of chronic bronchitis in yuan. We then incorporate this information into our national economic model of China. This provides a tool for a number of additional, unique analyses. Our team is able to consider not only how policy interventions may limit growth in emissions, reduce health damages, and affect the target sectors, but, just as important, also how those policies will reverberate throughout China's national economy. This permits a fuller understanding of the actual costs and benefits of such policies than most analyses to date have been able to accomplish. More specifically, our integrated analysis allows us to estimate the health benefits of environmental protection policies and, at the same time, estimate the costs in terms of lower output and consumption.

We ultimately examine two market-based policies to reduce air pollution and estimate their effects on economic performance, over both the short and long runs. We believe our results provide important insights into not only the ramifications of environmental policies, but also other important fiscal policy questions facing the government, such as tax reform. As an example, under some scenarios a pollution tax may lead to higher future economic growth that offsets some of the near-term costs.

The detailed results of our research program are reported in chapters 4–10 and the appendixes. They are written in considerable detail so that our methodologies and results, including the limitations and uncertainties noted previously and described further in this chapter, are clear and can inform future studies and other research teams. Nevertheless, the volume is intended for readers interested in pollution and environmental policies from across disciplinary backgrounds, with introductory sections for nonspecialists and limited disciplinary jargon.

This chapter summarizes the analysis and key results of the entire book with the same broad scope of the "summary for policy" of chapter 2, but this "summary for research" is a much richer presentation structured around the integrated methods of the project and linking specific results in sequence. It is intended as a transition between the general material of part I and the reports on original research in the rest of part II. As described in "How to Read This Book" in section 1.2 of chapter 1 (which we again encourage readers to keep in mind), this summary is still written in nonacademic language for both nonexpert and expert readers. For any nonexperts whose interest was raised by the previous chapter, this version will expand their understanding of assessment methods, assumptions, and uncertainties and help them judge for themselves how such research should inform policy decisions. For more expert readers, this chapter serves as an integrated introduction to the research core of book.

We help the general reader by explaining some methodological issues in side boxes. These are also written in lay terms, and we encourage readers to go through them to grasp important subtleties and uncertainties of the full assessment. It may be possible for those anxious for the policy discussion to skip these side explanations on their way to the final section on the health and economic assessment of our two green (or Pigovian) tax policies. Those readers might want to return to the boxes afterward, before moving on to the deeper investigation of the chapters that follow.

Scholarly conventions, including source citations, are therefore omitted in this chapter, but the discussion presented here is drawn entirely (including some passages) from the chapters that follow, where readers will find complete details and references. "We" and "our" refer to the entire team, not the authors of this chapter alone. As mentioned earlier, authors are responsible for their own chapters but do not have a collective responsibility for the entire assessment. That lies with the project leaders and editors of this volume.

In sections 3.2–3.6 we introduce the framework of our research program and sequentially present an overview of methods and key results from the rest of the book. Section 3.7 brings all of the foregoing results together in a single integrated approach to damage assessment. In section 3.8 we use the integrated model to assess the health and economic effects of two pollution-control policies—damage-weighted taxes on fuels and economic output.

3.2 Methodological Framework

We review the description in chapter 1 that our research strategy follows what is commonly referred to as the "pollution-impact pathway" or causal chain:

Component	Description	Chapter
1	From economic activity and energy use to pollutant emissions	1, 9
2	From emissions to concentrations (atmospheric transport)	4–7
3	From concentrations to human exposure	4–7
4	From exposures to health impact (dose-response)	4
5	Economic valuation of health impacts	8
6	National damage assessment by sector	9
7	Cost-benefit analysis of policies to reduce emissions	10

This program of research, graphically represented in figure 1.11 of chapter 1, makes primary contributions to components 1–3 and 5–7. Although we could not contribute primary research to component 4—air pollution epidemiology, which requires costly and time-consuming fieldwork—we do introduce the issues of defining and estimating functions that describe the human health response to pollution exposures. These are drawn from previously published studies, including several by colleagues with partial support from our program in prior years.

The structure of this summary chapter progresses in a logical sequence through each link in the above chain, under sections headed by the component descriptions. Each section may include background information, an introduction to assessment methods, and our research results. These various results are brought together in an integrated analysis toward the end, in component 6. It is then used to evaluate two possible tax-based pollution-control policies in component 7.

Some sections will be relatively lengthy. Not only must our research team analyze complicated processes from the individual perspectives of engineering, health science, risk analysis, economics, and policy, but we also then need to try to integrate them into one multifaceted assessment. There are no simple ways to do this.

3.3 Component 1: From Economic Activity to Energy Use to Emissions

We implement component 1—from economic activity to energy use to emissions—by using an economic model of China that features thirty-four distinct sectors, tracing the evolution of the economy across time. This model generates the output of each sector for each period and with it the use of fossil fuels. The combustion of fuels and the production of goods generate emissions of particulate matter (PM), sulfur dioxide (SO_2), and carbon dioxide (CO_2), and these levels are estimated by use of evolving emission factors. The model has been developed, updated, refined, and used in a number of analyses by Mun Ho and Dale Jorgenson, two economists on our team, with collaborators, over nearly a decade. It is described further in box 3.1.

We illustrate the structure and detail of the model with table 3.1, showing gross output, fuel use, and emissions of total suspended particulates (TSP) and SO_2 by sector for 1997. We chose this base year for our assessment because it accords with the most recent comprehensive interindustry (input-output) data available from the Chinese government at the time of our study.

In the next section, we will describe our assessment of the atmospheric transport of pollutants emitted by sectors and the resulting population exposures. Our research team cannot feasibly conduct this procedure on all sectors. Instead, we focus this detailed approach on five sectors that are dominant contributors to ambient TSP and SO_2: chemicals (sector 13 in the table), nonmetal mineral products (14, mainly cement), metals smelting and pressing (15, chiefly iron and steel), electricity generation (23), and transportation (26). The reader can see how large the emissions

Box 3.1
The Economic Model

The economic model is a standard, multisector growth model that is modified to recognize the two-tier plan-market nature of the Chinese economy. This version is based on the 1997 input-output table, where thirty-three production sectors are identified, including six energy industries, and one nonproduction sector, households. The largest sector in terms of employment and output is agriculture, the largest user of coal is electricity, and the largest emitter of TSP is the nonmetal mineral product (chiefly cement) sector.

The household sector maximizes a utility function that has all thirty-three sector commodities as arguments. Income is derived from labor and capital, supplemented by transfers. This is a so-called Solow (or dynamic recursive) model, where the private savings rate is set exogenously. Total national savings is made up of household savings and retained earnings of enterprises. These savings, plus allocations from the central plan, finance national investment, the government deficit, and the current account surplus. The investment in each period increases the stock of capital that is used for production in future periods.

Labor is supplied inelastically by households and is mobile across sectors. (By "inelastic labor supply," we mean that the total hours worked is a predetermined number not affected by economic events. The alternative seems to us to be too elaborate to implement sensibly for the current Chinese economy, with its large pool of underemployed workers.) The capital stock is partly owned by households and partly by the government. The plan part of the stock is immobile in any given period, whereas the market part responds to relative returns. Over time, plan capital is depreciated and the total stock becomes mobile across sectors.

The government imposes taxes on value added, sales, and imports and also derives revenue from a number of miscellaneous fees. On the expenditure side, it buys commodities, makes transfers to households, pays for plan investment, makes interest payments on the public debt, and provides various subsidies. The government deficit is set exogenously and projected for the duration of the simulation period. This exogenous target is met by making government spending on goods endogenous.

Finally, the rest of the world supplies imports and demands exports. World-relative prices are set to the data in the last year of the sample period. The current account balance is set exogenously in this one-country model, and endogenous terms of trade exchange rate clear this equation.

The rate of technology growth in each industry is projected exogenously. In the model, there are separate sectors for coal mining, crude petroleum, natural gas, petroleum refining, electric power, and gas (including coal gas) production. Nonfossil fuels, including hydropower and nuclear power, are included as part of the electric power sector.

Table 3.1
Gross output, fuel use, and combustion emissions in 1997

Sector	Gross Output (billion yuan)	Coal Use (million tons)	Oil Use (million tons)	Gas Use (million m^3)	TSP (kilotons)	SO_2 (kilotons)
1 Agriculture	2,467.7	12.7	9.61	0.0	159.7	367
2 Coal mining and processing	238.0	40.8	2.72	7.2	180.7	192
3 Crude petroleum mining	188.1	12.7	15.09	5.0	56.9	101
4 Natural gas mining	10.9	0.1	0.24	627.6	0.3	2
5 Metal ore mining	115.2	5.0	1.06	2.4	22.0	35
6 Non-ferrous mineral mining	222.4	9.9	3.32	3.1	44.0	72
7 Food products, tobacco	1,371.3	33.2	1.74	44.3	310.4	478
8 Textile goods	913.0	21.8	1.06	31.5	166.1	334
9 Apparel, leather	627.0	6.7	0.76	0.0	16.4	24
10 Sawmills and furniture	222.5	8.1	0.57	0.0	90.2	118
11 Paper products, printing	445.5	18.1	1.34	14.5	200.5	261
12 Petroleum refining and coking	308.5	20.0	8.75	421.7	71.6	163
13 *Chemicals*	*1,488.8*	*148.5*	*32.08*	*4,541.7*	*671.8*	*1,429*
14 *Nonmetal mineral products*	*874.9*	*206.7*	*10.34*	*1,442.4*	*2,548.8*	*2,178*
15 *Metals smelting and pressing*	*750.2*	*183.2*	*8.81*	*968.9*	*632.0*	*1,664*
16 Metal products	477.8	18.3	1.80	69.1	38.9	34
17 Machinery and equipment	882.1	36.2	3.81	366.8	125.4	189
18 Transport equipment	570.7	12.4	1.88	247.6	42.9	67
19 Electrical machinery	563.3	8.1	1.93	156.1	28.0	47
20 Electronic and telecommunication equipment	488.1	2.4	0.68	260.3	8.3	16
21 Instruments	96.0	1.7	0.26	1.6	6.0	10
22 Other manufacturing	286.9	13.7	1.08	6.4	82.8	235
23 *Electricity, steam, hot water*	*380.8*	*384.9*	*14.80*	*538.5*	*3,953.0*	*7,895*

Table 3.1
(continued)

Sector	Gross Output (billion yuan)	Coal Use (million tons)	Oil Use (million tons)	Gas Use (million m^3)	TSP (kilotons)	SO_2 (kilotons)
24 Gas production and supply	12.4	4.4	1.40	25.4	15.5	26
25 Construction	1,738.6	9.4	20.63	0.0	120.8	414
26 *Transport and warehousing*	*506.6*	*17.5*	*23.07*	2.9	*361.0*	*609*
27 Post and telecommunication	195.9	0.0	0.54	0.0	0.1	6
28 Commerce and restaurants	1,412.6	9.4	7.19	111.7	118.2	273
29 Finance and insurance	359.5	1.1	0.51	0.0	13.4	28
30 Real estate	185.5	5.8	0.19	0.0	72.3	124
31 Social services	564.5	17.8	6.55	36.3	221.8	441
32 Health, education, other services	658.5	33.5	1.15	38.5	416.0	714
33 Public administration	443.4	9.0	2.76	1.5	111.7	216
Households		37.1	2.11	1,990.8	461.7	778
Total	20,067.0	1,350.3	189.8	11,963.6	11,369.1	19,541

Note: The five major emitting sectors that are the core of our exposure analyses and that are discussed at length in the text are in italics.

from these sectors (which are listed in italics in table 3.1) are compared to others. For the other sectors, because they are lesser contributors to ambient pollution levels, we can use a simpler assessment procedure that is derived from the detailed work on the major sectors.

Before proceeding, we must make a few notes. Total emissions from transportation are lower than those of the other four target sectors and close to several others; we include it in part because of its swift growth into a major pollution source. Households also have high emissions and exposure risks, but, as described in chapter 1, this important sector has features that are distinct from production sectors (including combustion of large quantities of noncommercial, biomass fuels) and it requires a different sort of analysis that we must reserve for future research. The emissions of the health, education, and other services sector (32) also appear to be high, but this is due to the cruder methods used to estimate these widely dispersed and numerous sources. Finally, table 3.1 lists emissions from combustion only. Production of cement (i.e., nonmetal mineral products) also generates process emissions, pollutants caused by mechanical or chemical processes, on a scale almost equal to the total combustion TSP produced by the entire economy. Considerably smaller but significant quantities of process emissions are produced by another of our target sectors, metals smelting and pressing. We believe process TSP, however, is likely to exceed combustion TSP in its average particle size. This would have different health implications, because fine and ultrafine particulates have the strongest association with adverse health effects, as discussed in chapter 4. Process pollutants may also have substantially different chemical compositions, which would change health effects. This summary focuses on combustion emissions, but our full assessment also considers the results of including both combustion and process emissions. These issues are explained in more detail in the following chapters, notably chapter 9.

We will return to the economic model and its use in the health damage assessment and green tax policy simulations in the final two sections of this summary chapter.

3.4 Components 2 and 3: From Emissions to Concentrations to Human Exposures

To allocate total pollution damage to various sources, after identifying the emissions from each sector, one needs to relate how these emissions contribute to the ambient concentration of, and exposures to, pollutants. Although it is obvious that different sectors produce different levels of emissions, both in absolute terms as well as emissions per yuan of output, it may be less obvious that each ton of, for example, particulates emitted from different sectors may cause different amounts of health damage.

Why would this be the case? First, the location of the source of emissions can be important, because meteorological conditions and the proximity to population centers influence transport and exposure. Second, source characteristics such as stack height, emission temperature, and emission velocity matter, affecting how far emissions may be blown away from the source. Third, in the case of primary PM, particle size distribution is important because transportation rates and health impacts differ according to particle diameter. Other factors include those that affect the rate of chemical reactions that generate secondary particles.

The aim of components 2 and 3—which the reader will soon appreciate is a substantial research undertaking—is to translate emissions from a range of economic sectors into population exposures, including the intermediate stage of atmospheric transport and transformation. Past assessments of this sort have typically adopted a damage-function framework. In this approach, the emissions of relevant pollutants from all sources are first estimated, along with characteristics of the source (for example, stack height or pollutant exit temperature). Researchers use atmospheric models to estimate how these emissions influence ambient concentrations or pollutant deposition at a number of locations, validated or calibrated by field measurements. These models typically incorporate detailed meteorological characteristics, some also allowing for at least simple chemical interactions among pollutants. Finally, the geographical distribution of population is used to determine the population exposed to the pollutants.

Although this damage function is well established, it is simply impossible for a national-scale analysis in China because it requires an enormous amount of very detailed data, some of which do not yet exist in China and some of which are not publicly available, at least without great effort.

Nevertheless, national pollution-control decisions must be made, informed by some understanding of the quantitative trade-offs. We need to find techniques to make reasonable approximations of nationwide health impacts by using existing information, perhaps supplemented by substantial, but not overwhelming, collection of new data. With a framework of reasonable damage estimation in place, we can identify incremental improvements in the assessment approach and tackle them in turn. We can add refinements as more data become available and more sophisticated atmospheric models are developed for China. (As an example, colleagues in the Harvard China Project have a current research effort using a nested, high-resolution window over China within a global chemical tracer model, GEOS-Chem, to investigate air quality with more advanced chemistry, physical transport, and meteorology than is possible with the dispersion models employed by our team.) The aim is to develop a model to estimate human exposures and health impacts on the basis of fundamental physicochemical properties, but without the detailed input data that are now lacking.

To make this approximation, we apply the concept of "intake fraction," as developed by researchers at the Harvard School of Public Health and colleagues at other universities and institutes. This concept has been developed for precisely the problem at hand: to conduct informative, if approximate, health risk assessments with only limited available input information, as is the case in most developing countries.

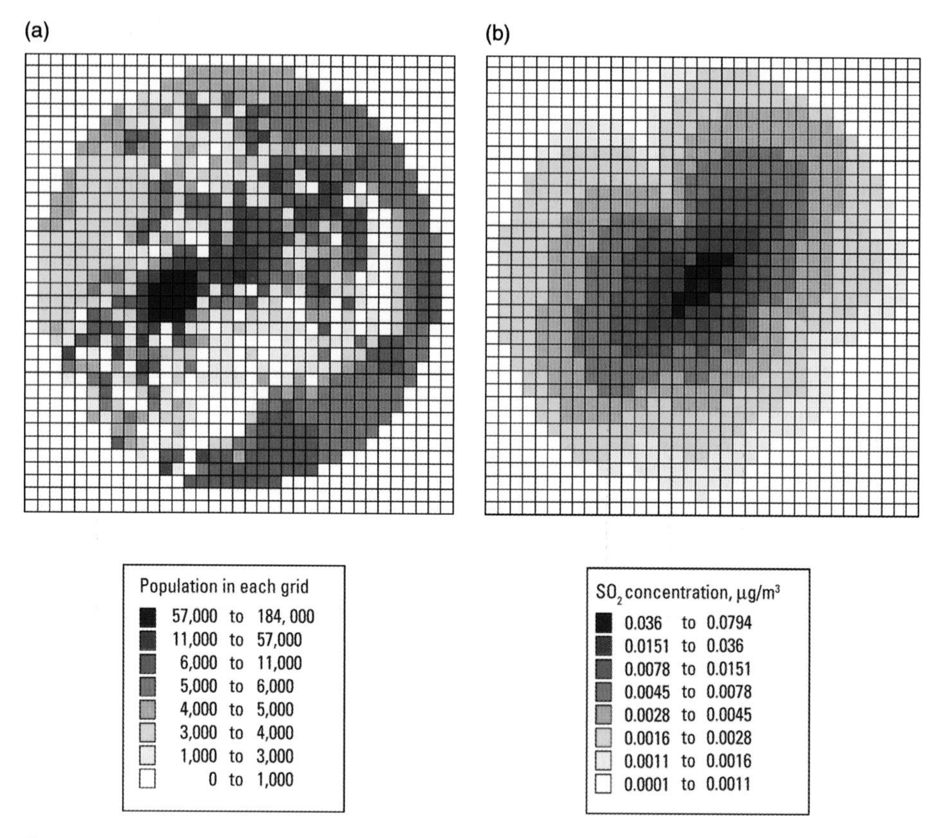

Figure 3.1
Estimating intake fraction for an example source, a cement stack in Jinan. (a) Population within 50 km of source. (b) Concentration within 50 km of source. (c) Population dose, multiplying concentration, population, and breathing rate at each grid cell.

3.4.1 Definitions of Intake Fraction

An intake fraction may well be the simplest way available to summarize the relationship between the emissions of a pollutant and the subsequent human exposure to that pollutant or a secondary by-product, including the atmospheric transport that happens in between. Informally, the intake fraction (*iF*) from a particular source is the amount of a pollutant emitted that is eventually inhaled by people before it dissipates from the atmosphere. A typical intake fraction value might be 0.000006 (or 6×10^{-6}), which can be interpreted simply as

> For every metric ton (10^6 grams) of a pollutant emitted, 6 grams will be inhaled by the population exposed to the source.

(c)

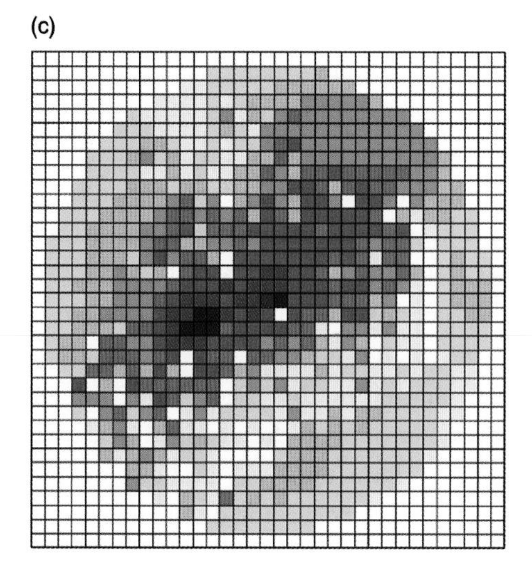

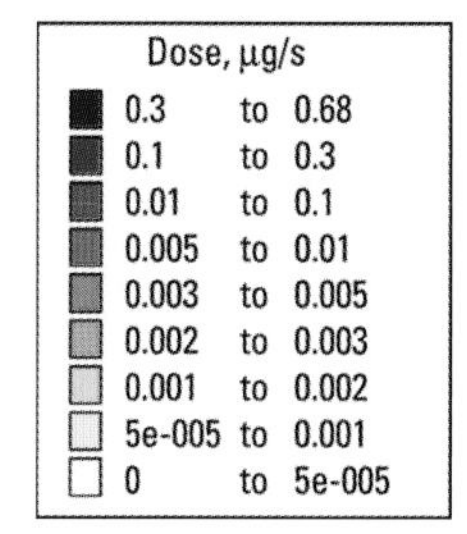

Figure 3.1
(continued)

Because intake fraction is both a central element and unique approach of the research reported in the book and summarized here, readers will want to keep this interpretation firmly in mind. The concept is more formally defined and explained in box 3.2.

A graphical representation of the determination of an intake fraction helps illustrate the idea. We use an example from our study of an actual cement plant in the city of Jinan, in Shandong province, illustrated in figure 3.1.

Determining the intake fraction for a pollution source requires the researcher first to set a geographical grid around the source. This grid will define a number of cells, perhaps 1 km × 1 km each, for which the researcher must then estimate two characteristics. First, she uses the best available information on population density to determine the number of people in each target cell. If such information is available only on a county level, for instance, she will assume that each cell in that county will have the same number of people, on the basis of the county's population density. If more detailed population information is available or can be estimated—for instance, by knowing land use distinctions such as residential versus agricultural areas—the population resolution can be finer. This determines the size of the at-risk population for each cell as shown in figure 3.1a. (In box 3.2, this is POP_d.)

Second, the researcher applies a pollution-dispersion model to the known emission quantity and other characteristics of the source, to estimate the average concentration of the pollution that results in each cell. (Pollution-dispersion models differ in their complexity and sophistication, depending in part on the detail of available input data—such as local meteorology and source characteristics—and also on the time and expenses required to model each source. See box 3.3 for discussion of the air-dispersion models employed in the study.) Figure 3.1b shows the pollution concentration (C_d) for each cell, reflecting the southwest and northeast prevailing winds of that particular area.

Each cell in figure 3.1c is the product of the values in figures 3.1a (population), 3.1b (concentration), and the average breathing rate for people. This is the estimated total quantity of the emitted pollutant inhaled by all of the people in that cell—or their dose, for a time period, in this case a day. Summing these for all of the cells in figure 3.1c gives the dose of the pollution from this source across the population of the entire domain.

This is not quite the intake fraction, but intake fraction follows easily from population dose. The summed pollutant dose across the entire domain is divided by the sum of the emissions for the given time period (i.e., a day). This is the portion of the pollutant emitted that is finally inhaled, the intake fraction. (Readers of box 3.2 will recognize these summations as simply the numerator and denominator in the formula that defines *iF*.)

A couple of important characteristics of the intake fraction concept deserve emphasis. First, this method takes into account not just where the air pollution goes, but also how people are distributed in its track to actually inhale it. This is a very important distinction over simpler assessments, in that intake fraction uses real information to distinguish pollutants that are likely to be inhaled from those that are less likely to be inhaled. Although the proximity of people to pollution seems an obviously central factor in health risk, exposure assessment is in fact one of the least developed and oft-neglected elements of pollution research. Very few national

Box 3.2
Formal Definition of Intake Fraction

For technically inclined readers, we provide a formal definition of *iF*. It is the integrated incremental intake of a pollutant released from a source or source category (such as mobile sources or power plants), summed over all exposed individuals during a given exposure time, per unit of emitted pollutant. Mathematically, it can be expressed as:

$$iF = \frac{\sum Population \times Concentration \times Breathing\ Rate}{\sum Emissions} = \frac{\sum_{d=1}^{n} POP_d \times C_d \times BR}{EM}$$

where *iF* is the intake fraction from a source, POP_d is the population at location d, C_d is the change in concentration at d, BR is a nominal population breathing rate, used to yield a unitless measure, and EM is the total emissions from a given source. The sigma (Σ) represents a summation, which means the quantity is computed for all locations d in a modeling domain and they are then added together.

Breathing rates vary substantially across individuals and their activities. Intake fraction is a population aggregate relationship, however, making it impractical to incorporate individual breathing rates in many situations. Given this, researchers generally use a nominal constant value of twenty cubic meters per day (20 m^3/d), which is higher than the average breathing rate in the population but is commonly used in risk assessment.

We emphasize a couple of characteristics of *iF* that are additional to the discussion in the main text. The reader should understand that an *iF* value is not an inherent property of a pollutant. Rather, it depends on a number of variables, including release location and conditions, atmospheric and meteorological conditions, pollutant properties, along with population patterns mentioned earlier. This implies that any attempt to extrapolate *iF*s to unstudied settings must adequately characterize these variables to ensure accurate estimates.

Also, there are several key assumptions built into our use of *iF* that require mention. For example, it assumes linearity both of the dose-response function and of the relationship between emissions and marginal concentration changes. We refer the reader to chapter 4 for a discussion of such important caveats.

How is *iF* used? To estimate health effects, we will multiply the *iF* of a source by the emissions of the source and a dose-response coefficient, and divide by *BR* (as shown in equation [9.15] of chapter 9). Some readers might notice that the assumed *BR* thus cancels out, and its value is ultimately immaterial to our assessment. It is included here to aid conceptualization. It also illustrates that if age-specific dose-response functions were available (based on doses rather than concentrations), one could refine such an assessment with age-specific *iF*s and *BR*s to differentiate health effects by age.

Box 3.3
Alternative Dispersion Models for Estimating Intake Fraction

In calculating exposure assessment for our five sectors, the team chiefly employed an air-dispersion model of relatively simple Gaussian form, an industrial source complex (ISC) model. These models, designed to support regulatory modeling programs of the U.S. Environmental Protection Agency (EPA), are recommended for evaluating the impacts of many types of emission sources on air quality and are a common and familiar choice of environmental modelers. ISC has two types—short term (ISCST) and long term (ISCLT)—reflecting different meteorological inputs and objectives. We used the latter. An important characteristic is that ISCLT is considered robust for domains up to 50 kilometers.

For the power sector, our program conducted a second assessment of *iF* and health risk that paralleled the ISCLT-based effort, using an alternative, more advanced air-dispersion model, CALPUFF. This work was initiated to consider its feasibility for our national assessment. CALPUFF is a Lagrangian puff model, simulating continuous puffs of pollutants released into the ambient wind flow. Puff diffusion is Gaussian, but the path of each puff changes according to evolving wind-flow direction. It is designed and recommended by the U.S. EPA to study long-range pollution transport and, in contrast to ISCLT, accordingly includes a chemistry module to estimate formation of secondary pollutants, such as sulfate and nitrate particulates. (Secondary pollutants are those formed chemically in the atmosphere, as the primary emissions from sources react with other compounds in the air.)

The time and computational requirements to apply CALPUFF to a full sample of our power-plant database (let alone to those of the other sectors) proved imposing and infeasible. For our integrated national assessment, we have to compromise, and so we try to capitalize on the relative strengths of both of our atmospheric modeling approaches.

Our national assessment is essentially rooted in the ISCLT-based results, because we are able to conduct it on far more extensive and representative samples of the dominant emission sources. More specifically, the assessment considers our *iF* estimates from ISCLT-based chapters 5 and 6 to reasonably capture the exposure characteristics for primary particulates and SO_2 within 50 kilometers of the sources, consistent with the design limits of the ISCLT model. By using multipliers based on our CALPUFF results and those of other studies, we extrapolate these primary pollutant *iF*s to cover a national domain.

Despite data limitations, we want to include in our national assessment exposures to secondary particulates, likely to be a significant additional factor in total health burden. To account for secondary particulates—largely sulfates formed from SO_2—we are able to use results from chapter 7, which is consistent with the design aims of the CALPUFF model. Specifically, the CALPUFF-derived intake fractions of sulfates from SO_2 from the power sector can be defensibly applied to SO_2 emitted from any sector, because this study also indicates that stack heights and other emission characteristics are relatively unimportant determinants of *iF*s of long-range pollutants such as sulfates. In this case, population is the dominant determinant.

Box 3.3
(continued)

We add that our CALPUFF-based study is additionally useful for validating the general intake fraction approach in China, because of its comparability with CALPUFF-based *iF* analyses that had already been conducted in the United States. It illustrates how much higher population densities in China may increase risks for a given quantity of emissions. The CALPUFF analysis, moreover, stands up as a valuable independent study, with additional scientific aims that are distinct from our integrated national analysis.

pollution studies in China have attempted to study exposures using the population resolution that our team attempts.

In the remainder of this section, we describe how this method is applied to a large sample of sources, and then, combining with national data, generates a "national" intake fraction. With both source-specific and national intake fractions, we are developing tools to estimate human exposure and health effects of air pollution. In our application in chapter 9, we begin estimating the national health impact, such as the number of cases of chronic bronchitis, by rearranging the intake fraction formula of box 3.2. The health effect of emissions from a particular sector is given by the product of that sector's emissions, the national intake fraction for that sector, and a "dose-response" coefficient from epidemiological literature, all divided by the same constant breathing rate. These are only approximations, but they will be informed by more real-world information than earlier, cruder estimations of the health impacts of air pollution sources in China.

3.4.2 Estimating Intake Fractions for Five Primary Polluting Sectors

We now describe our application of the intake fraction approach to major polluting sectors (i.e., to modeling the pollution causal chain) from pollution emissions to concentrations to population exposure (components 2 and 3 above). This section summarizes an enormous amount of research, both in the field and through computer modeling, conducted by a Tsinghua University—Harvard University collaborative team.

As mentioned earlier, our research team selected five key sectors: electric power, cement, chemicals, iron and steel, and transportation. Of these, we chose electric power for the most complete assessment, by Bingjiang Liu and Jiming Hao in chapter 6, with particular effort to compile a unique national database of emission

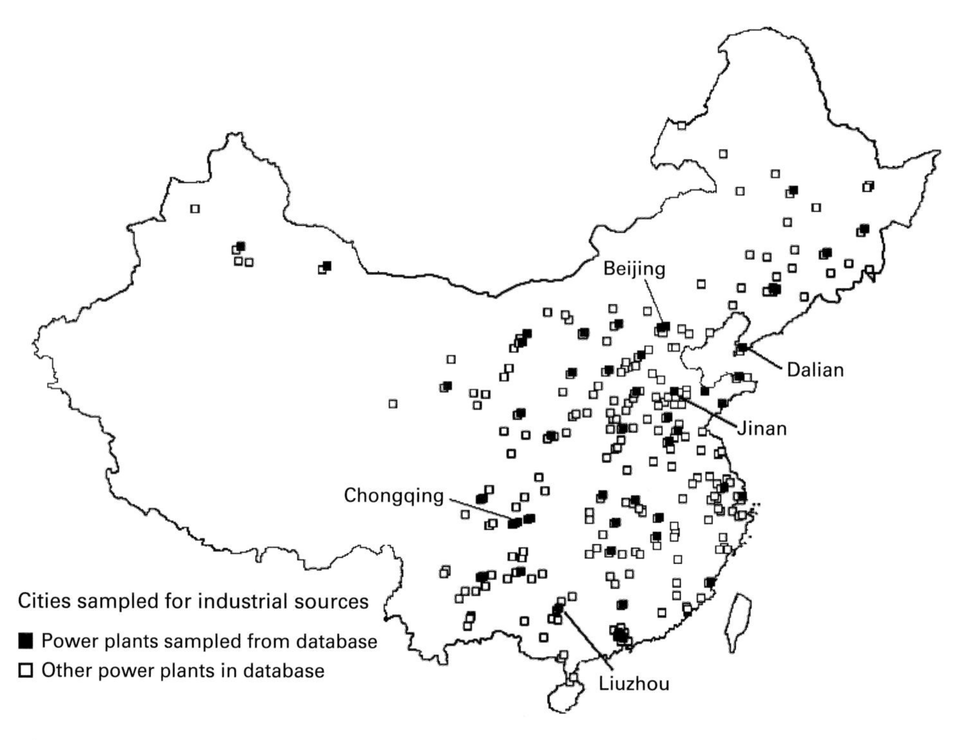

Figure 3.2
Location of sampled sources.

source characteristics of 686 smokestacks from 278 coal-fired power plants, largely relying on environmental impact assessment reports gathered for individual power plants. The locations of these plants are shown in figure 3.2. The plants account for 75% of all coal-fired power-generating capacity in China. The database contains voluminous information on each stack, including the location, local meteorological conditions, and source characteristics such as stack height, stack inner diameter, emission exit velocity and temperature, and emission rates of SO_2 and TSP. From this database, we select a representative fifty-two plants (the black squares in the figure) to calculate the intake fraction by using an air-dispersion model that is practical to apply to a large number of sources.

Our program includes a second, parallel assessment of intake fraction in the power sector, applying a more advanced atmospheric model to emission characteristics of one specific power plant, rather than a national plant sample. (The models are contrasted in box 3.3.) This study, chapter 7 by Ying Zhou, Jonathan Levy,

James Hammitt, and John Evans, then evaluates intake fraction of the power sector more generally by applying the model to those same plant characteristics in locations across China. The advantage of this effort is the much more advanced chemistry, meteorology, and physical transport in its dispersion model, though it proved infeasible for us to apply it to the many unique plants of our national sample. Nevertheless, this study provides our assessment with evidence to adjust prior intake fraction estimates from local to national scales, and with intake fractions for secondary pollutants, especially sulfate particles formed from SO_2, as described later in this section. These prove critical to making a comprehensive estimate of total health damages.

It was infeasible to compile a national database for the other sectors comparable to the one for electric power. Instead, we chose to investigate sample sources from five cities representing a range of economic development, air quality, and geographical location and for which data might be obtained by our field research teams: Beijing, Dalian, Jinan, Chongqing, and Liuzhou (see map in figure 3.2). Our nonpower sample for calculating intake fraction thus includes 169 smokestacks from cement plants, 185 from the chemicals industry, 187 from iron and steel plants, and 75 road segments for the transportation sector, for a total of more than 600 sources. This major research effort was conducted by Shuxiao Wang, Jiming Hao, Yongqi Lu, and Ji Li in chapter 5.

We illustrate the sources sampled in our example city, Jinan, in figure 3.3. Note that municipal jurisdictions in China include substantial rural areas in addition to urban centers. Of the 150 stacks selected in Jinan, 85 are classified as urban and 75 as rural. These sources cover more than 80% of emissions of TSP and SO_2 in Jinan.

The other key data ingredient in determining intake fractions is the population distribution. This too required a substantial research effort, in that digitized county-level population maps for China were available only on the basis of 1990 census data, well before our modeling year of 1997. Our team thus constructed an updated county-level population map for the entire country, shown in figure 3.4, from 1999 estimates available to us. To improve the population resolution for the nonpower industrial and transport sectors, in which emission heights are generally lower and thus local residents at greater risk, we developed more detailed population data for our five cities. The population distribution of our example city of Jinan is shown in figure 3.5, in which land use information was used to refine the population distribution to a 1 km $\times$ 1 km grid resolution.

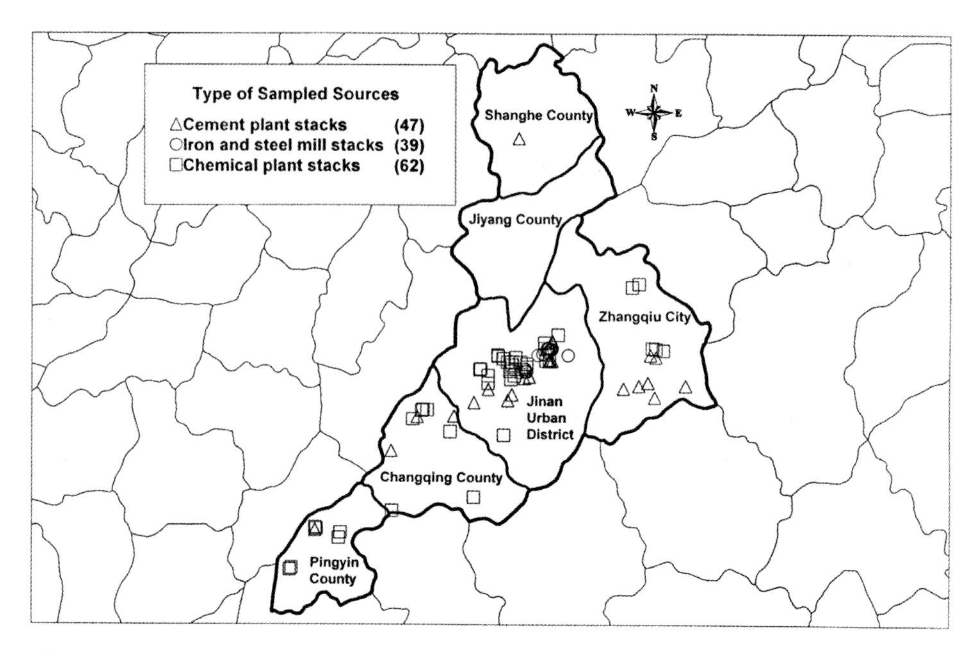

Figure 3.3
Selected sources for example city, Jinan.

We present now, in table 3.2, the results of all of the intake fraction modeling and calculations for PM emitted by our sampled sources for the five sectors. We also give the intake fraction for secondary sulfates from SO_2 emissions in the power sector. The purpose here is illustrative, and we do not also present the analogous results for primary SO_2 and nitrates, which can be consulted in chapters 5–7.

For our continuing example of cement plants, one can see that they have a calculated average intake fraction for TSP of 5.85×10^{-6}, which means that for every metric ton of TSP emitted from a cement plant, an average of 5.85 grams are inhaled by the population within 50 kilometers around the plant. Note that there are significant differences among the intake fraction estimates for different sectors. The intake fractions for nonpower industrial sectors are two to three times of that for the power sector. This is an immediately notable result because so much prior pollution research has focused on the power sector, the largest user of coal in China. Our results highlight the importance of considering other sectors that have much higher human exposures.

The mean intake fraction for mobile sources (i.e., the transport sector, analyzed by treating road segments as sources) is roughly an order of magnitude higher than

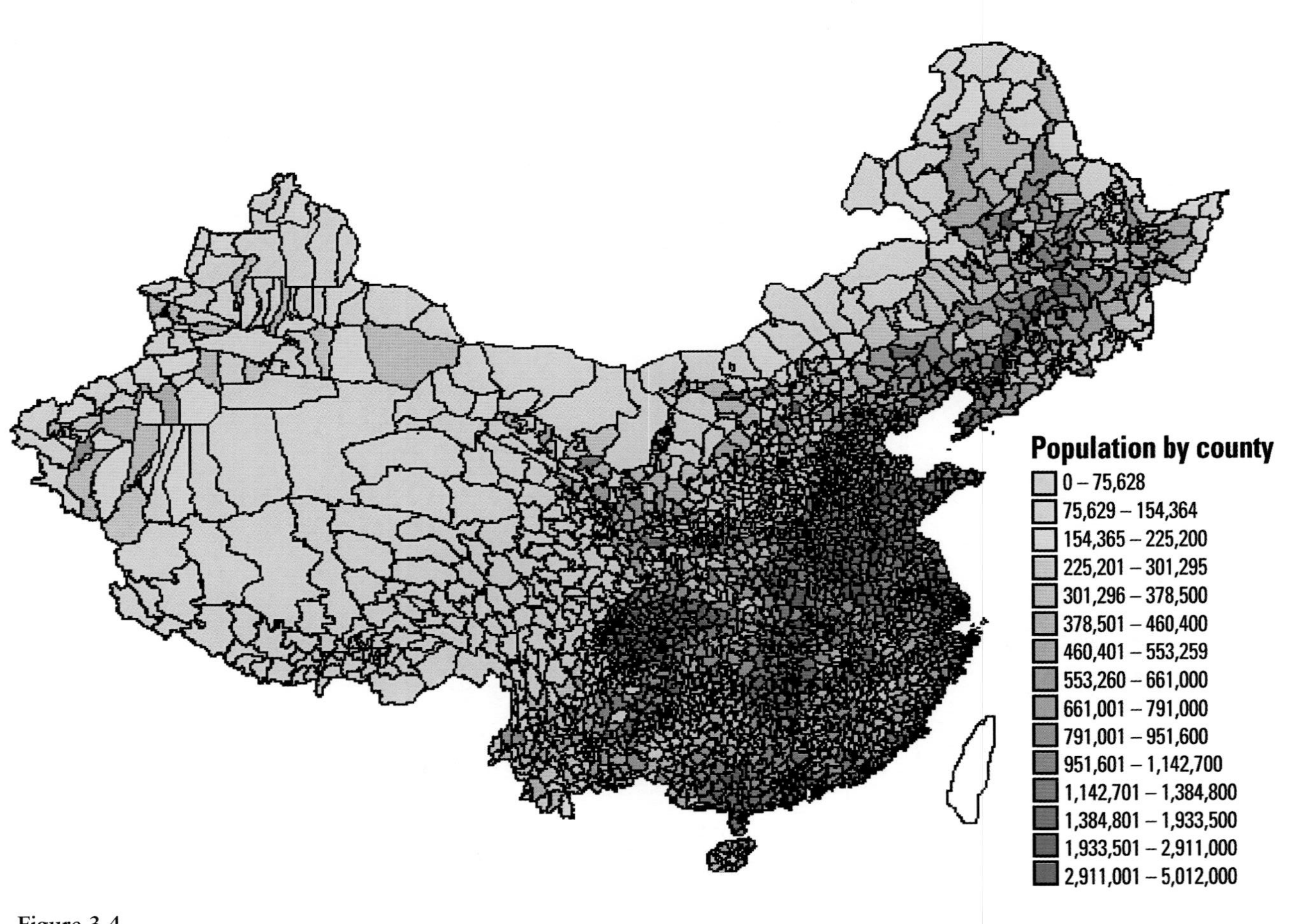

Figure 3.4
County-level population distribution, 1999.

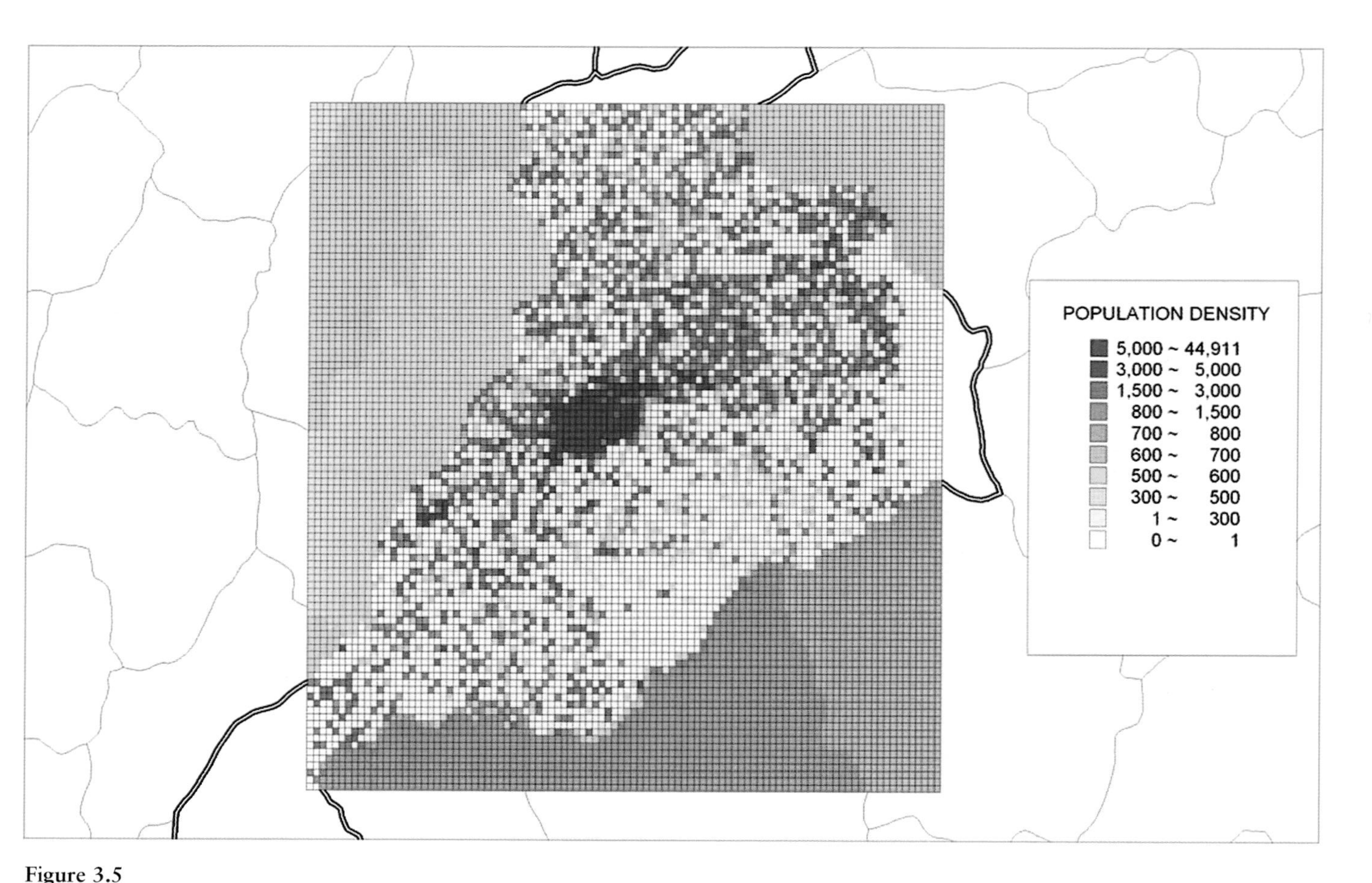

Figure 3.5
Population distribution for example city, Jinan.

Table 3.2
Sample intake fraction results for five-city sample

	Sample *iF* for Primary TSP (or PM_{10} for Roads)		
Sector	Mean	Maximum	Minimum
Cement plants	5.85×10^{-6}	3.07×10^{-5}	7.10×10^{-7}
Chemical plants	8.87×10^{-6}	5.05×10^{-5}	1.67×10^{-6}
Iron and steel plants	7.19×10^{-6}	5.32×10^{-5}	2.46×10^{-6}
Power plants	2.61×10^{-6}	1.23×10^{-5}	1.60×10^{-8}
Road segments	7.72×10^{-5}	1.54×10^{-4}	1.84×10^{-5}
	iF for Secondary Sulfate PM		
Sector	Mean	Maximum	Minimum
Power plants	4.40×10^{-6}	7.3×10^{-5}	7.3×10^{-6}

the industrial sectors. It may be that these levels result from closer proximity to people, because of both the ground-level height of vehicle exhaust pipes—contrasted with elevated stacks in the other sectors—and the high population densities around urban streets. We acknowledge here that our data for transportation are relatively thin, and improving the intake fraction estimations for this sector is the subject of an ongoing new phase of study by the research team. These intake fraction results, like other exposure estimates of this kind, involve substantial uncertainties. A brief summary of our sensitivity analyses of the assumptions is presented in box 3.4. It is important to understand these uncertainties if we are to emphasize to policy makers the degree of precision such research can offer. We believe these are useful estimates of the true exposures, even if they may be over- or understated by 100%.

Policy makers are better served by clearly qualified estimates than unexplained ones, which must be accepted at face value, or no systematic information at all. This also highlights the importance to well-informed policy of gathering basic data, making it objective, transparent, and available, and incrementally refining analyses such as those presented here.

3.4.3 Generalizing Intake Fraction Relationship to Nonmodeled Sources

The intake fractions reported in the previous section describe our sampled emission sources. As noted earlier, the advantage of the intake fraction approach is the ability to apply relationships derived from modeling a sample of sources, to give reasonable estimates for sources that cannot be individually modeled. This might be because

Box 3.4
Sensitivity Analysis of Intake Fraction

Every such study requires sensitivity analyses, by which researchers test the dependence of results on ranges of parameter inputs to gain a sense of the robustness of the methodology and the scale of uncertainties. In this case, we tested the results on alternative modeling domains, population resolution, terrain effects, half-life of SO_2, and aerodynamic diameter of PM. Our sensitivity analyses are explained fully in the following chapters, and we merely summarize here that these tests show effects ranging from a 49% decrease in results (from decreased population resolution) to as much as a threefold increase (from increased size of modeling domain). The most sensitive of the parameters tested was the size of the domain. We conclude that our method appears appropriate to evaluate *local* impacts on human health of primary SO_2 and TSP emissions but that estimates of total regional impacts have to be extrapolated very carefully. Our ISCLT-based estimates are conservative, in that a better accounting of exposures beyond the 50-kilometer limit of this dispersion model, of both primary and secondary pollutants, would increase the fraction of the pollutant (or its by-product) inhaled and the associated health impacts. Our CALPUFF-based estimates help us account for this.

there are too many of them—which is certainly the case when considering the entire Chinese economy—and/or because there are data limitations—which is often the case in developing countries, including China.

These relationships between intake fractions and source summary characteristics take the form of regression equations. In constructing a regression equation for intake fraction, we ask whether a short list of critical source characteristics may be sufficient to explain much of the variation in intake fraction values across different sources, or, in other words, whether a few variables are dominant determinants of human exposure. Examples of such source characteristics, or "explanatory variables," might be the total number of people near an emission source or the height of stacks, which might allow concentrated pollution plumes to diffuse before they reach human lungs. If this is the case, we will need relatively little information to produce useful, if approximate, population exposure estimates for all of the thousands of sources that are impossible to study individually.

As discussed in box 3.5, this proved to be the case with our intake fractions. We found that by using just stack height and two simple types of population values we could get intake fraction estimations that on average are only 15% off the values we get by doing the laborious work of running the dispersion model and assessing detailed population distributions. Such regression equations for each of our sectors and our two pollutants thus serve as the tools for generalizing our results more broadly.

Box 3.5
Regression of Intake Fractions

As described in chapter 4, prior studies in the United States have demonstrated the feasibility of developing *iF* regression equations based on sample data. We tested a number of regressions with each of our sector *iF*s on the left-hand side, and a number of descriptors on the right-hand side (noting that some variables are logarithms). Indeed, we found that relatively few right-hand variables explained, to use some statistical language, a high degree of variation of the *iF*.

We illustrate one of our regressions, on TSP, with our continuing example of the cement sector. To reflect the dispersion patterns of cement emissions, the explanatory variables include stack height, the population within a radius of 10 kilometers as the near-source population, and the population between 10 and 50 kilometers of the pollution source. The results of the regression are shown in the figure below, where the actual values are on the vertical axis and the fitted values from the regression are on the horizontal axis. We see that the fit is good (with an unadjusted R^2 of 0.87). The coefficient on near-source population is highly significant, as is the coefficient on stack height. We also ran regressions for the other sectors with the same or similar independent variables and found analogous significance of variables and R^2 values between 0.72 and 0.87. The full regression results are presented in chapters 5 and 6.

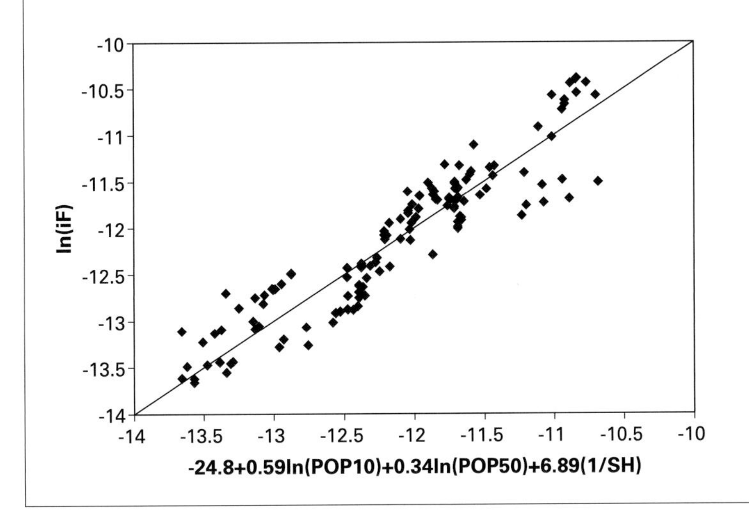

To estimate national intake fraction averages for the sectors, we need the average values of just these few explanatory variables for each sector: population exposed within given distances from the sources and the stack height. To find these for the electric power sector, we have the advantage of the national database that our research team originally constructed. For the other sectors, we use industry yearbooks that provide summary information for all enterprises in China and select a large sample. The summary information includes location, types of output, capacity, and so forth. From these data and general information about boiler sizes and stack heights, we estimate the population and stack heights for each source in the national sample. The intake fraction is then estimated for each source by using the summary regressions. The industry average *iF* is derived by averaging over the sample, weighting by production capacity.

3.4.4 Estimating Intake Fractions for Secondary Particles

As noted above, chapter 7 gives a separate assessment of the intake fractions in the electric power industry using a more sophisticated air dispersion model that includes estimates for different sizes of primary particulates, and secondary sulfates and nitrates formed from emissions of SO_2 and nitrogen oxides (NO_X). This considerably more elaborate calculation is performed for a smaller sample of sources. The mean intake fraction for sulfates (grams of secondary sulfates per gram of SO_2 emitted) is 4.4×10^{-6}, which is assumed for all industries (see box 3.3 for explanation). As we will describe below, in section 3.7, this translates into a large estimated contribution of secondary particulates to TSP concentration. Thus, although the estimates for the secondary effects are much less reliably established in this methodology, their role in health damages may be too large to simply exclude them from consideration.

With these national average intake fractions for the five most polluting sectors, we have information that we can combine with our national economic and emission data, along with information from component 4, to characterize the health risks due to TSP and SO_2 emitted from those sectors. Although the precise national intake fraction estimates may not be very meaningful to nonspecialists, having led the reader this far through the intake fraction assessment, we present the final estimates in table 3.3.

We note that all four industrial sectors have values for both TSP and SO_2 on the order of 10^{-6}, or, as we have interpreted previously, a range from 2.16 to 6.61 grams in every metric ton of the pollutant is estimated to be inhaled by the local population. Also, despite the steps to generalize the earlier sample results to national

Table 3.3
National average intake fractions

Sector	*iF*s		
	Primary TSP	Primary SO_2	Secondary PM (sulfate)
Cement plants	3.46×10^{-6}	3.99×10^{-6}	4.4×10^{-6}
Chemical plants	3.28×10^{-6}	6.61×10^{-6}	4.4×10^{-6}
Iron and steel plants	3.75×10^{-6}	6.03×10^{-6}	4.4×10^{-6}
Power plants	2.16×10^{-6}	2.38×10^{-6}	4.4×10^{-6}

averages, the power sector notably retains the lowest intake fraction for both pollutants by a considerable amount, presumably for the reasons described above.

3.5 Component 4: From Human Exposures to Health Effects

Once population exposure has been estimated, the next step in a health-damage assessment is to determine the effects of changes in exposure on health outcomes of concern—component 4. These effects include premature mortalities (deaths), morbidities (disease), and other health outcomes.

To determine the health impacts of particulate and SO_2 exposures in China, we have a methodological challenge of estimating the effects when relatively few studies have been conducted on Chinese populations. As mentioned above, this assessment does not offer any new air pollution epidemiology, the observational studies of human populations that identify relationships between exposure to measured pollutants and health effects. We must instead review scientific evidence of risks of health effects of air pollution from the limited Chinese studies that appear in the literature. (Several such studies were conducted and published by colleagues of our team in earlier phases of our program.) In chapter 4, Jonathan Levy and Susan Greco carefully consider how Chinese dose-response relationships—the derived coefficients relating pollution concentrations to health endpoints—compare with those determined in the much larger body of air pollution epidemiology in other countries. We also have to translate our particulate data, which in China is chiefly gathered for TSP, to accord to the relationships found through epidemiology, which chiefly investigate the narrower category of PM_{10}. To do this, we use a PM_{10}:TSP ratio derived from studies in China, as described in chapter 4.

At first, the application of dose-response functions to our exposure estimates appears to be a relatively simple step, because it relies on previously published

Box 3.6
Uncertainties in Air Pollution Epidemiology

The science of air pollution epidemiology is characterized by much uncertainty. A primary concern is that one must always control for possible confounding factors—the unobserved variables that can be independently associated with both the exposure and the health outcome. As a simple example of a confounder, an observed correlation between increased bronchitis and elevated carbon monoxide (CO) does not demonstrate a causal relationship if the researcher does not consider other pollutants—PM_{10}, for instance—that might be produced by the same sources producing the CO. The bronchitis may well be caused not by the CO, as the researcher assumes, but by the unobserved, confounding PM_{10}.

Confounding complicates the applicability of an epidemiological result for one population-pollution context to another. The United States and China, for instance, have different pollution mixes, particle-size distributions, threshold factors, and demographic and health characteristics of their populations. The problem is not chiefly that physiochemical processes of disease are likely to differ between peoples, but that it becomes harder to have confidence one has controlled for all possible confounders. This is one reason—others are discussed in chapter 4—why we strongly prefer studies conducted in China. Western dose-response coefficients should not be the primary source for studying Chinese health response if it can be avoided, at least without strong caveats about uncertainties. Unfortunately the literature for China is sparse enough that we do use some Western results for uncertainty bounds in our analysis.

In fact, the uncertainties of air pollution epidemiology are large enough to invite contentions even in a country with the voluminous supporting research of the United States. In the late-1990s, debate about interpretation of studies behind a ruling of the U.S. EPA to strengthen standards for fine particulates ended up in the federal court system.

findings. In fact, it is a step with great uncertainties and use of subjective judgment (see box 3.6). In the end we have little choice but to review scientific evidence of risks of mortality and morbidity from the limited Chinese studies that exist, Western studies, and the similar evaluations of dose-response relationships that other research teams have made. We then apply what we decide is the best available information, noting the uncertainties of our limited body of supporting research and of extrapolating evidence from one country to another.

After reviewing the literature, for the important premature acute mortality relationship, we use values from Chinese studies of an increase of 0.03% in the death rate per $\mu g/m^3$ for PM_{10} (particles less than 10 microns in diameter) and 0.03% for SO_2. With the current death rates in China these are approximately 1.95 excess deaths per million people per year due to 1 $\mu g/m^3$ increase in the concentration of

Table 3.4
Dose-response estimate for health effects, base case

Health Effect	Cases per Million People per $\mu g/m^3$ Increase
Due to PM_{10}	
Acute mortality (deaths)	1.95
Chronic mortality (deaths), for comparison	25
Respiratory hospital admissions (cases)	12
Emergency room visits (cases)	235
Restricted activity days (days)	57,500
Lower respiratory infection/child asthma (cases)	23
Asthma attacks (cases)	2,608
Chronic bronchitis (cases)	61
Respiratory symptoms (cases)	183,000
Due to SO_2	
Acute mortality (deaths)	1.95
Chest discomfort (cases)	10,000
Respiratory systems/child (cases)	5

PM_{10} or SO_2. For most of the other values in our base case, we adopt the dose-response relationships of the World Bank study mentioned earlier. A summary is reported in table 3.4. We also use differing results from other studies as upper and lower bounds for sensitivity analysis in the estimation of national health effects in chapter 7.

3.6 Component 5: Economic Valuation of Health Effects

Attributing the incremental morbidities and mortalities (e.g., the cases of actual respiratory disease and related death) to pollution emissions from particular industries provides a useful quantified basis for evaluating pollution control policies. Our assessment can evaluate and report the health effects alone as an output of the policy options that we will evaluate.

Most environmental policy makers in China and many in other countries, however, also ask that health impacts be translated into economic terms, to compare costs and benefits of possible policy interventions. Our economic model, as used in components 6 and 7, is naturally amenable to this interest. Because the policies we choose to evaluate are damage-weighted taxes, this also requires the economic valuation of the various health effects described in the prior section.

The monetization of health damages—component 5—is acknowledged even by its specialists as a quite imprecise and sometimes controversial science. It nevertheless offers useful, if approximate, information to help communicate and prioritize environmental risk in real-world policy processes. It is essential if policies are to be evaluated in a benefit-cost framework, once again with uncertainties carefully recognized.

A number of estimation methods, with associated terminologies, have been developed to value damages to health. These include assessments of "willingness to pay," determined by "revealed preference" observed from actual market transactions (such as higher wages for risky jobs), or by interview-based surveys ("contingent valuation"). They also include a more conservative approach sometimes favored by Chinese researchers, the "human capital" method that values health outcomes only by their effect on expected earnings. A brief overview of valuation methods is presented in box 3.7.

Our assessment includes a new contribution to this field, one of the first contingent valuation studies conducted in China, chapter 8 by Ying Zhou and James Hammitt. We studied three different locations in China to value adverse health effects associated with air pollution, as well as to study the regional differences in willingness to pay within China. We included three health endpoints—colds, chronic bronchitis, and premature mortality—to represent a range of severity of health endpoints related to air pollution and for their comparability to similar estimates from other countries. The survey was conducted in three locations to represent the preferences of residents of big cities, small cities, and the rural areas of China.

We cannot present the full results of the contingent valuation study here. We provide just one important example of the results. Depending on location, the mean value of a "statistical life" indicated by the survey was 230,000 to 506,000 yuan, and we use a simple average of this for our base-case value. These values are lower than those used in many studies (including the World Bank's *Clear Water, Blue Skies*) that transfer U.S.-based valuations to China assuming a proportional income effect ("unit income elasticity," in the jargon of this literature). They are, however, quite comparable to results from one other China willingness-to-pay study, by Hong Wang and colleagues, initiated in an earlier project of the Harvard China Project and described further in chapters 1 and 9. They are also consistent with estimates from richer countries transferred to China using higher income elasticities.

Box 3.7
Valuing Health

Although there are many studies valuating health in industrialized countries, there are relatively few such estimates for developing countries. As a result, values are often approximated by transferring estimates from industrialized countries. These benefit transfers attempt to account for differences in income, though the rate at which benefits vary with income is highly uncertain. Because of the lack of estimates for developing countries, it is impossible to account for any differences in preferences that may result from disparities in economic opportunities, health care systems, and cultural characteristics.

The monetary value of a reduction in health risk is defined as the affected individual's willingness to pay (WTP) to obtain it. WTP describes the amount of money that the individual views as equally desirable as the reduction in health risk.

There are two methods for estimating WTP for risk reduction—"revealed preference" and "contingent valuation." Revealed-preference methods observe the choices people make in settings where they are choosing between different levels of risk and money, from which an analysts determines their WTP for risk reduction. Most of the revealed-preference estimates of the value of mortality risk are based on comparing wages and occupational fatality risks. Among the jobs for which an individual is qualified, the more dangerous jobs offer higher wages. (If not, employers could not attract workers to take these jobs.) From these revealed preferences, the value of an unidentified statistical life—a social value, different from a value of a given individual's life—can be estimated.

The alternative method, contingent valuation (CV), is based on asking survey respondents how they would act in hypothetical settings involving choices between alternatives that differ in risk and financial consequences. Compared with revealed-preference methods, CV is much more flexible, because respondents can be asked about their choices in a range of hypothetical settings. In contrast, revealed-preference estimates can only be obtained in cases where the analyst can determine not only what option each individual chose, but also what alternatives were rejected. CV also offers the advantage that respondents can be informed of the risks and monetary consequences associated with alternative choices. With revealed preference, the analyst does not know how well the individuals knew the consequences of alternative choices. It is important to note, however, that revealed-preference methods offer the advantage of relying on decisions with real consequences, unlike CV, where there is less incentive for the individual to respond carefully to hypothetical scenarios.

Table 3.5
Range of values of a statistical life

	Valuation per Case (in constant 1997 yuan)		
Health Effect	Base Case	Lower Bound	Upper Bound
Premature mortality	370,000	130,000	950,000

In view of the large uncertainties, it is common practice to apply a range of valuations from different studies, rather than a single estimate, to the assessed health impacts. Our contingent valuation results are thus one of several monetization inputs to the economic damage assessment and policy analyses that follow. We use them for a base case and lower bound on premature mortality from PM_{10} exposure, using an upper bound derived from estimates from other countries, adjusted to China. These are shown in table 3.5. Valuations are also applied to nine morbidity and other health endpoints, and the reader is referred to chapters 8 and 9 for details.

3.7 Component 6: National Damage Assessment by Sector

We now bring all of the foregoing results together to estimate the health damage due to air pollution from each sector of the economy, returning to the economic framework and research of Mun Ho and Dale Jorgenson in chapter 9. These estimates are needed to identify the major sources of damage, including the sector share of total damages, and the damages per unit output. This information is directly related to our thirty-four-sector environment-economic model of China. We are thus able to estimate the change in health damages over time. This will also allow us to examine the effects of pollution-control policies in our last component, as described in the final section below and in chapter 10. We present only key results here; the reader is referred to the final two chapters for a full discussion of the health and economic results of the integrated assessment and the policy implications.

3.7.1 Damage by Sector

We begin with a return to the estimates of emissions of TSP and SO_2 for each sector, as described in component 1 and represented in table 3.1. We combine these emissions with the intake fraction results from components 2 and 3 to get human exposures (i.e., the dosages of primary TSP, SO_2, and secondary sulfates and nitrates due to our five key polluting sectors). For the less critical, cleaner sectors of the economy, we attribute population exposures by use of a simplified approach. By applying the dose-response relationships described in component 4, we translate these sector population exposures into health outcomes, and using valuation results of component 5, the result is a valuation of the marginal health damage from air pollution for each and every sector. This will enable us to derive the marginal damage per unit output for each sector, in terms of yuan of damages per yuan of output. These are presented for our base year in the third column of table 3.6. Most of the damage is from TSP (primary and secondary) and very little from primary SO_2.

Electric power alone is estimated to cause more than 26% of all health damages from ambient air pollution from combustion. Perhaps unexpectedly, the nonmetal mineral products (cement) sector ranks second, responsible for more than 12% of damages, but still less than half of the electric power share. Transportation ranks third, at nearly 10%, and this was for a base year well before personal vehicle ownership in China skyrocketed. The other sectors that we studied carefully in components 2 and 3, chemical production and metals smelting, contribute less to population health risk, at less than 7% and 5%, respectively, though still more than most remaining sectors.

We can draw some conclusions from our assessment thus far. First, our more careful consideration of population exposure in assessing health damage appears to affect recommended sector priorities. Electric-power generation, which consumes the most coal, is likely the largest source of health damage from ambient air pollution, as is often assumed. Its primary particulates cause much health damage, but the power sector ranks so high because it also emits large quantities of SO_2. These emissions pose a direct threat to health, but, more important, engender enormous secondary particle formation. This illustrates the difference between analyses that focus on local damages versus those that consider regional damages. The local damage for electric power stations is often estimated at low values because of their very high stack heights. Secondary particles, however, are spread over great distances, and thus the regional and national damages are very high.

Second, cement production may be underemphasized as a source of environmental health risk from fossil fuel combustion. It appears to far outrank all industrial sectors except power generation, and its impact could be even larger if its enormous noncombustion PM emissions are shown to imperil health substantially. Third, although our estimates for the transport sector are weak, because of a limited database, damages from this sector appear very large even in our base year of 1997, before the vehicle population of China began its especially swift growth. As noted earlier, this embodies the higher *iF*s for transport, where emissions occur both at ground level and predominantly in the midst of population concentrations (cities).

Table 3.6
Total health damages by sector share, 1997 (manufacturing and transportation)

Sector	Damage Rate (yuan damage per yuan of output)	Share (%)
7 Food products, tobacco	0.00262	2.62
8 Textile goods	0.00255	1.70
9 Apparel, leather	0.00030	0.14
10 Sawmills and furniture	0.00420	0.68
11 Paper products, printing	0.00467	1.52
12 Petroleum refining and coking	0.00276	0.62
13 *Chemicals*	*0.00607*	*6.60*
14 *Nonmetal mineral products (cement)*	*0.01960*	*12.52*
15 *Metals smelting and pressing (iron and steel)*	*0.00875*	*4.79*
16 Metal products	0.00060	0.21
17 Machinery and equipment	0.00162	1.04
18 Transport equipment	0.00087	0.36
19 Electrical machinery	0.00059	0.24
20 Electronic and telecom. equip	0.00020	0.07
21 Instruments	0.00073	0.05
22 Other manufacturing	0.00432	0.90
23 *Electricity, steam, hot water*	*0.09362*	*26.02*
24 Gas production and supply	0.01467	0.13
25 Construction	0.00275	3.49
26 *Transport and warehousing*	*0.02668*	*9.87*
Other sectors		26.41
Total		100.00

Note: The five major emitting sectors that are the core of our exposure analyses and that are discussed at length in the text are in italics.

In addition, this does not yet incorporate impacts of the more complex secondary mobile-source pollutant, ozone. This suggests that pollution-control efforts give increased priority to transportation and that researchers try to improve data and estimates of mobile-source damages (as our own program is currently doing in a next phase of research).

We must emphasize one critical qualification about the results in table 3.6. As noted in our discussion of component 1, these are for pollutants generated by combustion only and do not consider process emissions. If one assumes that particulates from mechanical and chemical processes are similar to combustion particulates in size, composition, and thus health effects, the rank order of sector damage share would be quite different. The enormous process emissions of the cement industry would vault it far above all other sectors, including electric power, as a contributor to total damages. As reported in chapter 9, in this case cement would cause nearly 32% of all damages, with electric power falling below 18%, and iron and steel rising to 8%. An assumption of similar particulate characteristics from industrial processes and combustion, however, is a large one. We draw a more qualified conclusion for now that the cement industry in China, already second ranking in damages for its combustion emissions, could possibly be an even greater concern. Policy makers should take note. Before we can be more conclusive, it is essential to analyze the nature of process particulates from the cement and iron and steel industries.

3.7.2 Damage per Unit Fossil Fuel

We also estimate the damage due to primary fuels, to provide guidance for possible fuel tax policies. Although different sectors produce different emissions per ton of fuel burned and we could produce damage estimates accordingly, we do not differentiate damages by both fuel and sector. Taxing fuels has proven politically feasible in some countries, but setting different fuel taxes for different sectors is impractical, whether in China or elsewhere. Instead, we estimate the average national damage per ton of coal, ton of oil, or cubic meter of gas. Our figures here are based on impacts on urban populations only.

The values of the average damage are given in table 3.7. The estimated damage from coal is very high, 93.84 yuan per ton in 1997, or 0.58 yuan of health damage per yuan of coal. This means that the damages nearly match the market cost. The damage from oil is much less, only 0.025 yuan per yuan of oil, while the damage from gas is negligible. We will return to these results in one of our policy simulations in the next section.

Table 3.7
Health damage from fuels, 1997

Fuel	Average Marginal Damage National Fuel Use	Unit	Yuan of Damage per Yuan of Fuel Burned
Coal	93.84	yuan/ton	0.5751
Oil	54.28	yuan/ton	0.0249
Gas	0.08	yuan/1000 m^3	0.0002

3.7.3 Aggregate National Damages

Policy makers who want to make use of pollution-damage research often have great interest in an aggregate national damage value, typically expressed as a percentage of GDP. This is simple for us to estimate, in light of the research described so far. Readers may be surprised that we report this number here for an unexpected reason: to warn against regarding this value as a particularly useful result of such assessments, whether by us or by other teams. Although we believe ours may be one of the most detailed analyses to date, aggregating all of the foregoing research into a single number also aggregates all of the many uncertainties. In view of the limits of data in China and the assumptions that are implicit in the varied modeling steps, such percentage-of-GDP figures should be interpreted very cautiously. From our discussion of the intake fraction sensitivity analysis, the dose-response uncertainty, and the valuation uncertainty, the reader may rightly gather that we can legitimately report only a broad range of possible values. The true value may be off by 100% or more from our base-case estimation, and this applies to other similar assessments, and not just of China. In view of the magnitude of the damages, however, even a large underestimate is worrisome enough to warrant an active pollution-control policy. More data and information would sharpen the total damage estimates, but we caution that the degree of uncertainty will remain unavoidably large.

Readers disappointed by this warning should understand that such a figure was never the objective of our assessment in the first place. Note that there is little that policy makers can do with such a value, as top officials of the State Environmental Protection Agency (or SEPA) have themselves noted to us, except to advertise the seriousness of the air pollution problem as a whole. Even if such a figure were strictly accurate, it would by itself recommend little about how to target control of pollutants.

An analysis like ours is both more robust and more useful for its indications of relative damages within the Chinese economy. The absolute numbers are only rough

Table 3.8
Total health damages in 1997 using central values

Health Effect	Number of Cases	Value (million yuan)
Due to PM_{10}		
Acute mortality	43,200	16,000
Respiratory hospital admissions	265,000	464
Emergency room visits	5,200,000	738
Restricted activity days	1,270,000,000	18,200
Lower respiratory infection/child asthma	507,000	41
Asthma attacks	57,600,000	1,420
Chronic bronchitis	1,350,000	64,900
Respiratory symptoms	4,040,000,000	15,000
Due to SO_2		
Acute mortality	50,600	18,700
Chest discomfort	259,000,000	1,600
Respiratory systems/child	129,000	1
Total		137,000

estimates, but, because our assessment methods are consistent across the sectors, we are on firmer ground drawing conclusions about how they relate to each other. Indeed this is far more useful in a policy sense. Decision makers need to know not just how much air pollution impairs public health or the economy, but also how to focus policies on those parts of the economy that are most responsible, such as power, cement, and transport sectors. As we will see in the final component, which follows, differentiating more harmful industries (or fuels) from less harmful ones enables us to investigate pollution-control policies that are better targeted or, in economic parlance, that internalize the environmental externalities with greater efficiency.

With this strong caveat about the utility of the following numbers, our estimated value of national health damages due to air pollution using the central parameter values is 137 billion yuan for 1997, or 1.84% of GDP. Of this figure, primary particulates only account for 56 billion yuan, with the remainder due to secondary particulates and primary SO_2. As shown in table 3.8, premature mortality accounts for 34 billion yuan and chronic bronchitis 65 billion yuan.

Our final note on this component is that, as previously in the intake fraction analysis, we investigate the uncertainties in our approach and test the sensitivity of our results to key parameter assumptions. As explained in more detail in chapter 9, when low-end parameter values are used for dose-response and valuation, the national damages come to 0.65% of GDP, whereas upper-end values yield 4.7%.

3.8 Component 7: Benefit-Cost Analysis of Policies to Reduce Emissions

In light of the severe damages caused by air pollution in China, there have been many studies examining options to control or reduce emissions. Unlike most of these studies, we make an integrated estimate of the health benefits and costs of pollution control policies throughout the entire economy. In chapter 10 by Ho and Jorgenson, we focus on economywide tax policies and examine how they affect fuel use and, hence, emissions and health damages. At the same time, we estimate how these taxes affect output, allocation of resources, existing taxes, and, over time, economic growth. We also consider the effects on GHG emissions.

Why analyze market-based policy instruments when more traditional pollution control approaches, such as the technology mandates familiar to most officials and researchers, are more the norm in China? Market-based instruments such as the taxes we analyze are increasingly preferred for pollution control in Western countries, in those circumstances where the nature of the hazard suits the approach and where institutional and legal capacities are in place to support them. Such conditions are often lacking in developing countries, particularly for emissions trading, including to some degree in China as described by Morgenstern et al., Alford and Shen, and others, referred to in chapter 1. Chinese officials have strong interest in environmental policy innovations of the West, however, and a track record of trying them in carefully limited real-world policy experiments. In some cases, market-based approaches are increasingly workable in China now. In others, analyses of them may point to the future as market, institutional, and legal reforms proceed.

The full effects of such policies extend beyond environment. Ideally, for instance, taxes can rationalize energy inputs throughout the economy, including in sectors with few emissions at all, and can establish a revenue source that itself may benefit the economy if it displaces more distortionary taxes. Most traditional pollution-control assessments—including nearly all conducted to date in China, many of which are described in chapter 1—are not designed to evaluate such secondary economic impacts of a policy intervention. These aggregate impacts, however, can be enormous. An appropriate way to estimate the full effects of such policy alternatives is with a dynamic national economic model, such as the one employed in the policy simulations of chapter 10.

Some green tax studies consider the possibility of a "double dividend"—reduced environmental damage and improved economic performance. Although this is possible in a system with many existing distortions such as that in China, we do not explicitly address this issue. We limit ourselves to asking a simpler question: what are the effects on prices, output, consumption, and economic growth of taxes that are proportional to environmental damage? Our answer is subject to the same uncertainties that underlie the damage estimates discussed previously. Nevertheless, we believe our results are instructive. They give us a sense of the relative magnitudes of the costs and benefits involved that would help prioritize mitigation efforts.

3.8.1 Methodology for Analyzing Pollution Taxes

Our approach is to begin with the estimates of damages due to the current patterns of output and fuel use in each industry, as described in the previous component. Our framework also allowed us to calculate the health and economic damages per unit of coal or oil used. The economic model is dynamic and allows us to estimate the value of damages each year as the projected economy expands and changes in structure. If we do this assuming no new policy interventions, we have a base case. The assumptions imbedded in our base case and its results over a thirty-year period are described in box 3.8.

In light of the negative externalities from production of goods and use of fuels, it would be useful to impose taxes to force all actors in the economy to consider them in their market decisions. This is to internalize the externalities (i.e., to correct for an imposition of pollution costs on the entire society, rather than properly on the consumers of pollution-creating products).

We examine two sets of policies. The first is a tax on sector output, in which the tax rate is proportional to the health damage caused by the production of the commodity. This tax will cause the buyers of goods to face a price that reflects the pollution damages. For example, users of cement will pay much higher prices in the pollution tax regime compared to users of apparel. The second policy is a tax on primary fuels, in which the tax rate is proportional to the average damage per unit of fuel, causing producers to internalize the damages caused by their choice of fuels.

The most efficient policy would be a direct tax on emissions, not on output or fuels; however, emissions of total TSP are not comprehensively measured but are derived from sample surveys and data on fuel inputs. Universal TSP measurement would be prohibitively expensive. SO_2 from large sources may eventually prove amenable to a national control policy such as the U.S. trading program (as China

Box 3.8
Base-Case Economic Simulation

In our base-case projection, it is not our aim to give the most sophisticated forecast possible. The projections are made in a relatively simple manner, involving many assumptions regarding population growth, technical progress, changes in preferences, and changes in the world economy. We briefly describe our assumptions.

We start the simulation in 1997, initializing the economy with the capital stocks and labor supply of that year. The economic model then calculates, for this period, the output of all commodities, the purchases of intermediate inputs including energy, the consumption by households and government, exports, and the savings available for investment. This investment augments the capital stock for the next period, and, using the projected population for the next period, we solve the model again. This exercise is repeated for the subsequent periods. The level of output (of specific commodities and total GDP) thus calculated depends on our projections of the population, savings behavior, changes in spending patterns as incomes rise, the ability to borrow from abroad, improvements in technology, and so forth.

Our base-case assumptions produce a 5.1% growth rate of GDP over the next thirty years, slightly less optimistic than the 6.7% growth rate projected several years ago for China by the World Bank, but still a very rapid growth in per capita income.

Our assumptions for energy-use improvements are fairly optimistic and, together with changes in the structure of the economy, result in an energy-GDP ratio in 2026 that is about 60% that of 1997. There is a large decline in the use of coal per unit output, but there is a rise in oil and electricity use about equal to that of GDP growth and an even greater increase in the use of gas. These changes are due to our assumptions about changes in transportation demand, electricity-generation technologies, space-heating technologies, and improvements in energy efficiency. This shift from coal to oil and gas results in the rate of growth of CO_2 emissions slightly slower than the growth in energy use.

With the industry output and input requirements calculated for each period, we calculate the total emission of pollutants, their urban concentration, and their health effects.

Total PM_{10} emissions are projected to fall at a 1.44% annual rate despite the increase in energy use. This is due to the sharp difference in the assumed emission coefficients for new and old capital and the shift from coal to oil. Most manufacturing sources of PM_{10} fall dramatically, whereas those from transportation, construction, and services rise. Emissions from electric-power generation fall and then rise due to the opposing trends of lower emissions per unit output and rapidly rising total electricity use. As a result of these opposing trends, the decline in urban concentrations of PM is short lived and, over thirty years, it rises back to current levels. Projected SO_2 emissions rise much faster than particulates, because of a less optimistic estimate of the improvement in emission coefficients. There is still a substantial projected improvement, however; whereas coal use rises at 2.3% per year, SO_2 emissions rise only at 1.85% per year over this thirty-year period.

Box 3.8
(continued)

Our base-case estimate of premature mortality in 1997 is about 94,000 deaths. The growth rate of health effects from our more optimistic assumptions of energy trends are lower than that projected by other studies, with estimated excess deaths at 2.2 times the first-year level after twenty-five years, compared to a 3.7-fold increase in the comparable World Bank study. Premature mortality first falls and then rises due to the initial reductions in emission coefficients being bigger than the increased urbanization and fuel use, but over time the higher economic output and greater urbanization become more dominant.

is experimenting with in some localities), but this depends on monitoring and implementation mechanisms that are only gradually developing in China. Our interest is in policies that both have broad impact and may have other practical advantages for implementation—possibly including net economic benefits—and hence we consider taxes on output and fuels.

In light of the size of the estimated health damages, these pollution taxes can be large. To maintain revenue neutrality, we reduce existing taxes. The choice of which taxes to cut, or which sectors to compensate, affects both the mix of "winners" and "losers" in a given period, as well as the mix over time.

For our benefit-cost analyses of policies—which are discussed in full in chapter 10—we begin with a base-case growth path of the economy that includes no pollution taxes. We then estimate the alternative case with the green taxes and compare to the base case. In the base case, or "business-as-usual" scenario, the model projects GDP to grow at 5.1% per year over the next thirty years, while energy use only grows at a 3.3% rate. The growth of coal use is projected at 2.3%, while for oil use this figure is 5.2%, consistent with the expectations of improvements in energy efficiency and the rapid growth of motor vehicles.

3.8.2 Damage-Based Output Taxes

This policy imposes green taxes on industry output at rates given in table 3.6. This is a substantial tax on electricity, a modest one on cement, and a very small one on most others. The economywide effects of this policy are given in columns 2 and 3 of table 3.9, for the first and twentieth years, corresponding to 1997 and 2016. The initial effect of the taxes is to raise the price of electricity by almost 6% and the prices of the other dirty commodities by about 1.5–2.5%. This change in prices

Table 3.9
Effects of damage-based taxes on outputs or fuels

	Output Tax		Fuel Tax	
Variable	Effect in First Year	Effect in Twentieth Year	Effect in First Year	Effect in Twentieth Year
GDP	−0.04	+0.18	−0.04	+0.05
Consumption	−0.06	−0.01	−0.06	−0.02
Investment	+0.06	+0.55	+0.07	+0.20
Coal use	−3.95	−3.36	−13.2	−18.2
Primary particulate emissions	−3.08	−2.30	−6.0	−9.0
Combustion TSP	−4.25	−4.00	−11.8	−16.3
From electricity	−9.21	−7.42	−12.4	−16.3
From low-height sources	−1.17	−1.21	−11.0	−15.3
SO_2 emissions	−4.56	−4.00	−10.2	−14.0
CO_2 emissions	−3.41	−2.41	−10.7	−12.5
Premature deaths	−3.21	−2.88	−10.7	−14.3
Value of health damages	−3.53	−3.01	−10.7	−14.0
Change in other tax rates	−8.5	−6.1	−2.8	−2.1
Reduction in damages/GDP	0.07	0.09	0.20	0.43
Pollution tax/total tax revenue	9.1	6.0	2.72	1.99

Note: The entries are percentage changes between the counterfactual tax case and the base case, except for the last two rows, which are changes in percentage shares.

lead to a very modest reduction in aggregate GDP, but a shift in consumption to investment.

In the first year the output of electricity falls by 6.7%, whereas the other dirty sectors—nonmetal mineral products (cement), transportation, and health-education-services—fall by 1–3%, leading to a reduction in the demand for electricity and coal. Inputs released from these shrinking sectors go to the cleaner industries, leading to an expansion in finance, food products, and apparel by about 0.2–0.8%.

These changes lead to a reduction in total primary particulate emissions of 3.1%, with the power sector having the biggest reduction, 9.2%, whereas emissions for sectors with low stack heights only fall 1.2% in aggregate. Industries have different emission factors, and changes in relative industry outputs cause a greater fall in PM emissions than coal use, 4.2% versus 3.9%. For SO_2, total emissions fall by a large 4.6%.

The effect of these pollution reductions is to lower the value of health damages by 3.5%. The number of cases of premature mortality falls by 3.2%, whereas those of chronic bronchitis fall 3.7%. The value of this reduction in damages in the first year comes to about 0.07% of GDP of that year. This is a very modest reduction compared to the base-case estimated total health damages of 1.9% of GDP.

Turning to the effects on government finances, the revenue raised from this broad-based tax is substantial, 9.1% of total revenues in the first year. With our requirement that government revenues and spending be kept fixed at base-case levels, this allows a reduction in value-added and capital income taxes of 8.5%. This tax cut eventually leads to higher retained earnings and resultant investment. The higher rate of investment leads to a higher stock of capital and a greater productive capacity for the whole economy. As shown in table 3.9, by the twentieth year GDP is 0.18% higher than the base case.

The emissions of the GHG CO_2 are related to the local pollutants. In the first year, CO_2 emissions from fossil fuels fall by 3.4%, a little less than the fall in coal use because there is some switching toward oil and gas via reallocation of industry activity. This is a relatively inefficient policy to reduce emissions of both PM and CO_2, and we next consider another policy.

3.8.3 Damage-Based Fuel Taxes

Emissions are a function of output levels, fuel choice, energy efficiency, and control strategies. We now consider a more sharply targeted air pollution–control policy, a tax on fossil fuels in proportion to the health damage that result from current combustion technologies.

The heavy tax on coal reduces its use by 13% initially, whereas the modest tax on oil reduces refining output by 1.3%. Heavy users of these fuels have to raise their output prices to compensate, causing a reduction in demand for these goods. The electricity sector is the biggest user of coal. Its price rises by 2.4% and output falls by 2.7%. Metals smelting (iron and steel) and nonmetal mineral products (cement) are the next biggest consumers of coal, and their outputs fall by 0.8 and 0.6%, respectively.

The total additional tax burden is relatively small. Pollution tax revenue comes to only 2.7% of total revenue, and the offsetting cuts are correspondingly small. The small additional retained earnings of enterprises allow aggregate investment to rise by 0.07% in the first year. This is accompanied by a fall in real consumption of 0.06% as households face mostly higher prices of goods. The changes in composition of aggregate output lead GDP to fall by a minor 0.04%.

These changes in industry structure and fuel switching reduce total primary PM emissions by 6.0%, with those from electricity falling the most, 12%, and those from manufacturing (in aggregate) the least, 3.9%. Large reductions in SO_2 and primary particle emissions reduce total PM, which in turn generates an 11% reduction in health damages. The value of reduced damage comes to a substantial 0.20% of GDP, compared to our base-case total damage of about 1.9% of GDP.

Because only the three fossil fuels are taxed, the revenue is much smaller than the thirteen collected in the output tax policy, which taxes all commodities; however, the output tax case only generates a 3.5% reduction in health damages compared to the 11% here. Thus, a narrowly based but well-targeted tax that raises only a modest amount of revenue can lead to a sizable reduction of pollution and related health damages.

The patterns over time are also very different from the output tax case. Here the reduction in emissions and damages rise over time, comparing the first- and twentieth-year columns in table 3.9. By the twentieth year, even though GDP is 0.05% higher than the base case, the reduction in primary PM emissions is 9.0%, greater than the 6.0% reduction in the first year, when GDP fell. Similarly, the reduction in SO_2 emissions grows over time. Health damages, driven by these emissions and the growing urban population, fall even more. By the twentieth year, the total health damages are down 14% from the base path.

In light of the effects on CO_2, we see that the reduction in emissions is smaller than the fall in coal use due to the switch to other fossil fuels. In the first year, when coal falls by 13%, CO_2 falls only by 11%. Over time, as the tax rises, the CO_2 share from oil rises even more. The reduction of CO_2 in the twentieth year is only 12%, whereas coal use falls 18%. Although this may seem an inefficient instrument to reduce CO_2, it is actually an effective "second-best" instrument, even ignoring the very large health benefits. In our view, one should see substantial reductions in CO_2 emissions as a critical side-benefit of dealing with the urgent issues of local air quality and public health.

3.9 Conclusion

Our assessment consists of an integrated series of studies, exploring and linking many aspects of the health and economic damages of ambient air pollution in China. Understanding the causes, effects, and policy options in air quality is a complex analytical challenge, not to be underestimated. This is partly because the problem is inherently multifaceted, encompassing economics, engineering, atmospheric science, health science, and public policy. It is certainly more difficult (and the results less precise) if underlying information and data are limited.

We summarize the integrated assessment by spotlighting several key conclusions:

- The damage to health of air pollution in China is considerable, even under conservative choices in the parameter ranges of our assessment.
- It is feasible and valuable to focus on damages, and not on energy use or emissions per se, to properly understand the adverse effects of air pollution. Industries that generate the most emissions are not necessarily those that cause the most damages. Our assessment finds the electric-power sector to be the largest source of damages, followed by the cement industry and transportation.
- The intake fraction methodology, and our specific intake fraction results, can produce a simple, quick, but reasonable approximation of exposures to TSP and SO_2 from emission sources, if data, time, or other resources to run dispersion models are lacking.
- The regional dimensions of air quality are critical and should not be overlooked, because damages from primary and secondary pollutants far from the source of emissions are considerable. They are of similar magnitude as the local damages that are most often (and most easily) calculated.
- The benefits of pollution control in China likely far exceed the costs, if market instruments are used. This high benefit-cost ratio is robust to the many parameter uncertainties. Short-run adjustment costs can be mitigated by gradual implementation.
- A green tax policy could generate substantial new revenues for the government that could be used for tax reform (i.e., the reduction of other, distortionary taxes). This tax rationalization over the long run could produce higher retained earnings and investment, thus leading to higher capital stock and greater productive capacity for the whole economy. Such policies in principle could not only reduce health damages of air pollution, but also improve economic performance over time.
- In light of the high level of damages, the whole range of pollution-reduction policies should be analyzed, beyond those considered here. Wealthier countries generally used end-of-pipe regulations to reduce emissions of particulates and SO_2, before developing and beginning to adopt market-based policy instruments more recently. The former mandate particular types of scrubbers or emission standards. Analysis of such policies would give the authorities a wider range of options, and should be high on the priority list of action. A careful analysis of these regulations would require data on costs of operating various scrubbers, washing coal, using low sulfur coal, and so forth.
- We strongly encourage efforts to analyze global and local pollution in concert, which give a more comprehensive accounting of total costs and benefits. The pollution control policies that we examined produced concurrent reductions in GHG emissions. The magnitude of the gain to global welfare should spur serious

consideration by wealthier nations concerned with climate change to aid China in these control efforts.

Surprise at the wide range of uncertainty described at many steps in our assessment is understandable. We emphasize two points in response. First, assessing environmental damage anywhere in the world is by nature an uncertain science, and the difficulties grow as the scope of an assessment grows. Analyses at the municipal or provincial levels can be more precise, but other systematic studies of damages at the national scale in China include approximating steps (and thus uncertainties) like our own. It was the challenge of improving on the best known of these, *Clear Water, Blue Skies* by the World Bank, that inspired the more academic research venture reported here. The goal of damage estimation is to inform the design of efficient pollution control policies, and here we go further than *Clear Water, Blue Skies* by analyzing the effects of two market-based policies on both damage reduction and economic performance. The rest of this book describes the research in complete, transparent, and replicable detail so that others can critique, apply, adapt, or improve any and all elements of our methods.

Second, at every step of this and similar assessments, uncertainty results from limited availability of data and underlying research. This is understandable and hardly unique for a developing country like China. Nevertheless, we emphasize that for any society to efficiently and successfully confront its environmental hazards, an essential first step is to build understanding of the true nature of the physical, economic, and institutional challenges. A major commitment to enhance collection and dissemination of systematic, objective, and transparent data and other information could be one of the most important steps that China can take to address the risks of air pollution to the health and welfare of the public over the long run.

We note key areas where effort could be directed to reduce the uncertainty, to fill gaps in the damage estimates, and to improve subsequent policy analyses, by our research team and by colleagues in the same field:

- Increased study of the epidemiology of air pollution in Chinese populations, both indoors and outdoors;
- More analysis of household energy use, fuel choice, heating and cooking technologies, and indoor air quality;
- Enhanced air-quality modeling, to cover larger areas, more industries (notably the transportation sector), and more complex chemistry and meteorology, including additional pollutants (e.g., nitrogen oxides and ozone);
- More complete and detailed emission inventories by industry or source types, including characteristics like quantity, particle size distribution, and toxicity;

- More health valuation studies to cover a greater sample of the population and apply alternative valuation methodologies; and
- Estimates of enterprise behavior, to understand how easily different industries switch fuels or substitute capital for energy.

It is clear that such a long and ambitious list of data and research needs cannot be addressed all at once. The challenge for government, and for scholars, is to work on them incrementally, collectively building a comprehensive body of applicable research that can better inform China's choices on pollution control and environmental health for the future.

4

Estimating Health Effects of Air Pollution in China: An Introduction to Intake Fraction and the Epidemiology

Jonathan I. Levy and Susan L. Greco

4.1 Introduction

To determine the health impacts of air pollution in developing countries like China, as outlined in chapter 1, there are considerable methodological challenges. In this chapter, we focus on two specific questions:

1. How can one model the effect of emissions on exposure using limited resources in a way that is interpretable from a health perspective?
2. How does one determine the effects of air pollution on health when limited studies have been conducted on the population of concern?

We introduce the concept of an intake fraction (*iF*) as a solution to the first question, focusing largely on the implications of the concept and past study findings for a comprehensive analysis. We devote the remainder of the text to consideration of the second question, presenting an overview of the major evidence of mortality and selected morbidity risks from air pollution found in the worldwide literature and in Chinese studies and discussing potential uncertainties in extrapolating evidence from one country to another. We conclude with some preliminary recommendations for investigation of the health benefits of air pollution control in China now and in the future.

4.1.1 Air Pollutants of Concern

Before estimating the health impacts of air pollution, it is worthwhile to give an overview of common air pollutants that are regulated in the United States, China, and elsewhere. In these countries, six major "criteria" pollutants have regulated ambient concentrations, under standards set by the U.S. Environmental Protection Agency (U.S. EPA) and the State Environmental Protection Administration (SEPA) in China. These pollutants are particulate matter (PM), sulfur dioxide (SO_2), carbon monoxide (CO), nitrogen oxides (NO_X), ozone (O_3), and lead (Pb). Hazardous air pollutants (HAPs), on the other hand, are regulated in the United States by technologically based emission limits set forth in the National Emission Standards for Hazardous Air Pollutants by the EPA. The pollutants are briefly described below, reproduced in part verbatim from the concise summary by Rubin and Davidson (2001).

Particulate Matter The health effects associated with particulate air pollution include respiratory and cardiovascular disease, damage to lung tissue, and (potentially) carcinogenesis and premature death. Particles may also act as carriers that adsorb other pollutants on particle surfaces. The size of particulate matter strongly affects the scope and severity of its impacts. Fine particle sizes such as those smaller than 2.5 microns ($PM_{2.5}$) appear to be most closely linked to adverse health effects. Particles smaller than 10 microns (PM_{10}) are fine enough to enter the respiratory tract, with $PM_{2.5}$ able to penetrate deeply into the lungs, releasing pollutants on moist lung surfaces. Total suspended particulates (TSP) include particles that are larger than 10 microns, which are generally unable to enter the respiratory tract. The chemical nature of particulate matter also is important in determining health and environmental impacts; for example, heavy metals or pesticide residues are of greater concern than less toxic materials.

Sulfur Dioxide Exposure to high concentration of SO_2 can lead to respiratory illnesses, alterations in the lung's defenses, and aggravation of existing cardiovascular or chronic lung disease. Asthmatics and individuals with diseases such as bronchitis or emphysema, as well as children and the elderly, are most sensitive to elevated levels of SO_2.

Carbon Monoxide CO is a colorless, odorless gas that is produced when fossil fuels or other carbon-containing materials are not completely combusted. When inhaled, carbon monoxide is absorbed by blood hemoglobin, which normally carries oxygen to the body. Exposure to elevated levels of CO in the atmosphere can produce a spectrum of adverse health effects ranging from shortness of breath and dizziness as the body's oxygen delivery system is choked off to death.

Oxides of Nitrogen At the lower concentrations typical of polluted urban air, NO_X can irritate the respiratory system and produce respiratory illnesses such as bronchitis. Children may be particularly affected by elevated NO_X levels. Like SO_2, NO_X also can cause acidification of soil and water, which generate serious effects on certain species of plants and animals.

Ozone The air pollutant O_3 found at ground level can be thought of as "bad ozone," in contrast to the protective layer of "good ozone" found in the stratosphere high above the earth. Ground-level O_3 causes health problems because it attacks lung tissue, reduces lung function, and sensitizes the lungs to other irritants. Studies have shown that ambient levels of O_3 affect not only people with impaired respiratory systems, such as asthmatics, but healthy people as well.

Lead Lead (Pb) is a heavy metal that can cause neurological damage and adverse effects on organs such as the liver and kidneys. Children exposed to Pb are particularly vulnerable to a range of effects that can impair normal development. Once ingested via inhalation or other means, Pb tends to bioaccumulate in blood, bone, and soft tissues, so that its effects are not easily reversible.

Hazardous air pollutants These chemicals are emitted in much smaller quantities than criteria air pollutants, but their effects can be severe, even in small doses. Carcinogenic substances like asbestos and benzene are of particular concern, as are heavy metals and other chemicals that may cause neurological, immunological, mutagenic, and other serious health effects.

This volume will focus on PM and SO_2 in determining health effects in China, because these pollutants would likely dominate a health impact assessment in China and because data for estimating their effects are relatively available. For example, a recent meta-analysis of exposure-response functions for health effects in China exclusively considered PM and SO_2 (Aunan and Pan 2004); however, as automobiles become more prevalent in China, we would expect NO_X and O_3 to become greater problems. Fortunately, China has already made progress against a third mobile-source pollutant, Pb, with a national ban on production and sale of leaded gasoline that became effective in 2000 (World Bank 2001). The other pollutants are mentioned as additional concerns for future policy action and supporting research.

4.2 Exposure Modeling

4.2.1 Definition of Intake Fraction

An intake fraction (*iF*) is perhaps the simplest way available to summarize the relationship between the emissions of a pollutant and the subsequent exposure to that pollutant or a secondary by-product. It has been defined as the integrated incremental intake of a pollutant released from a source or source category (such as mobile sources, power plants, or refineries) and summed over all exposed individuals during a given exposure time, per unit of emitted pollutant (Bennett et al. 2002). Essentially, it is the amount of a pollutant or its precursor emitted that is eventually inhaled by someone.

Mathematically, it can be expressed as

$$iF = \frac{\sum Population \times Concentration \times Breathing\ Rate}{\sum Emissions} = \frac{\sum_{d=1}^{n} POP_d \times C_d \times BR}{EM},$$

where iF is the intake fraction from a source, POP_d is the population at location d, C_d is the change in concentration at d, BR is a nominal population-average breathing rate, used to yield a unitless measure, and EM is the total emissions from a given source. The summation is over all locations. This concept has been discussed in the scientific literature for decades, though with an array of different names (including exposure efficiency, committed dose, exposure commitment, exposure factor, exposure effectiveness, inhalation transfer factor, exposure constant, potential intake, population-based potential dose, and fate factor). The impetus behind the concept was to find straightforward ways to organize scientific information in a manner that informs risk-based environmental policy and to allow findings from exposure studies to be extrapolated to other settings.

A few characteristics of an *iF* are important to keep in mind for any policy application:

- An *iF* is a unitless value, which enhances comparability across sources and settings. It provides population-weighted exposures per unit emissions, with a nominal population-average breathing rate used to yield a unitless measure to facilitate comparability of values. An *iF* calculation incorporates variability in exposures but integrates them to provide an aggregate exposure value.
- An *iF* is not an inherent property of a pollutant. Rather, it depends on a number of variables, including release location and conditions, atmospheric and meteorological conditions, pollutant properties, and population patterns. This implies that any attempt to extrapolate *iF*s to unstudied settings must adequately characterize these variables to ensure accurate estimates.
- As defined above, an *iF* implicitly assumes that the health effect in question has a linear dose-response function, is dose-rate independent, and exhibits no threshold. The above formula is based on the premise that an incremental concentration change is equally important regardless of the current level of exposure; if this assumption is not correct, an *iF* cannot be directly linked with health risks. The *iF* concept, however, could be applied for pollutants that do not display these characteristics, only with modifications to the definition and formula. Similarly, the *iF* assumes a linear relationship between emissions and total population intake,

which will not be true for many secondary pollutants if the changes in emissions are large enough.

- Ultimately, we want to know the health impact of polluted air (i.e., the number of cases of chronic bronchitis, premature mortality, or other health endpoint). In many studies, this is represented by multiplying the concentration change by a "concentration-response coefficient" (the number of cases per additional microgram per cubic meter in increased pollutant concentration, or the relative risk multiplied by the baseline disease rate). Denoting the slope of this concentration-response curve β and assuming linearity, the population risk for an identified population with a given level of exposure is $\beta \times concentration \times population$. Given this fact and the above assumptions about linearity, it is simple to estimate population risk given an intake fraction. Rearranging the above equation,

$$\sum_{d=1}^{n} C_d \times POP_d = EM \times iF/BR.$$

Thus, population risk can be estimated if one knows the *iF*, emission rate, nominal average breathing rate in the population (generally assumed to be 20 m^3/day), and β, the slope of the concentration-response curve. Namely, the population risk is $(EM \times iF/BR) \times \beta$.

As with any summary measure, the *iF* has some obvious drawbacks. It requires some modeling to be conducted in a relevant geographic area, or else the incremental concentration changes per unit emissions cannot be determined. Thus, in a setting with no resources or available meteorological data, it would be difficult to determine appropriate *iF* values, as it would require extrapolation from other geographic areas. In addition, it is only relevant for aggregate population calculations and is not as suitable for estimation of individual risk patterns. That being said, it provides a useful mechanism to estimate the population health benefits of air pollution control in settings where atmospheric dispersion modeling is implausible or limited.

4.2.2 Past Studies of Intake Fractions: United States and China

There have been two important arguments in favor of using *iF*s for a large-scale analysis in China. The first is that it would allow for dispersion modeling across a limited number of sources, with the findings then extrapolated to other sources across the country. This is the approach used in this volume, with intake fractions calculated in chapters 5, 6, and 7 (see also Zhou 2002; Zhou et al. 2003; Zhou et al. 2006), applied nationally in chapter 9, and used to analyze policy options in chapter 10. This is also the approach of similar analyses in the U.S. (Smith 1993;

Wolff 2000; Evans et al. 2002). The second is that it would allow findings from studies in other countries to be applied to China, in situations where available information was too limited to construct dispersion models in China. As mentioned above, however, extrapolating across settings requires one to adequately characterize all factors that could influence *iF* values to ensure comparability.

Although the past and ongoing *iF* work in China is detailed through the rest of this volume, we give a brief summary of relevant past work and use some key findings from the China power sector analysis that was completed first for this volume (chapter 7). This will provide context for the health evidence and foster discussion of similarities and differences between estimates in China and elsewhere. This is intended to be a historical look at this literature rather than an exhaustive summary of publications to date.

One of the earlier studies in the *iF* field (Smith 1993) estimated *iF*s for primary particles from a variety of sources, including coal-fired power plants in the United States and in developing countries. Smith estimated a value of 1×10^{-6} for the United States, compared to 1×10^{-5} in less-developed countries. The difference was based on assumed differences in population density. He also estimated a higher *iF* for motor vehicles of 2×10^{-5}. The interpretation of the last *iF* is as follows: for every metric ton (10^6 grams) of primary particles emitted, 20 grams will be inhaled by the population.

Another study (Phonboon 1996) computed *iF* estimates for particles, SO_2, and benzene from an oil refinery in Thailand at both local (within 50 km) and regional scales (within 1,000 km). Phonboon reported estimates of 1.4×10^{-5} for particles and SO_2 and 4.3×10^{-5} for benzene and found these values not to increase significantly beyond 50 km. This was attributed to the short half-lives of the pollutants in question, the extremely high population densities close to the source, and the low stack heights at the refinery.

A U.S. study (Wolff 2000; Evans et al. 2002) estimated *iF*s for both primary and secondary particulate matter for coal-fired power plants and mobile sources in the United States. Wolff used CALPUFF, a long-range dispersion model, to capture the intake fractions across the entire United States. The power plant and mobile source results are summarized in tables 4.1 and 4.2 below.

To summarize these results, intake fractions for primary $PM_{2.5}$ (particles less than 2.5 microns in diameter) from power plants were on the order of 2×10^{-6}, similar to the earlier developed country estimate by Smith (1993). Much of this total was due to long-range transport of particulate matter (PM), with about half of the total *iF* occurring within 500 km of the source (related to long half-lives and tall stacks).

Table 4.1
Intake fractions for fine particles from U.S. coal-fired power plants (Wolff 2000), based on stratified random sample of forty large power plants

Pollutant (Emissions/Exposure)	Mean	S.E.M.	Minimum	Maximum
Primary $PM_{2.5}$/primary $PM_{2.5}$	2.2E-06	1.9E-07	2.5E-07	6.3E-06
SO_2/sulfate	1.6E-07	5.9E-09	6.0E-08	2.2E-07
NO_x/nitrate	2.7E-08	1.0E-09	7.4E-09	5.8E-08

Notes: S.E.M. = standard error of the mean. NO_x/nitrate values were computed by dividing CALPUFF estimates by a factor of four to account for the role of low temperature in formation of particulate nitrate. It is no longer clear that such a correction is necessary.

Table 4.2
Intake fractions for fine particles from U.S. mobile sources (Wolff 2000), based on stratified random sample of twenty urban and twenty rural highway segments

Site and Pollutant (Emissions/Exposure)	Mean	S.E.M.	Minimum	Maximum
Urban				
Primary $PM_{2.5}$/primary $PM_{2.5}$	9.4E-06	8.0E-07	3.0E-06	1.8E-05
SO_2/sulfate	1.2E-07	9.3E-09	3.5E-08	2.0E-07
NO_x/nitrate	2.3E-08	2.0E-09	4.1E-09	4.6E-08
Rural				
Primary $PM_{2.5}$/primary $PM_{2.5}$	8.8E-06	1.0E-06	1.2E-06	1.8E-05
SO_2/sulfate	1.4E-07	9.0E-09	4.0E-08	2.2E-07
NO_x/nitrate	2.6E-08	2.3E-09	1.2E-08	5.1E-08

Notes: S.E.M. = standard error of the mean. NO_x/nitrate values were computed by dividing CALPUFF estimates by a factor of four to account for the role of low temperature in formation of particulate nitrate. It is no longer clear that such a correction is necessary.

The values for sulfates from SO_2 emissions are approximately an order of magnitude lower than the primary $PM_{2.5}$ intake fractions, with the nitrate values from NO_X emissions even lower. Secondary PM *iF*s appear similar for mobile sources and power plants, but the primary $PM_{2.5}$ *iF* is somewhat higher for both urban and rural mobile sources.

A final set of U.S. power-plant values can be taken from work conducted for a recent analysis of more than 500 power plants in the United States (Wilson 2003). The investigators applied a source-receptor (S-R) matrix to estimate *iF*s for both primary and secondary particulate matter. This model incorporated nitrate increases

Table 4.3
Intake fractions for fine particles from U.S. power plants (Wilson 2003), based on S-R matrix application to 507 power plants across the United States

Pollutant (Emissions/Exposure)	Mean	S.E.M.	Minimum	Maximum
Primary $PM_{2.5}$/primary $PM_{2.5}$	1.2E-06	3.8E-08	3.6E-07	8.7E-06
SO_2/sulfate	3.4E-07	5.3E-09	7.5E-08	7.3E-07
NO_x/nitrate	5.1E-08	1.2E-09	4.4E-09	1.7E-07

Note: S.E.M. = standard error of the mean.

associated with SO_2 emission controls due to the increase in available ammonium. The mean and range of values are reported in table 4.3 and are generally comparable with the findings from Wolff (2000).

In the power-plant application in China described in chapter 7 as well as in Zhou (2002) and Zhou et al. (2006), *iF*s are estimated by use of CALPUFF on a random sample of twenty-nine power plants, one in each of twenty-nine provinces. Mean values include 1×10^{-5} for primary $PM_{2.5}$, 4×10^{-6} for secondary sulfates, and 4×10^{-6} for secondary nitrates. These estimates are roughly an order of magnitude larger than corresponding U.S. values, as predicted by the earlier work by Smith, but with higher values for secondary pollutants relative to primary pollutants. The difference between U.S. and China *iF*s appears to be explained by the higher population density in China and proximity of power plants to urban centers, but meteorological factors may also be partly responsible.

An advantage of using the *iF* approach is the ability to construct regression models that can then be applied in settings for which there are limited data. Regression models are functions of factors, like population and meteorology variables, that would influence *iF*. Zhou (2002) and Zhou et al. (2006) developed a predictive regression equation for *iF*s in China, based solely on population and an indicator variable for region. When these regressions were used to predict *iF*s in the United States, they performed quite well for primary $PM_{2.5}$ but overpredicted sulfate and nitrate intake fractions by approximately a factor of six. This demonstrates that a regression-based approach can potentially address differences in population patterns but that differences in atmospheric or meteorological conditions might make secondary pollutant estimates more difficult to extrapolate. Nevertheless, the robustness of the atmospheric modeling coupled with the strength of the regression equations (involving population predictors) suggests that the model could predict total population exposure to power-plant pollution in China.

4.3 Health Effects

Once population exposure has been estimated, the next step in a health-impact assessment is to determine the effects of changes in population exposure on health outcomes of concern. At first, this appears to be the simplest step in the analysis, because it involves the application of previously published findings rather than development of complex dispersion models; however, it is the step with the greatest uncertainties and use of subjective judgment.

We can highlight these issues by focusing on premature mortality instead of the whole range of health effects. In most past assessments where economic values were placed on health outcomes (e.g., U.S. EPA 1999), mortality reductions dominate total benefits, in light of the relative economic values typically placed on mortality and morbidity outcomes. Nevertheless, the issues raised when addressing mortality are representative of those found in the morbidity literature. A more detailed assessment of the mortality, as well as the morbidity literature, can be found in Aunan and Pan (2004) and HEI (2004). It is not our intent to replicate this work, but rather to highlight some of the dimensions that complicate the development of concentration-response functions in China.

In this section, we first describe the types of evidence used to quantify the relationship between pollution exposure and mortality risk, summarizing the strengths and weaknesses of each study type. We then briefly summarize the worldwide literature on mortality risk from air pollution, focusing on the most important pollutants and the magnitude of the effect. We discuss the state of evidence within China and compare the findings to the worldwide literature, discussing reasons why similar findings might or might not be anticipated. We conclude by suggesting approaches for determining an appropriate mortality concentration-response function for China.

4.3.1 Types of Mortality Studies

The two major sources of evidence for any assessment of mortality risks from air pollution are toxicological and epidemiological studies. Toxicological studies generally involve administering high doses of chemicals to laboratory animals and measuring the health effects in a controlled laboratory setting. In contrast, epidemiological studies are observational studies of human populations designed to show relationships between exposure to actual levels of pollution and the development and exacerbation of disease.

Although toxicological studies provide much of the evidence for cancer risk assessment, they are rarely used to determine dose-response functions for criteria pollutants (e.g., PM, O_3, SO_2, NO_X, or CO) or for respiratory or cardiovascular mortality. This is because extrapolating from the high doses needed to detect effects in a small number of animals to the low, long-term exposures that humans experience contains significant uncertainties. Thus, for premature mortality from criteria pollutants, dose-response functions are derived almost exclusively from epidemiological studies, with toxicology providing supporting evidence of biological plausibility.

Although epidemiological studies avoid some of the problems of applying toxicology results to humans, their observational nature can make it difficult to determine the causal agent. The primary issue in interpretation of epidemiological findings is whether there might be an unobserved variable that is independently associated with both exposure and disease. This unobserved variable is known as a confounder.

Three major types of epidemiological studies have been used to infer relationships between pollution exposure and mortality. The first are cross-sectional studies, which compare death rates in different cities with pollution rates in those cities. These studies provide some indication of the role of air pollution on mortality but can suffer from confounding and the ecological fallacy. The ecological fallacy is the incorrect assumption that relationships evaluated from groups are applicable to individuals. It is possible to observe a difference between groups that does not hold for individuals—a classic example is a study that showed that states with more foreign-born residents had higher literacy rates but that foreign-born individuals have lower literacy rates than native-born individuals. Because cross-sectional studies cannot incorporate individual behaviors and risk factors into the model, they provide limited information about the risk of air pollution for individual mortality (Pope, Bates, and Raizenne 1995). Thus, although many of the early studies of air pollution and mortality were cross-sectional studies, these studies are rarely conducted today and are generally considered less useful than other study types.

A second type of epidemiological study is the time-series study, in which changes in air pollution levels on a daily basis are compared with daily changes in numbers of deaths. These studies are sometimes referred to as "acute mortality" studies, because they provide only information on the effect of today's pollution levels on deaths over the next few days (or, at most, a couple of months). In addition to being relatively quick and inexpensive to conduct, time-series studies have the advantage of a limited number of potential confounding variables. A confounder must be correlated with both the exposure and the outcome, so behavioral variables like cigarette smoking are not plausible confounders (because they would have to change on a daily basis, for example, increased smoking on more-polluted days). Potential confounders for time-series studies therefore include weather-related variables and other air pollutants.

The final type of epidemiological study used to evaluate mortality risks from air pollution is the cohort study, in which researchers follow individuals for a significant period of time to evaluate whether long-term pollution exposure is linked to mortality rates. Unlike cross-sectional studies, cohort studies use subject-specific information about covariates such as smoking status or occupation and can reasonably determine the exposure history of each person. They are therefore better able to deal with confounding variables, though a greater number of variables can plausibly confound a cohort study versus a time-series study. Because cohort studies capture the effects of long-term pollution exposure, they would be expected to yield greater mortality risks (called "chronic mortality") than time-series studies.

4.3.2 Worldwide Literature on Air Pollution Mortality

Time-series studies In recent years, hundreds of time-series mortality studies have been conducted across the world, largely because they can be done more quickly and at lower cost than cohort mortality studies. Studies have evaluated all criteria air pollutants (including PM, O_3, SO_2, NO_2, and CO), but the most substantial evidence has been associated with PM exposure.

Two recent studies provide some indication of the magnitude of this effect. A meta-analysis of twenty-eight time-series studies in the worldwide literature (Levy et al. 2000) estimated that daily deaths increased by approximately 0.6% per 10 $\mu g/m^3$ increase in daily PM_{10} concentrations, independent of the effects of other pollutants. Another study considered the ninety largest cities in the United States and pooled the findings across all cities (Samet et al. 2000). The latter study originally reported an estimate of 0.5% per 10 $\mu g/m^3$ increase in daily PM_{10} concentrations, but researchers have revised their estimate to approximately 0.3% given statistical flaws in the original analysis (NMMAPS 2002).

Many investigators have attempted to determine which component(s) of PM_{10} might be responsible for the mortality effects. Although this literature is too large and varied to summarize in this document, a few salient points have emerged:

- Most studies that have considered both fine ($PM_{2.5}$) and coarse ($PM_{10-2.5}$) particulate matter have found a greater effect associated with the smaller particles, consistent with the known deposition patterns of PM. Most estimates fall in the range of a 1–2% increase in daily deaths per 10 $\mu g/m^3$ increase in daily $PM_{2.5}$ concentrations (Schwartz, Dockery, and Neas 1996; Burnett et al. 2000).

- Studies that have tried to determine whether the time-series deaths were simply cases of people dying who would have died a few days later (sometimes referred to as the harvesting effect) have concluded that this is not the case and that at least two months of life expectancy are lost (Schwartz 2001).
- Although it is unclear which PM constituents are most strongly linked to health outcomes, the literature has demonstrated that combustion-related particles appear more toxic than noncombustion particles (Laden et al. 2000).

In addition, it is important to realize that some positive evidence exists for pollutants other than PM. For example, although O_3 did not significantly influence the effect of particulate matter on mortality in the above studies, recent meta-analyses (Bell, Dominici, and Samet 2005; Ito, De Leon, and Lippman 2005; Levy, Chemerynski, and Sarnat 2005) have concluded that there may be an independent effect of O_3 on mortality. Since O_3 has not been the focus to date in China, we do not discuss this evidence further. Finally, although there has been limited evidence supporting the role of SO_2 or NO_2 on daily mortality in the United States at present, higher levels of ambient pollution in China could imply that effects would exist in China in spite of negative findings in the United States. For example, a recent study investigating daily mortality in a district of Chongqing (Venners et al. 2003, supported in part by the same Harvard China Project) found significant associations for SO_2, but not $PM_{2.5}$. The relative risk of respiratory mortality with a two-day lag after a 100 $\mu g/m^3$ increase in mean daily SO_2 was 1.11 (95% confidence interval [CI]: 1.02–1.22). For cardiovascular mortality with a three-day lag, it was 1.20 (95% CI: 1.11–1.30). These effects may be due to the high SO_2 concentrations in China per se, or SO_2 might be a marker for particulate sulfate, despite the fact that the SO_2 effects remained after controlling for $PM_{2.5}$. These issues are discussed in more detail below, in reference to the Chinese literature on air pollution mortality (section 4.3.3).

Cohort studies Many fewer cohort mortality studies exist in the literature, because they require individuals to be followed for decades with detailed information about personal risk factors. Two major studies have figured most prominently in the literature: the Six Cities study (Dockery et al. 1993) and the American Cancer Society study (Pope et al. 1995, 2002). Other recent studies, such as the Adventist Health Study of Smog (McDonnell et al. 2000) and the Veterans' Administration—Washington University cohort study (Lipfert et al. 2000), did not evaluate representative populations and are less useful for national assessments. In addition, the Six Cities and American Cancer Society (ACS) studies have undergone extensive reanalysis by an independent research group to confirm their conclusions (Krewski et al. 2000). We describe the Six Cities and ACS studies in more detail below.

The Six Cities study followed more than 8,000 white adults in six cities in the eastern half of the United States for approximately 15–17 years. Concentrations of both PM and gaseous pollutants were measured at centrally located monitoring stations in each community. After controlling for potential confounders (including smoking, education, obesity, and occupational exposures), investigators found statistically significant associations with mortality rates for multiple pollutants, with the strongest relationships found for three different measures of particulate matter: PM_{10}, $PM_{2.5}$, and sulfates. For every microgram per cubic meter increase in annual average $PM_{2.5}$ concentrations, mortality rates increased by approximately 1.2%. Air pollution was associated with deaths from cardiopulmonary disease and with lung cancer (though the lung cancer effect was not statistically significant), but not with deaths from other causes. Of note, the range of concentrations across study cities was 11.0–29.6 $\mu g/m^3$ of $PM_{2.5}$ and 18.2–46.5 $\mu g/m^3$ of PM_{10}.

Primary limitations of this study include the relatively small number of cities evaluated, the limited set of confounders analyzed (addressed to a significant degree by the Krewski et al. 2000 reanalysis), and the assumption that measured concentrations during the study period reflect the relevant exposure period (Lipfert and Wyzga 1995). In addition, the three most-polluted cities also had the oldest populations, the least-educated cohorts, and the highest smoking rates, which implies that the estimated magnitude of the air pollution effect could depend on the methodology for addressing potential confounders (U.S. EPA 1996).

The ACS study was conducted to follow the Six Cities study and to determine whether the conclusions were robust over a larger population. Pope and colleagues considered more than 500,000 subjects who lived in all 50 U.S. states, drawn from an established cohort that was being followed to investigate cancer development. Air pollution exposures were estimated from a database organized for an earlier cross-sectional mortality study in the original ACS publication but included recent monitoring data in a follow-up analysis (Pope et al. 2002). This recent publication expanded on the original work by considering additional years, refining the statistical approach to account for spatial correlations, and evaluating a number of potential confounders (including alcohol consumption, passive smoking, occupational exposure, diet, and obesity). In this analysis, concentrations of $PM_{2.5}$ were approximately 10–30 $\mu g/m^3$ at the start of the study period, and fell to 5–20 $\mu g/m^3$ by the end of the study period.

For $PM_{2.5}$, the estimated mortality effect was approximately half the effect measured in the Six Cities study, with a 0.6% increase in mortality rates per $\mu g/m^3$ increase in annual average $PM_{2.5}$ concentrations (using the average pollution concentrations across the study period). The relative risk was slightly higher for cardiopulmonary mortality and for lung cancer, both of which were statistically significant. In single-pollutant models, there was no relationship between total suspended particles (TSP) and mortality, with a slight relationship for PM_{10}. All gaseous pollutants were insignificant, with the exception of SO_2, which was significant for all causes of mortality.

The primary critiques of the ACS study are that the population studied was somewhat older, better educated, and contained more nonsmokers than the U.S. average (U.S. EPA 1996) and that some possible confounders (such as migration and sedentary lifestyle) were omitted. The ACS study, however, is the most comprehensive analysis of long-term mortality risks from air pollution published to date, which is why it is the basis of most air pollution health-impact assessments.

If the findings from the cohort studies are valid and the concentration-response function is linear, the levels of risk that they suggest are substantially higher than the levels indicated by the time-series mortality studies. Furthermore, although most of the deaths from the time-series studies may involve persons who are already quite ill (and therefore who have short life expectancies), the deaths seen in the cohort studies may reflect greater losses of life expectancy and possibly induction of disease in otherwise healthy individuals. Thus, although the cohort mortality evidence may be more uncertain than the time-series evidence, in light of the smaller number of studies and greater number of potential confounders, the cohort evidence should be central in any health impact assessment.

4.3.3 Chinese Literature on Air Pollution Mortality

Although substantial evidence exists for air pollution mortality in the United States and in numerous other countries, information is more limited on air pollution mortality in China. As described below, one meta-analysis (Aunan and Pan 2004) included six time-series studies of all-cause mortality in China, along with one cross-sectional study of long-term mortality and no cohort studies. A second assessment (HEI 2004) considered six Chinese time-series studies to be appropriate for a meta-analysis, with additional time-series or cross-sectional studies excluded because of various study limitations. These limited studies measured a variety of different pollutants and included both Hong Kong and mainland China. The critical questions are therefore whether the epidemiological studies in China are sufficient for a quantitative analysis, whether the findings from the worldwide literature can inform an analysis in China, and whether one could not infer air pollution health effects in China without additional epidemiological studies conducted in China.

A few barriers could cause difficulties in applying findings directly from the worldwide literature to China or in interpreting the Chinese literature in reference to the worldwide literature. First, as described below, although the majority of evidence points toward fine particles ($PM_{2.5}$) as the causal agent behind air pollution mortality effects, most monitoring efforts and Chinese epidemiological studies have focused on TSP or SO_2. Because it is thought that $PM_{2.5}$ is the fraction of TSP that penetrates deep into the lungs and is responsible for health impacts, conversions are necessary to compare findings directly, which contributes uncertainty, and any exposure misclassification may obscure the true effect. This uncertainty is illustrated by the fact that TSP was not a significant predictor of cohort mortality in the Six Cities study, in spite of the significance for $PM_{2.5}$. The ACS study did not consider TSP extensively but drew similar conclusions in limited analyses. In addition, because particulate matter is a mixture of a number of different compounds that might differ in their toxicity, any differences in particle composition between China and other countries would cause differences in the magnitude of health impacts.

Another important issue concerns the ambient concentrations in China and the countries primarily studied to date, such as the United States. If the concentration-response function were nonlinear, the slope would differ in the two countries, in light of large differences in pollution patterns. If there were a threshold for health effects from a given pollutant, health effects could be seen in China but not in the United States. Furthermore, because air pollution has different effects on different causes of death, differences in the underlying mortality patterns and age structures of the population would contribute to differences in concentration-response functions. A refinement to the health effects estimate could incorporate age-specific exposure-response relationships to determine separate estimates for children or the elderly as has been done recently in a Norwegian report (Aunan et al. 2002).

To provide a basis for comparison, we first summarize available evidence on ambient concentrations of air pollution in China. We then summarize the time-series mortality evidence in China and make some first-order comparisons with the worldwide literature. Although no cohort mortality studies have been conducted in China, to date, we compare the results from one cross-sectional mortality study with the cohort literature in the United States to determine whether patterns are similar.

Ambient pollutant levels in China In China, the air pollutants TSP and SO_2 have been the focus of monitoring, in light of the importance of coal combustion for energy production. Beyond the discussion in chapter 1, a helpful published study measured particulate matter in three size categories (TSP, PM_{10}, and $PM_{2.5}$) as well as SO_2 and NO_X from 1993 to 1996 in four cities in China: Lanzhou, Chongqing, Wuhan, and Guangzhou (Qian et al. 2001). Four-year average concentrations of SO_2 in this study ranged from 15 to 331 mg/m^3, whereas the study cited in chapter 1 (PRCEE et al. 2001) reported declining average concentrations that by 1999 ranged from 47 to 61 mg/m^3. In any case, these far exceed the corresponding measure in the ACS cohort study, 7 mg/m^3.

For PM_{10}, the range of annual means for two years (1995–1996) in the Chinese cities investigated by Qian et al. (2001) was 81–232 mg/m^3. Although PM_{10} measurements by municipal environmental protection bureaus (EPBs) have not become routine throughout China, they are now made in some cities, including Beijing, Guangzhou, Harbin, Shanghai, Shenzhen, and Tianjin. Annual means reported on official websites of EPBs in these six cities for 2001 range from 63 to 165 mg/m^3, somewhat lower values than Qian et al. but consistent with progress in particulate control since the mid-1990s (see table 4.4). Concentrations in the four cities of Qian et al. and the six cities of the EPB websites appear consistent with those in an earlier study (Xu 1998) and in PRCEE et al. (2001). These are all substantially higher than U.S. values, notably the mean value of 29 mg/m^3 cited for the ACS cities.

The data from the six municipal EPBs allow us to estimate current ratios between PM_{10} and TSP, which is crucial in comparing epidemiological studies. On the basis

Table 4.4
Annual mean particulate concentrations (in mg/m^3) and PM_{10}/TSP ratio in six cities for 2001

City	TSP	PM_{10}	PM_{10}/TSP
Beijing	370	165	0.45
Guangzhou	151	73	0.48
Harbin	219	135	0.62
Shanghai	162	100	0.62
Shenzhen	122	63	0.52
Tianjin	283	167	0.59
Six-city average	218	117	0.54

Sources: Beijing EPB (2001); Guangzhou EPB (2001); Harbin EPB (2002); Shanghai EPB (2001); Shenzhen EPB (2001); Tianjin EPB (2001).

of 2001 data for these six cities, the ratio of PM_{10}/TSP ranged from 0.45 to 0.62, with a mean of 0.54. The somewhat older data of Qian et al. 2001 includes not just TSP and PM_{10}, but also $PM_{2.5}$ (of which no other systematically measured annual averages have been found, recent or otherwise). This allows at least one estimate of the ratio of $PM_{2.5}$ to PM_{10} on the basis of Chinese data, for eight sites in four cities in 1995–1996. This ratio ranged from 0.52 to 0.73, with a mean of 0.61. In the United States, default ratios that are generally used are 0.55 for PM_{10}/TSP and 0.6 for $PM_{2.5}/PM_{10}$. Thus, on the basis of the very limited data available for China, the conversions between $PM_{2.5}$, PM_{10}, and TSP appear almost identical to those of Western studies. In the approximations below, we apply a PM_{10}/TSP ratio of 0.54 and a $PM_{2.5}/PM_{10}$ ratio of 0.61.

Acute effects on mortality in China Six studies of the relationship between daily air pollution levels and all-cause mortality in China were summarized in Aunan and Pan 2004—in Beijing (Xu et al. 1994), Benxi (Jin et al. 1999), Shenyang (Xu et al. 2000), Hong Kong (Wong et al. 2001), Shanghai (Kan and Chen 2003), and Chongqing (Venners et al. 2003). The HEI (2004) meta-analysis also considered a second study in Beijing (Gao et al. 1993). On the basis of the studies included in the meta-analyses, Aunan and Pan report pooled all-cause mortality concentration-response functions of 0.3% for a daily 10 $\mu g/m^3$ increase of PM_{10} and 0.4% for a daily 10 $\mu g/m^3$ increase of SO_2. Similarly, HEI reports an all-cause mortality concentration-response function of 0.4% for a daily 10 $\mu g/m^3$ increase of PM_{10} in China. No comparable value is reported for SO_2, but a pooled estimate across all Asian cities considered was an approximate 0.5% increase in mortality for a daily 10 $\mu g/m^3$ increase of SO_2.

Although these studies provide some important insight, there are some significant limitations in developing a pooled concentration-response function from the current literature. For example, the Kan and Chen analysis used case-crossover methods with logistic regression, and the Jin et al. analysis used linear regression rather than the time-series design using Poisson regression found in other studies (Aunan and Pan 2004). In addition, both the air pollution levels and sociodemographic factors are likely significantly different between Hong Kong in the late 1990s and mainland China. We discuss two of the remaining studies in greater detail below, to provide additional insight about the strengths and weaknesses of the literature and the assumptions needed to develop concentration-response functions as presented above.

In Beijing, the association between daily mortality and air pollution was examined in 1989 in two residential areas (Xu et al. 1994). Annual mean concentrations were 102 μg/m^3 of SO_2 (with a maximum daily concentration more than six times that) and 375 μg/m^3 of TSP (with a maximum daily concentration almost three times that). After controlling for temperature, humidity, and day-of-week effects, a significant association was found between ln(SO_2) and daily mortality: an 11% increase with each doubling in SO_2 concentration [95% CI: 5–16]. The central estimate for the association of ln(TSP) and mortality was positive but not statistically significant at the 5% level (4% increase in mortality with each doubling in TSP [95% CI: –2%, 11%]). In multipollutant regressions, the coefficient for TSP was reduced substantially, whereas that of SO_2 was largely unchanged, which indicates that SO_2 is a more robust indicator of air pollution mortality.

The relationship between air pollution and daily mortality was also examined in Shenyang by use of data from 1992 (Xu et al. 2000). Pollution levels were somewhat higher than those in Beijing, with mean levels of 197 μg/m^3 of SO_2 and 430 μg/m^3 of TSP. The statistical approach was generally similar, though the authors reported the results on a mass basis rather than as a function of doubled air pollution levels. In univariate regressions, significant associations with mortality were found for both SO_2 and TSP (a 2.4% and 1.7% increase, respectively, for a 100 μg/m^3 increase in pollution levels). When both pollutants were included in the model, unlike the findings in Beijing, only the TSP relationship remained statistically significant. It is interesting that the dose-response curve for SO_2 appeared to have a threshold near 150 μg/m^3, which potentially explains its insignificance in U.S. studies. No threshold was apparent for TSP.

If we take the Beijing findings for TSP (in spite of the fact that the relationship with SO_2 was more robust), a doubling from the mean TSP concentration would be an increase of 375 μg/m^3. Using the PM_{10}/TSP ratio of 0.54 reported earlier for six Chinese cities, 375 μg/m^3 of TSP is approximately equal to 200 μg/m^3 of PM_{10}. Thus, this study implies that for every 10 μg/m^3 increase of PM_{10}, daily total mortality increases by about 0.2% (4% divided by 20). Similarly, the Shenyang study can be translated into a 1.7% increase in daily total mortality for a 54 μg/m^3 increase of PM_{10}, or a 0.3% increase in daily total mortality for a 10 μg/m^3 increase of PM_{10}.

Noting the numerous simplifying assumptions, these estimates (and those from the meta-analyses) are close to the slopes seen in the worldwide literature for PM_{10}, though at the lower end of the values reported (many of which ranged from 0.3% to 1.0%). This would be the case if the concentration-response function for particulate matter were linear or had a slightly decreasing slope as concentrations increased; however, in light of the uncertainties in translating TSP measures into

PM_{10} or $PM_{2.5}$, the potential significance of SO_2 in China, and differences in statistical methods, it is impossible to draw conclusions other than that the time-series concentration-response functions in the United States and China appear of a similar order of magnitude.

Chronic effects on mortality in China As mentioned earlier, there are no cohort mortality studies currently available in China. Some limited cross-sectional mortality evidence exists, however, which can be compared with the literature at large to draw some preliminary conclusions about potential long-term mortality effects.

An ecological air pollution mortality study (Xu 1993) that was conducted in Shenyang was described in Xu (1998). This study simply categorized the population of Shenyang into low, medium, and high pollution groups. Comparing the lowest and the highest polluted areas, an increase in TSP concentrations of 207 $\mu g/m^3$ was associated with a 70% increase in age-adjusted mortality rates (from 409 to 697 per 100,000 people). If we use the conversions described earlier, 207 $\mu g/m^3$ of TSP is approximately equivalent to 68 $\mu g/m^3$ of $PM_{2.5}$. This implies an increase of approximately 1% in mortality per $\mu g/m^3$ of $PM_{2.5}$ (assuming that the entire TSP effect is due to fine particle exposure).

This estimate is difficult to interpret, as it is a cross-sectional rather than a cohort study and therefore has no information about other potential confounders (such as smoking status, diet, socioeconomic status, or indoor air pollution). In any case, it provides some descriptive indication of an air pollution effect that could be greater than the effect found in the time-series literature and of a similar magnitude as the cohort mortality studies in the United States.

There are some significant issues in applying this value or a U.S. cohort mortality concentration-response function to China. First, if the function were assumed to be linear down to zero concentration, the results would be implausible. Given $PM_{2.5}$ levels on the order of 50–100 $\mu g/m^3$, even the lowest cohort mortality concentration-response functions would imply that perhaps half of all deaths in China are due to air pollution exposure. It is important to remember, however, that there is no plausible near-term policy that would eliminate all air pollution from China. Thus, the relevant question for policy analysis is not what the slope of the concentration-response curve is from 100 to 0 $\mu g/m^3$, but rather what it is for an incremental reduction in ambient concentrations. Within this framework, it becomes somewhat more reasonable to include a cohort mortality effect.

Nevertheless, directly applying the U.S. cohort mortality evidence to China poses some significant difficulties. As noted, pollution concentrations were significantly

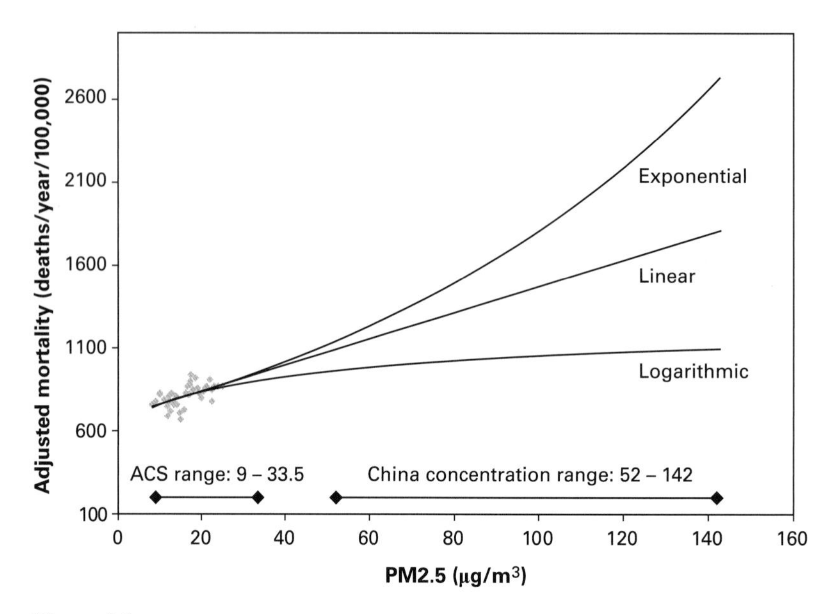

Figure 4.1
Relationship between mortality rate and $PM_{2.5}$ concentrations from ACS study, with three dose-response curves fit to the data. The China concentration range is from Qian et al. 2001.

lower in the U.S. studies than in China. The cities included in the ACS study had annual average $PM_{2.5}$ concentrations between roughly 10 and 30 μg/m^3, while levels in China according to the limited available data (Qian et al. 2001) are closer to 100 μg/m^3. Although the ACS study was consistent with a linear concentration-response curve, figure 4.1 shows that deviations from linearity might have a minimal effect within the concentration range of the ACS study but would significantly influence the slope at pollution levels in China.

Summary of health effects We have now described both acute and chronic mortality and seen reasonably similar concentration-response functions in studies from developed countries and in China, albeit with limited evidence and significant assumptions. To summarize our comparisons, commonly used concentration-response functions applied in the United States and China are included in table 4.5. For comparative purposes, some morbidity endpoints have been included even though mortality impacts tend to dominate the quantification of health benefits from reduced air pollution. From the table below, one sees that the concentration-response functions from studies conducted in China tend to fall in the range of values applied for

Table 4.5
Sampling of dose-response functions from epidemiological studies in China and the United States for PM and SO_2

Health Endpoint (increase per 10 μg/m^3 increase in pollutant)	Dose-Response (China)	Source	Dose-Response (U.S.)	Source
Time-series mortality (daily PM_{10})	0.3–0.4%[a]	Aunan and Pan (2004), HEI (2004)	0.3–0.6%[b]	Samet et al. (2000),[c] Levy et al. (2000)
Time-series mortality (daily SO_2)	0.4–0.5%	Aunan and Pan (2004), HEI (2004)	—[d]	
Chronic mortality (annual average PM_{10})	6%[a,e]	Xu (1993)	4–7%[f,g]	Pope, Bates et al. (1995), Pope et al. (2002), Dockery et al. (1993)
Respiratory or cardiovascular hospital admissions (daily PM_{10})	0.3–1.6%	Xu et al. (1995),[h] Aunan and Pan (2004)	1.1–1.4%[i]	Samet et al. (2000)
Chronic bronchitis (annual average PM_{10})	—[j]		8%[k]	Abbey et al. (1995)

[a] These studies measured TSP, not PM_{10}. A conversion ratio of $PM_{10}/TSP = 0.54$ was used.
[b] This is not the full range of values seen in individual cities, but a range of pooled national average estimates.
[c] The NMMAPS study (Samet et al. 2000) revised its original estimate of 0.5% after correcting for statistical flaws (Dominici et al. 2002).
[d] SO_2 is rarely statistically significant in multivariate time-series studies in the United States.
[e] This was an ecological, or cross-sectional, study rather than a cohort study.
[f] The lower value is used in calculations for the EPA's benefit cost analysis of the Clean Air Act, because it is thought to capture the long-term mortality and the study was more comprehensive than the Dockery et al. 1993 study.
[g] These studies measured and reported effects for $PM_{2.5}$. A $PM_{2.5}/PM_{10}$ ratio of 0.6 is used for the conversion to PM_{10}. Note that chronic mortality dose-response estimates are about an order of magnitude greater than acute mortality.
[h] This study found a relationship between nonsurgery outpatient visits and the square root of TSP. The 0.3% is found using the slope (0.015% per μg/m^3 TSP) from the portion of the curve near the mean study concentration (388 μg/m^3 TSP) and a PM_{10}/TSP ratio of 0.54.
[i] Range for one day mean (lag 0) increase in hospital admissions for cardiovascular disease and COPD.
[j] No comparable epidemiological studies have been located in China that quantify dose-response functions for chronic bronchitis.
[k] Effect for chronic bronchitis in those over 25 years of age.

acute and chronic mortality in the U.S., with chronic mortality about an order of magnitude greater than acute in both countries. (Regardless of how the study was conducted, all PM dose-response functions above are reported as the effect per 10 mg/m^3 of PM_{10} to facilitate comparison between effects.) For SO_2 mortality, however, there is no corresponding U.S. value, because SO_2 is often an insignificant predictor of air pollution mortality in the United States.

Two morbidity endpoints, hospital admissions and chronic bronchitis, are provided to illustrate some key points. In terms of hospital admissions, comparability is severely impaired by differences in the subpopulations and health outcomes considered. More broadly, differences in access to health care and health care systems between the United States and China might make international comparisons of these values complicated. For example, there may be vast differences in hospital usage rates or how diseases are coded. That being said, the concentration-response functions are similar, though the higher values for China were seen in Hong Kong, with significantly lower values in mainland China. For chronic bronchitis, the only estimates available from China are from cross-sectional studies of prevalence with limited exposure information or from occupational exposure studies, so no studies are included in the table. This complicates comparison with a prospective cohort study of chronic bronchitis incidence in the United States. Again, morbidity endpoints are difficult to compare between any two countries because of differences in study type and disease definition.

4.3.4 Health Evidence Conclusions

At first glance, the findings from epidemiological studies in China appear roughly comparable with the U.S. literature, though significant assumptions were embedded in our comparisons. It is also possible that the estimates happen to be similar to one another because of numerous differences that cancel each other out but that extrapolating findings from the United States to China may be inappropriate in light of the array of differences.

What is clear is that there is extremely strong evidence of an effect of PM on mortality from time-series studies in the United States but that the magnitude of the cohort mortality effect is significantly greater. A health impact assessment in the United States that neglects cohort mortality (assuming that the observed relationship is causal) would significantly understate health impacts. Similar significance has been observed for the effect of particulate matter and sulfur dioxide on mortality in time-series studies in China, and it is reasonable to assume that cohort mortality would be similarly important in China. It is also clear that conducting a new prospective cohort study in China would be expensive and take decades and that policy action is required long before such a study could be completed. Although investigating air pollution effects within existing health cohorts in China (similar to the ACS approach) might address this issue in a more timely fashion, near-term decisions will need to be based in part on studies outside of China.

That the American time-series literature appears to show similar concentration-response values as the Chinese studies lends support to extrapolating the cohort mortality findings from the U.S. to China; however, in light of the significant differences in pollution levels, this should only be done for incremental concentration changes and with appropriate characterization of uncertainty. This uncertainty currently may be quite large, in view of inherent uncertainties in the cohort mortality literature as well as numerous extrapolations. As additional epidemiological or exposure assessment studies in China become available, the concentration-response function (including both the central estimate and uncertainty bounds) for mortality should be iteratively updated, because it is likely that the magnitude of this relationship will strongly influence assessments used in policy decisions.

Acknowledgment

The research of this chapter was funded by a generous grant to the China Project of the Harvard University Center for the Environment from the V. Kann Rasmussen Foundation.

References

Abbey, D. E., B. E. Ostro, F. Peterson, and R. J. Burchette. 1995. Chronic respiratory symptoms associated with estimated long-term ambient concentrations of fine particulates less than 2.5 microns in aerodynamic diameter ($PM_{2.5}$) and other air pollutants. *Journal of Exposure Analysis and Environmental Epidemiology* 5 (2):137–159.

Aunan, K., J. Fang, H. Vennemo, K. Oye, and H. M. Seip. 2002. Co-benefits of climate policy: Lessons learned from a study in Shanxi, China. Center for International Climate and Environmental Research-Oslo (CICERO). Oslo, Norway.

Aunan, K., and X.-C. Pan. 2004. Exposure-response functions for health effects of ambient air pollution applicable for China: A meta-analysis. *Science of the Total Environment* 329:3–16.

Beijing Environmental Protection Bureau (EPB). 2001. Beijing environmental status report 2001. Available at http://www.bjepb.gov.cn/newhb/file/filelist3.asp?path=html/hjg. In Chinese.

Bell, M. L., F. Dominici, and J. M. Samet. 2005. A meta-analysis of time-series studies of ozone and mortality with comparison to the National Morbidity, Mortality, and Air Pollution Study. *Epidemiology* 16 (4):436–445.

Bennett, D. H., T. E. McKone, J. S. Evans, W. W. Nazaroff, M. D. Margni, O. Jolliet, and K. R. Smith. 2002. Defining intake fraction. *Environmental Science & Technology* 36 (9):207A–211A.

Burnett, R., J. Brook, T. Dann, C. Delocla, O. Philips, S. Cakmak, R. Vincent, M. Goldberg, and D. Krewski. 2000. Association between particulate- and gas-phase components of urban air pollution and daily mortality in eight Canadian cities. *Inhalation Toxicology* 12 (Suppl. 4):15–39.

Dockery, D. W., C. A. Pope, X. Xu, J. D. Spengler, J. H. Ware, M. E. Fay, B. G. Ferris, and F. E. Speizer. 1993. An association between air pollution and mortality in six U.S. cities. *New England Journal of Medicine* 329 (24):1753–1759.

Dominici, F., A. McDermott, S. L. Zeger, and J. M. Samet. 2002. On the use of generalized additive models in time-series studies of air pollution and health. *American Journal of Epidemiology* 156 (3):193–203.

Evans, J. S., S. Wolff, K. Phonboon, J. I. Levy, and K. Smith. 2002. Exposure efficiency: An idea whose time has come? *Chemosphere* 49:1075–1091.

Gao, J., X. P. Xu, Y. D. Chen, D. W. Dockery, D. H. Long, H. W. Liu, and J. Y. Jiang. 1993. Relationship between air pollution and mortality in Dongcheng and Xicheng Districts, Beijing. *Zhonghua Yu Fang Yi Xue Za Zhi* 27:340–343. In Chinese.

Guangzhou Environmental Protection Bureau (EPB). 2001. Guangzhou environmental status report 2001. Available at http://www.gzepb.gov.cn/hjgb/200312090003.htm. In Chinese.

Harbin Environmental Protection Bureau (EPB). 2002. Harbin environmental status report 2002. Available at http://hrb-hlsep.gov.cn/hjgb/hjgb.htm. In Chinese.

Health Effects Institute (HEI). 2004. Health effects of outdoor air pollution in developing countries of Asia: A literature review. Special report no. 15, International Scientific Oversight Committee, HEI. Boston.

Ito, K., S. F. De Leon, and M. Lippman. 2005. Associations between ozone and daily mortality: Analysis and meta-analysis. *Epidemiology* 16 (4):446–457.

Jin, L. B., Y. Qin, Z. Xu, and B. Chen. 1999. Association between air pollution and mortality in Benxi, China. *Chinese Journal of Public Health* 15 (3):211–212. In Chinese.

Kan, H., and B. Chen. 2003. A case-crossover analysis of air pollution and daily mortality in Shanghai. *Journal of Occupational Health* 45:119–124.

Krewski, D., R. Burnett, M. Goldberg, K. Hoover, J. Siemiatycki, M. Jarrett, M. Abrahamowicz, and W. White. 2000. *Particle epidemiology reanalysis project. Part II. Sensitivity analyses*. Boston: Health Effects Institute.

Laden, F., L. M. Neas, D. W. Dockery, and J. Schwartz. 2000. Association of fine particulate matter from different sources with daily mortality in six U.S. cities. *Environmental Health Perspectives* 108 (10):941–947.

Levy, J. I., J. K. Hammitt, and J. D. Spengler. 2000. Estimating the mortality impacts of particulate matter: What can be learned from between-study variability? *Environmental Health Perspectives* 108 (2):109–117.

Levy, J. I., S. M. Chemerynski, and J. A. Sarnat. 2005. Ozone exposure and mortality: An empiric Bayes metaregression analysis. *Epidemiology* 16 (4):458–468.

Levy, J. I., A. M. Wilson, J. S. Evans, and J. D. Spengler. 2003. Estimation of primary and secondary particulate matter intake fractions for power plants in Georgia. *Environmental Science and Technology* 37:5528–5536.

Lipfert, F. W., H. M. J. Perry, J. P. Miller, J. D. Baty, R. E. Wyzga, and S. E. Carmody. 2000. The Washington University-EPRI Veterans' Cohort Mortality Study: Preliminary results. *Inhalation Toxicology* 12 (Suppl. 4):41–73.

Lipfert, F. W., and R. E. Wyzga. 1995. Air pollution and mortality: Issues and uncertainties. *Journal of the Air and Waste Management Association* 45 (12):949–966.

McDonnell, W. F., N. Nishino-Ishikawa, F. F. Petersen, L. Hong Chen, and D. E. Abbey. 2000. Relationships of mortality with the fine and coarse fractions of long-term ambient PM_{10} concentrations in nonsmokers. *Journal of Exposure Analysis and Environmental Epidemiology* 10 (5):427–436.

National Morbidity and Mortality Air Pollution Study (NMMAPS). 2002. Frequently-asked questions. Johns Hopkins University, Bloomberg School of Public Health. Available at http://www.biostat.jhsph.edu/biostat/research/nmmaps_faq.htm.

Phonboon, K. 1996. Risk assessment of environmental effects in developing countries. Ph.D. diss., Harvard School of Public Health.

Policy Research Center of Environment and Economy (PRCEE) of SEPA, China National Environmental Monitoring Center (CNEMC), and Chinese Academy of Environmental Sciences (CRAES). 2001. New countermeasures for air pollution control in China. Background report for World Bank.

Pope, C. A., III, D. V. Bates, and M. E. Raizenne. 1995. Health effects of particulate air pollution: Time for reassessment? *Environmental Health Perspectives* 103 (5):472–480.

Pope, C. A., III, R. T. Burnett, M. J. Thun, E. E. Calle, D. Krewski, K. Ito, and G. D. Thurston. 2002. Lung cancer, cardiopulmonary mortality, and long-term exposure to fine particulate air pollution. *Journal of the American Medical Association* 287 (9):1132–1141.

Pope, C. A., III, M. J. Thun, M. M. Namboodiri, D. W. Dockery, J. S. Evans, F. E. Speizer, and C. W. Heath, Jr. 1995. Particulate air pollution as a predictor of mortality in a prospective study of U.S. adults. *American Journal of Respiratory and Critical Care Medicine* 151 (3 part 1):669–674.

Qian, Z., J. Zhang, F. Wei, W. E. Wilson, and R. S. Chapman. 2001. Long-term ambient air pollution levels in four Chinese cities: Inter-city and intra-city concentration gradients for epidemiological studies. *Journal of Exposure Analysis & Environmental Epidemiology* 11 (5):341–351.

Rubin, E. S., and C. I. Davidson. 2001. *Introduction to engineering and the environment.* Boston: McGraw-Hill.

Samet, J. M., S. L. Zeger, F. Dominici, F. Curriero, I. Coursac, D. W. Dockery, J. Schwartz, and A. Zanobetti. 2000. *The National Morbidity, Mortality, and Air Pollution Study. Part II. Morbidity, mortality, and air pollution in the United States.* Boston: Health Effects Institute.

Schwartz, J. 2001. Is there harvesting in the association of airborne particles with daily deaths and hospital admissions? *Epidemiology* 12 (1):55–61.

Schwartz, J., D. W. Dockery, and L. M. Neas. 1996. Is daily mortality associated specifically with fine particles? *Journal of the Air and Waste Management Association* 46 (10): 927–939.

Shanghai Environmental Protection Bureau (EPB). 2001. Shanghai environmental status report 2001. Available at http://www.sepb.gov.cn/gongbao/2001.asp. In Chinese.

Shenzhen Environmental Protection Bureau (EPB). 2001. Shenzhen environmental status report 2001. Available at http://www.szepb.gov.cn/web/hjzl/gb/gb_2001.htm. In Chinese.

Smith, K. R. 1993. Fuel combustion, air pollution, and health: The situation in developing countries. *Annual Review of Energy and the Environment* 18:529–566.

Tianjin Environmental Protection Bureau (EPB). 2001. Tianjin environmental status report 2001. Available at http://www.tjhb.gov.cn/kqzl/hjzb/2001gb.htm#13. In Chinese.

Thurston, G. D., and K. Ito. 2001. Epidemiological studies of acute ozone exposure and mortality. *Journal of Exposure Analysis & Environmental Epidemiology* 11 (4):286–294.

U.S. Environmental Protection Agency (U.S. EPA). 1996. *Air quality criteria for particulate matter*. Volume III. Washington, D.C.: Office of Research and Development.

U.S. Environmental Protection Agency (U.S. EPA). 1999. *The benefits and costs of the Clean Air Act: 1990 to 2010*. Washington, D.C.: Office of Air and Radiation.

Venners, S., B. Wang, Z. Peng, Y. Xu, L. Wang, and X. Xu. 2003. Particulate matter, sulfur dioxide and daily mortality in Chongqing, China. *Environmental Health Perspectives* 111 (4):562–567.

Wilson, A. 2003. Improved characterization of fine particle intake fraction for air pollution control and research decision-making. Ph.D. diss., Harvard School of Public Health.

Wolff, S. K. 2000. Evaluation of fine particle exposures, health risks and control options. Ph.D. diss., Harvard School of Public Health.

Wong, C. M., S. Ma, A. J. Hedley, and T. H. Lam. 2001. Effect of air pollution on daily mortality in Hong Kong. *Environmental Health Perspectives* 109 (4):335–340.

World Bank. 2001. *China—Air, land, and water: Environmental priorities for a new millennium*. Washington, D.C.: World Bank.

Xu, X., D. W. Dockery, D. C. Christiani, B. Li, and H. Huang. 1995. Association of air pollution with hospital outpatient visits in Beijing. *Archives of Environmental Health* 50 (3):214–220.

Xu, X. 1998. Air pollution and its health effects in urban China. In *Energizing China: Reconciling Environmental Protection and Economic Growth*, edited by M. B. McElroy, C. P. Nielsen, and P. Lydon. Cambridge, Mass.: Harvard University Committee on Environment/Harvard University Press.

Xu, X., J. Gao, D. W. Dockery, and Y. Chen. 1994. Air pollution and daily mortality in residential areas of Beijing, China. *Archives of Environmental Health* 49 (4):216–222.

Xu, Z., D. Yu, L. Jing, and X. Xu. 2000. Air pollution and daily mortality in Shenyang, China. *Archives of Environmental Health* 55 (2):115–120.

Xu, Z. Y. 1993. *Report of the World Health Organization consultation on air pollution and health*. Washington, D.C.: World Bank, China and Mongolia Department.

Zhou, Y. 2002. Evaluating power plant emissions in China: Human exposure and valuation. Ph.D. diss., Harvard School of Public Health.

Zhou, Y., J. I. Levy, J. K. Hammitt, and J. S. Evans. 2003. Estimating population exposure to power plant emissions using CALPUFF: A case study in Beijing, China. *Atmospheric Environment* 37 (6):815–826.

Zhou, Y., J. I. Levy, J. S. Evans, and J. K. Hammitt. 2006. The influence of geographic location on population exposure to emissions from power plants throughout China. *Environment International* 32 (3):365–373.

5

Local Population Exposure to Pollutants from Major Industrial Sectors and Transportation

Shuxiao Wang, Jiming Hao, Yongqi Lu, and Ji Li

5.1 Introduction

Until recent decades, heavy industry was considered the foundation of China's economy. Most major cities had large iron and steel mills and chemical-manufacturing plants located in their city limits as a result of government policy, so that they might serve as bases of the local and regional economy and provide employment. These industrial plants often expanded into very large complexes. Many of them have remained located near city centers even as populations have grown dense around them because of urbanization. Cement manufacturing has a somewhat different history, because simple cement kilns can be built in small sizes. As a result China has thousands of such kilns spread throughout the country, along with more industrial-scale cement manufacturing. Many know that the transport sector has undergone a huge transition in recent years, as motor vehicle ownership has soared, modern highway systems and urban roads have been built, and freight hauling by trucks using the new roads has quickly expanded.

As chapter 4 has introduced and previous chapters have mentioned, human exposure is a key step in quantifying health damages of pollution emissions. It can depend on a variety of factors. It is a help that the four sectors mentioned above seem to offer a mix of these basic characteristics, which gives a better opportunity to estimate and compare their effects.

It is logical to assume that the proximity of sources to people would be a central factor in exposure. That small cement kilns are sometimes located in less-settled outlying areas indicates one scale of population proximity, and that iron and steel mills and chemical complexes are often near city centers indicates another. Urban transport corridors probably put this source type in the closest proximity to population of all. The four sectors also differ somewhat in terms of the height of their emission outlets, which logically would also have an effect on exposure. The large urban steel mills and chemical factories will often have tall smokestacks, whereas cement kilns perhaps somewhat shorter ones. The exhaust pipes of vehicles, of course, put transport emissions at ground level.

What is common about these sources is that they are all among the highest-emitting sectors. It was for this reason that we first selected them to examine the health damages due to their emissions of air pollutants. The first three are manufacturing industries and together contribute 17% of total industrial sulfur dioxide (SO_2) emissions, and 49% of total suspended particulate (TSP) emissions from industry in 1999.[1] We included the transportation sector additionally because it is the most rapidly increasing source of air pollution and hence an increasing concern to the government. In many urban environments road traffic is now responsible for more than 70% of atmospheric PM_{10} (Künzli et al. 2000). We might note that the greatest consumer of coal and generator of combustion pollution emissions is the electric power sector, the subject of chapters 6 and 7.

We estimate health exposures to TSP and SO_2 emissions from the three industrial sectors, and to PM_{10} emissions from urban road traffic, for the year 1999. In this and the next two chapters, we analyze the emission-exposure relationship by using the concept of intake fraction (*iF*) introduced in chapter 4. These *iF* estimates serve three purposes. The first is to estimate exposure risk for these sectors on a national basis so they may be used for national analyses like those done in chapters 9 and 10. Second, they will provide a reduced-form relationship between exposure risk and easily observed source characteristics so that exposures may be quickly approximated for other sources without laborious air-dispersion modeling for each. The third is to provide the basis for future analyses of air pollution health risk in detailed case studies of the five cities from which samples are drawn.

The sources of emissions (stacks) in these sectors number in the tens of thousands. There is no national database of plant emissions to call upon or develop that, as we will see in chapter 6, is feasible for the power sector. Therefore, a different methodology is required. After considering various options, we conclude that using data from sample plants in five distinctly located cities would provide a good sample at reasonable cost and time. We use an air-dispersion model to estimate *iF*s for hundreds of sources in this sample. The next step of our procedure is to estimate reduced-form exposure-emission relationships by running regressions of *iF*s on sample source characteristics. We then compile summary data from national lists of enterprises in these three sectors and combine them with the reduced-form coefficients to estimate national exposures to air pollution.

On the local scale (i.e., less than 50 kilometers [km]), primary pollutants emitted directly in the combustion process dominate secondary pollutants formed by chemical reactions in the atmosphere. Because modeling secondary pollution formation requires much more complex calculations, we concentrate on primary pollutants and the local scale. Particulate matter and SO_2 are major primary pollutants affecting urban air quality in most cities of China, and official agencies collect data for industrial TSP and SO_2 emissions. We examine TSP and SO_2 from the three industrial sectors and primary PM_{10} for urban road transportation.

In the next section we describe the data from the five-city sample and the calculation of the intake fractions from each source in this sample. In section 5.3, we report the *iF* results and the reduced-form of emission-exposure relationships. In section 5.4, we examine the sensitivity of the results to assumed parameter values in the air-dispersion model. These five-city estimates are then combined with national data to obtain national intake fractions in section 5.5. In the final section, we discuss our results and compare them with other estimates.

5.2 Calculation of Intake Fractions from Five-City Sample

Our methodology is first to obtain a reasonable sample of the tens of thousands of emission sources in the cement, chemical, iron and steel, and transportation sectors. We then calculate the fraction of emissions inhaled by the local population for each source in the sample using the methods described in chapter 4. For ease of reference, the methodology is briefly repeated here.

We can refine the expression of intake fraction of chapter 4, defining it for a given pollutant x from a given source i:

$$iF_{ix} = \sum_{d=1}^{n}(POP_d \times C_{idx} \times BR)/EM_{ix}, \tag{5.1}$$

where POP_d is the population at location d; C_{id} is the incremental pollutant concentration at location d in grams per cubic meter (g/m^3) caused by emissions of source i; BR is the breathing rate; and EM_{ix} is the emission rate of x from i in grams per second (g/s).

For an exhaustive estimate of *iF*, the summation should be over any area where the concentration is positive; however, we will concentrate on local effects and consider only a domain of 50 km around the source. For all of our calculations here, we use an average breathing rate of 20 cubic meters per day (m^3/day) that is commonly adopted by other researchers. We now describe how we select the samples and calculate the concentrations and exposed populations.

5.2.1 Emission Source Sampling

The detailed data on source characteristics that we seek are not available in any national database but can be collected for individual cities. We begin by selecting five of the ninety-three cities in China that have more than 500,000 people in their urban areas by the end of 2000 (noting that city administrative jurisdictions include both urban and rural areas). We are unable to select them randomly because such data are not easily available for each city. Instead, we divide the country into regions and select one from each where we could enlist the support of the local Environmental Protection Bureau (EPB). The cities chosen are Beijing, Dalian, Jinan, Chongqing, and Liuzhou, located, respectively, in the north, northeast, east, central, and south, as shown in figure 5.1.

The five cities also represent different air pollution conditions in China, from highly polluted Chongqing to relatively clean Dalian. Detailed information on these five cities is given in appendix C. With the help from each city's EPB, we obtain data on population distribution and enterprises in the selected industries. These include

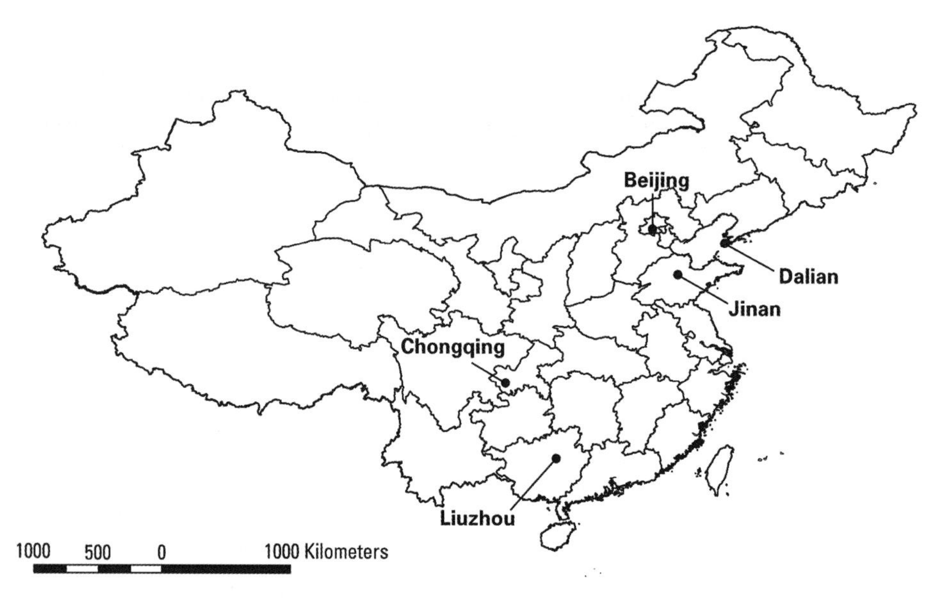

Figure 5.1
Geographic locations of five selected cities.

the location, production, and emission source characteristics. From the list of enterprises in each city, we randomly select a sample, giving us both urban and rural pollution sources. All data are updated to the year 1999. Table 5.1 lists the number of emission sources selected for each sector from each city.

Industrial sectors For these sectors, the sample consists of a total of 541 stacks, of which 169 stacks are from 67 cement plants, 185 stacks from 110 chemical plants, and 187 stacks from 5 steel production complexes. For each stack, we include in our database the location, stack height, stack inner diameter, emission exit velocity and temperature, and emission rates of SO_2 and TSP.

Our focus on the local human exposures means that the stack height is the most important emission feature. The complete distribution of stack heights in our sample is given in appendix C, figure C.5. The range of stack heights is quite large; the average stack heights for the cement, chemical, and steel sectors are, respectively, 34 meters (m), 41 m, and 44 m, with standard deviations (s.d.'s) of 17.95, 25.02, and

Table 5.1
Summary of samples in selected cities

	Industrial Sectors				
City	Cement Stack (plant)	Chemical Stack (plant)	Iron and Steel Stack (plant)	All Industrial Stacks	Transport Road Segments
Number of samples:					
Beijing	25 (8)	34 (17)	52 (1)	111	30
Jinan	47 (20)	64 (52)	39 (1)	150	36
Dalian	78 (28)	23 (12)	34 (1)	135	9
Chongqing	10 (5)	12 (6)	43 (1)	65	n.a.
Liuzhou	9 (6)	52 (23)	19 (1)	80	n.a.
Total	169 (67)	185 (110)	187 (5)	541	75
Stack height (m):					
Average	34	41	44	40	n.a.
Standard error	17.95	25.02	26.94	24.12	n.a.
Population within 10 km (million):					
Average	0.307	0.784	0.685	0.601	2.303
Standard error	0.344	0.813	0.540	0.636	1.545

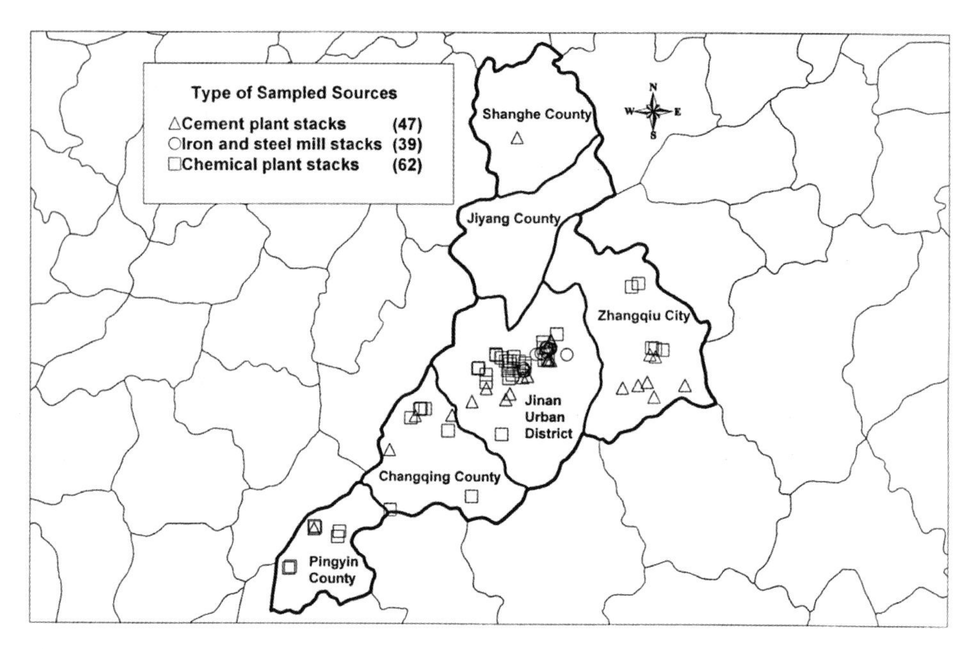

Figure 5.2
Location of selected samples in Jinan.

26.94. The highest one reaches 165 m, but more than 80% of the sampled stacks are lower than 50 m.

The next item important for health impact analysis is the particle size distribution of TSP emitted. This is characterized using information from two sources. A study conducted in Beijing provided measurements for fifteen stacks in 2001. We supplement that with particle size distributions given in the U.S. Environmental Protection Agency report AP-42 (U.S. EPA 1996; Florig and Song 2000). Details are given in appendix C.

To give a sense of the location of the selected pollution sources relative to the population centers, we present a map of one of the cities, Jinan, in figure 5.2. Similar maps of the other cities are in appendix C. Typically, cement plants are scattered in both urban and rural areas, whereas there is usually only one iron and steel complex with stacks concentrated in a single district. To illustrate the difference, we calculated the population within 10 km of each source, and average them for each sector. This is given in the bottom of table 5.1. The average population within a 10-km radius is 0.78 million (with s.d. of 0.81) for chemicals plants, 0.3 million (s.d. 0.34) for cement plants, and 0.69 million (s.d. 0.54) for iron and steel complexes.

Urban transportation The data for traffic pollution is much more limited than those of the industrial sectors. We are able to sample only three cities, getting a total of seventy-five road segments where we have information on traffic flows. The vehicle numbers were video recorded at each intersection and counted by our collaborators from the local EPBs. Of these, thirty are from Beijing, thirty-six from Jinan, and nine from Dalian. This is not an ideal sample because there is no systematic traffic flow information for all roads. Figures in appendix D show that most of the sampled road segments are located in the urban areas. The average population within a 10-km radius of the selected road segments is a high 2.3 million, with s.d. of 1.54. A second limitation is that we do not have information on emission factors and vehicle speeds. We simply assume that emissions from mobile sources are proportional to the traffic flows.

5.2.2 Air-Dispersion Modeling

In order to implement equation (5.1), we need to estimate the incremental concentration due to emissions from each sampled stack in each area *d* in the vicinity of the source. To do this we employ the Industrial Source Complex Long-Term (ISCLT3) air-dispersion model that will also be used in chapter 6. The details of this steady-state Gaussian plume dispersion model are given in appendix A.

To calculate the annual average concentration of pollutants, ISCLT3 requires inputs of source emission characteristics and receptor and meteorological parameters. The meteorological data and receptor locations that we use are specific to each city. In contrast, the following ISCLT3 model parameters are held constant for all samples. First, the terrain of both sources and receptors are set as "flat," because complex terrain simulations result in air concentrations that are highly dependent on site-specific topography. Simulations from flat terrain produce values that are more broadly applicable. Second, the half-life for exponential decay of SO_2 is set as 4 hours (14,400 seconds), the default value in the ISCLT3 model. Third, for TSP, we set the cause of plume depletion as dry removal mechanisms. Other minor assumptions are described in appendix A.

Meteorological data Meteorological information is collected for each of the five cities for the year of 1999 and applied to all emission sources within a given city. This includes the joint frequency distribution of wind speed, wind direction, and stability class (stability classes are described in appendix A). The ISCLT3 model also requires the average wind speed, average temperature, average mixing height, and average wind profiles. These data are obtained from local EPBs in the five cities.[2]

The meteorological characteristics of the five cities are summarized in table 5.2. They range from the very wet Liuzhou, with 1,489 millimeters (mm) of rain per year, to the dry Dalian, with 284 mm of precipitation.

Receptor system and population information To implement equation (5.1), we need to design a suitable grid system for the location index *d* for each source of emissions. In light of the pattern of source locations in figure 5.2, we design one system for each city. Our goal is to include exposures within 50 km of a given source. Taking Jinan as an example, after examining the locations of all the sampled sources, we decide that the grid should cover a 240 km × 240 km area. The fineness of the grid spacing is determined by the quality of the population data available. The smallest administrative unit for population data in the urban areas is the city block, whereas for rural areas it is the village- or town-level unit.

In the case of Jinan, we are able to use a 3 km × 3 km grid for the outer regions and a 1 km × 1 km one for the densely populated areas. This is shown in figure 5.3a. The domain is divided into 80 grid cells in both x and y directions, giving a total of 6,400 cells, each 3 km × 3 km. For the 100 km × 100 km area of densest population, we have detailed block data and use grid cells that are 1 km × 1 km, as shown in figure 5.3b.

Our procedures to allocate the population data to the chosen grid system are described in appendix B. The mapping operations are done using ArcView, version 5.2. The population is assumed to be distributed evenly within a given grid cell. In each of the smallest available administrative units, the population is distributed based on the land use. No population is assigned to lakes, grasslands, or mountain areas, which are shown as blank grid cells in figure 5.3b. For those counties without detailed population information, the county-level population data are used. In this case we have to assume that the population is evenly distributed across the entire county.

5.2.3 Estimating Intake Fraction

With the population mapped into fine grid cells as described above and meteorological information, we are now ready to calculate the *iF* according to equation (5.1). We illustrate with a cement stack in Jinan as an example. Figure 5.4b gives the annual-average incremental concentration for SO_2 estimated by the ISCLT3 air-dispersion model within a circle of 50-km radius; darker shades indicate higher concentrations. (For a color version of this figure, see figure 3.1 in chapter 3.) The

Table 5.2
Summary of meteorological information in selected cities

City	Wind			Mixing Height (meters) by Stability Class						Average Ambient Temperature (°K)	Rain Precipitation (mm/year)
	Prevailing Wind Direction	Prevailing Wind Frequency (%)	Wind Speed (meters /sec)	A	B	C	D	E	F		
Beijing	SW–NE	20.8	2.2	1,047	1,038	1,106	977	803	754	286.3	650
Chongqing	NNE–N	29.0	1.2	1,758	1,156	747	530	258	109	291.6	1,142
Dalian	N–S	20.7	3.5	900	600	600	600	400	400	285.1	234
Jinan	SW–NE	26.0	2.7	1,300	1,288	900	600	405	223	288.3	640
Liuzhou	NNW–NE	35.4	1.6	827	827	650	650	928	928	293.7	1,489

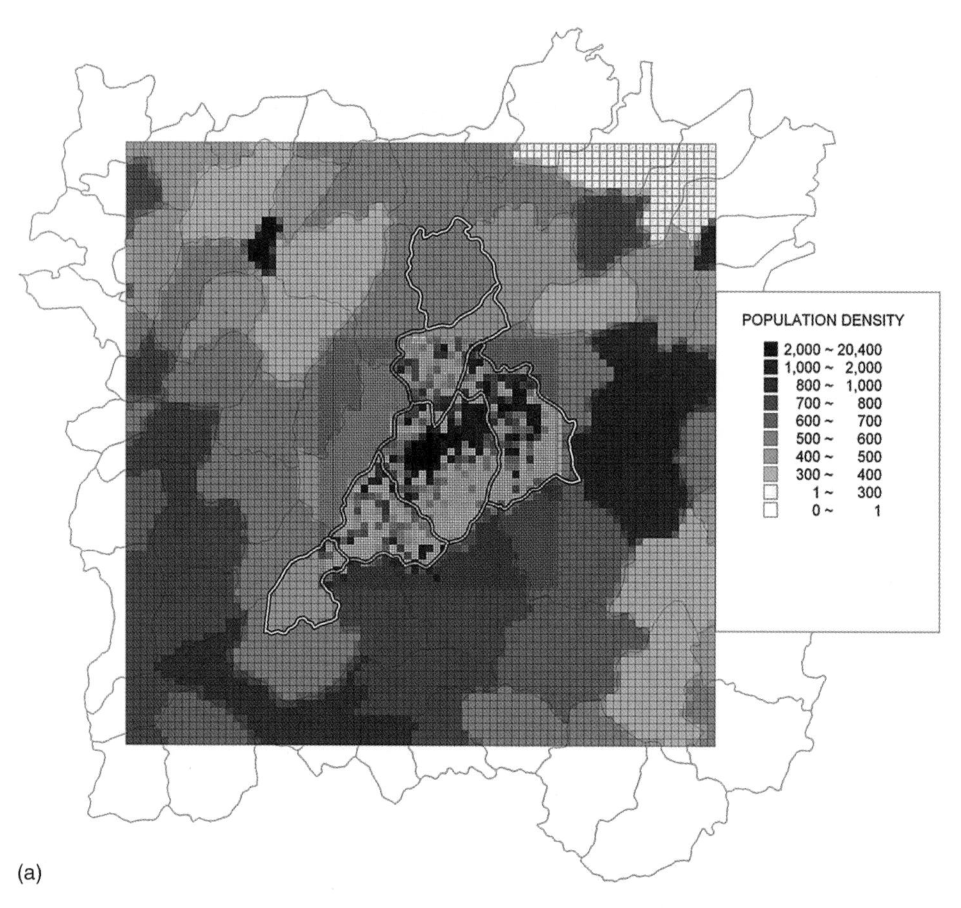

Figure 5.3
Gridded receptor domain and population distribution of Jinan City. (a) Entire domain. (b) Urban area. (See figure 3.5 in chapter 3 for a color version of figure 5.3b.)

prevailing wind directions for Jinan are southwesterly and northeasterly, which are clearly reflected in the concentration profile. In the adjacent figure 5.4a, we graph the population densities, with the darker cells indicating denser areas. The total quantity of pollutant inhaled is given as the product of the population and the concentration within each grid cell, also multiplied by the breathing rate; this is shown in figure 5.4c, where the darker grid cells indicate higher total human exposures. Summing over every grid cell in the domain, we obtain the total dosage. Dividing by the emissions of the given source i, we obtain the intake fraction, iF_{ix}. We do this for both x = TSP and SO_2 and repeat it for every source in that city.

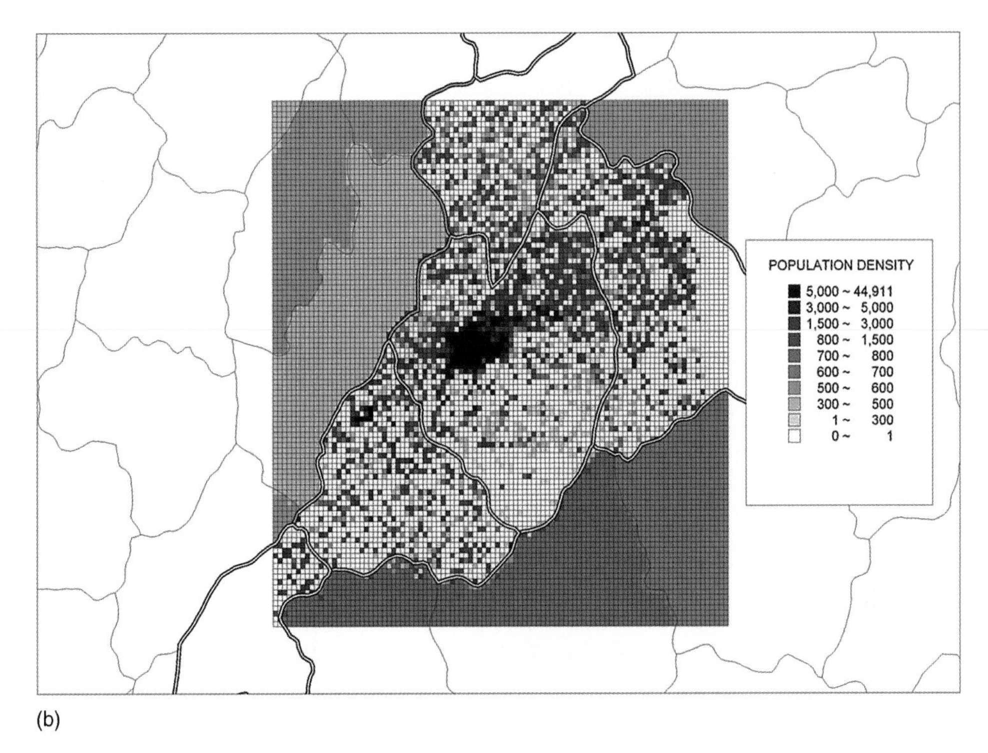

(b)

Figure 5.3
(continued)

This exercise is repeated for all the cities: examining the location of the sources; designing a grid system; mapping the population using ArcView; entering the source and meteorological data into ISCLT3; calculating the concentrations of TSP and SO_2; and, finally, calculating the *iF*s.

5.3 Estimates and Regression of Intake Fractions

5.3.1 Intake Fractions for Industrial Sources

The *iF* estimates of the 541 sample sources from the three industrial sectors are summarized in table 5.3. For SO_2 they range from 2.4×10^{-7} for a rural cement plant in Liuzhou, to 4.8×10^{-5} for a steel stack in Beijing. That is, of every metric ton of SO_2 emitted from that steel plant, 48 g are inhaled by the population within 50 km around the plant. The pollution loading of each source varies significantly, from a small cement plant that emits 0.1 tons of SO_2 per year to a large steel plant stack

(a)

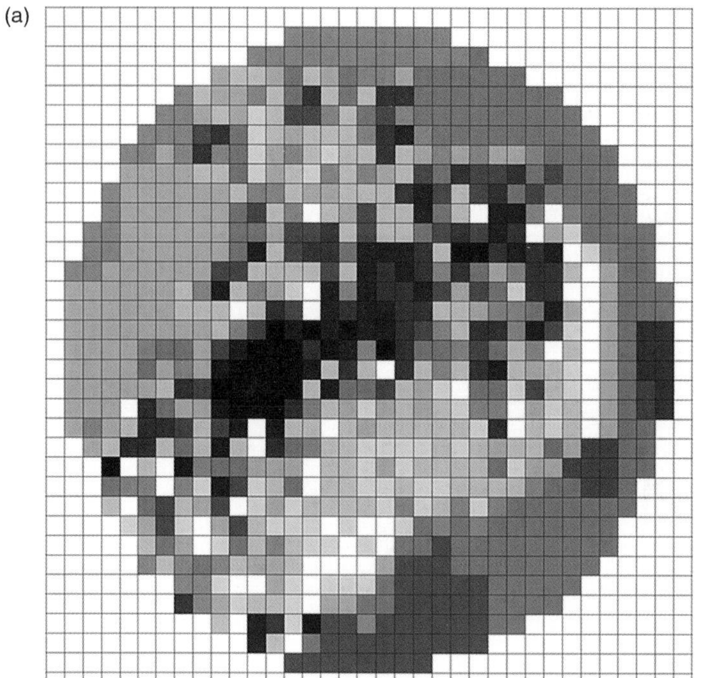
POPULATION
20,000 ~ 184,000
10,000 ~ 20,000
7,000 ~ 10,000
6,000 ~ 7,000
5,000 ~ 6,000
4,000 ~ 5,000
3,000 ~ 4,000
1,000 ~ 3,000
1 ~ 1,000
0 ~ 1

(b)

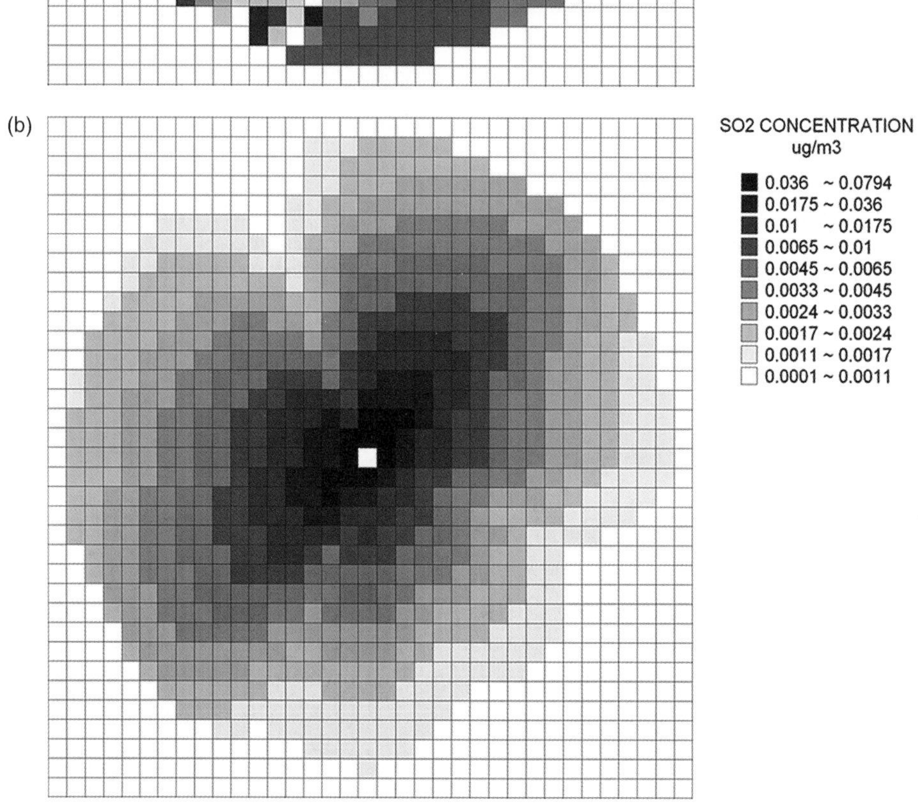
SO2 CONCENTRATION
ug/m3
0.036 ~ 0.0794
0.0175 ~ 0.036
0.01 ~ 0.0175
0.0065 ~ 0.01
0.0045 ~ 0.0065
0.0033 ~ 0.0045
0.0024 ~ 0.0033
0.0017 ~ 0.0024
0.0011 ~ 0.0017
0.0001 ~ 0.0011

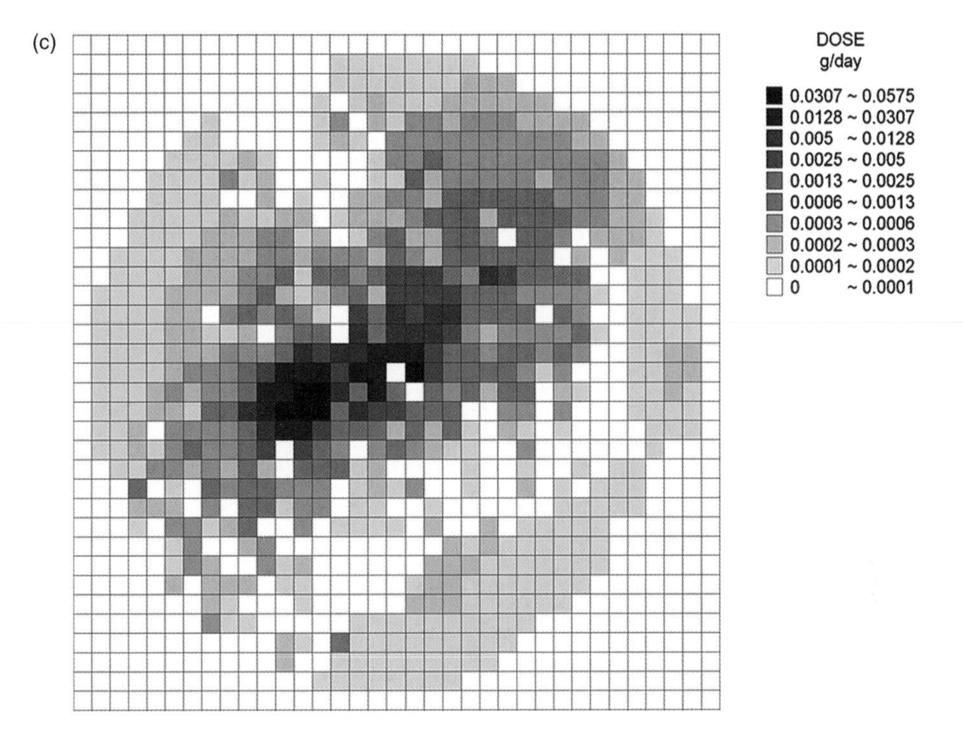

Figure 5.4
Estimating intake fraction for a cement stack in Jinan. (a) Population within 50 km from source, in total individuals. (b) Concentration within 50 km from source, in μg/m^3. (c) Multiplying concentration with population and breathing rate at each grid cell. (See figure 3.1 in chapter 3 for a color version.)

emitting 2.8×10^3 tons. We therefore calculate the emission-weighted average *iF* for each sector and for the whole sample. The weighted average iF_{TSP} for the whole sample is 7.2×10^{-6}, and weighted average iF_{SO2} is 6.7×10^{-6}, with s.d.'s of 8.3×10^{-6} and 1.1×10^{-5}, respectively. If the distribution of characteristics in the five-city sample is the same as the national distribution, the 90% confidence interval for the mean *iF* is (5.9×10^{-6}, 7.4×10^{-6}) for SO_2, and (6.5×10^{-6}, 7.8×10^{-6}) for TSP.

There are significant differences among the *iF* for the different sectors. The average *iF* for SO_2 from cement sources is $(4.4 \pm 0.9) \times 10^{-6}$, which is about half of the $(9.1 \pm 1.1) \times 10^{-6}$ value for iron and steel sources. In addition, the *iF*s also vary considerably across each of the three sectors. The range for SO_2 from the 169 stacks from cement plants is 2.4×10^{-7} to 3.9×10^{-5}, whereas the range for steel plants

Table 5.3
Summary of intake fractions for industrial samples

	iF for SO_2				*iF* for TSP			
Sector	Mean[a]	Maximum	Minimum	Standard Deviation	Mean[a]	Maximum	Minimum	Standard Deviation
Cement	4.41×10^{-6}	3.86×10^{-5}	2.38×10^{-7}	6.73×10^{-6}	5.85×10^{-6}	3.07×10^{-5}	7.10×10^{-7}	6.07×10^{-6}
Chemical	7.60×10^{-6}	4.52×10^{-5}	6.20×10^{-7}	1.32×10^{-5}	8.87×10^{-6}	5.05×10^{-5}	1.67×10^{-6}	1.04×10^{-5}
Steel	9.14×10^{-6}	4.75×10^{-5}	2.21×10^{-6}	8.86×10^{-6}	7.19×10^{-6}	5.32×10^{-5}	2.46×10^{-6}	7.56×10^{-6}
Average	6.65×10^{-6}	4.75×10^{-5}	2.38×10^{-7}	1.06×10^{-5}	7.15×10^{-6}	5.32×10^{-5}	7.10×10^{-7}	8.27×10^{-6}

[a] Emission weighted average values.

Table 5.4
SO_2 Intake fractions versus spatial variation of population

Source Location	Population within 1 km	5 km	10 km	50 km	Intake Fraction[a]
City center	46,170	678,140	1,542,255	5,403,672	1.59×10^{-5}
City fringe	10,529	136,227	709,245	5,391,970	0.98×10^{-5}
Inner suburb	2,797	50,975	149,040	5,411,307	0.47×10^{-5}
Rural area	936	23,961	147,913	4,604,674	0.33×10^{-5}

[a] Estimates for a cement plant with 50-m high stack.

is a much narrower 2.7×10^{-6} to 4.8×10^{-5}. The iF_{TSP} estimates for sampled chemical and steel plants both vary by one order of magnitude.

As we can see from the definition in equation (5.1), many factors contribute to the intake fraction value. We examine them carefully in section 5.3.3 below, but let us first examine the partial contributions with simple figures. It is clear that the main source of variability in the *iF* values is the population distribution around the emission source. For given source characteristics and meteorological conditions, the *iF*s from industrial emissions are found to vary by a factor of five between the densely populated urban centers and less-inhabited rural areas. An example of the relation between *iF* and population densities, ignoring the other factors, is shown in table 5.4.

Many cement kilns are scattered in rural areas with low population densities, and, despite having the lowest average stacks, the sector has the lowest average *iF*s, for both TSP and SO_2. In contrast, most of the big chemical sources examined are located in urban areas and are surrounded by many more people, as described earlier in reference to figure 5.2. Their emission-weighted average *iF*s are almost twice that of cement plants. The samples from the five steelworks have the highest average *iF* for SO_2, which is mostly due to the contribution of the large Shougang Group in Beijing.[3] The situation for TSP, however, is different because advanced dust collectors are installed at Shougang Group. The TSP emissions there are only 17% of total TSP emissions of the five iron and steel complexes, and hence the sample-average *iF* of the sector is much lower than that for not only SO_2, but also chemicals.

The next set of important factors determining *iF*s are the emission characteristics of stacks, which influence how pollutants are emitted into the air and thus how they

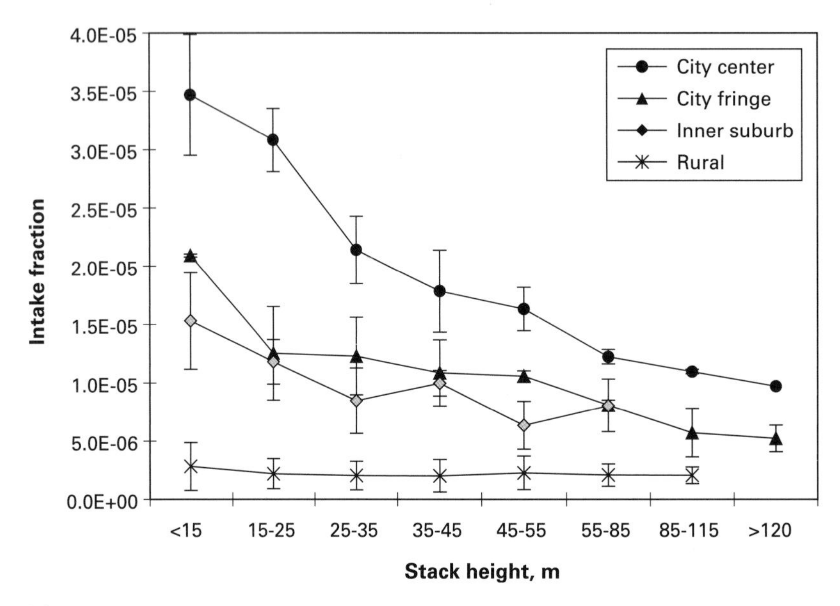

Figure 5.5
Sulfur dioxide intake fractions versus stack heights.

behave in the atmosphere. In light of how the ISCLT3 air-dispersion model works, stack height is one of the most important factors because it is the main determinant of the maximum ground-level concentration as well as the distance to reach that maximum. The stack heights in our sample vary from 10 to 165 meters and represent almost the entire range for industrial sources in China (not including power plants, described in chapters 6 and 7). To get a quick impression of the effect of stack height, we first divide the sample into four types of population densities—city centers, city fringes, inner suburbs, and rural areas—and then for each type calculate the average iF_{SO2} of a number of height categories.[4] These average *iF*s are plotted against stack height in figure 5.5.

The inverse relationship between iF_{SO2} and stack height is very clear. We can also see that the *iF*s of urban stacks are much more sensitive to emission height compared to those of rural sources. This is because as stack height rises, the ground-level concentration within the 10-km vicinity of the emission source falls rapidly. This relationship has a significant influence on urban sources in densely populated areas. As shown in figure 5.5, a reduction of stack height by a factor of 10 (from 100 to 10 meters) increases the *iF* by a factor of 3.2 for sources at city centers, whereas the factor is only 1.4 for rural sources with much lower population density. Increasing

Table 5.5
Summary of intake fractions for mobile-source samples

Primary PM_{10}				Secondary $PM_{2.5}$ (nitrate/NO_X)[a]		
Mean	Maximum	Minimum	Standard deviation	Mean	Maximum	Minimum
7.72×10^{-5}	1.54×10^{-4}	1.84×10^{-5}	3.75×10^{-5}	3.10×10^{-6}	6.25×10^{-6}	4.42×10^{-5}

[a] Estimated from literature values (see description in 5.5.2 and appendix D).

the stack height, however, diminishes the effect of population densities on *iF*s. Moving a stack from a rural area to a city center raises the *iF* by a factor of 12 at a stack height of 10 m but raises it by a factor of only 5.3 at 100 meters.

The relationship between iF_{TSP} and stack height is similar to that plotted in figure 5.5 for iF_{SO2}.

5.3.2 Intake Fractions for Urban Mobile Sources

For emissions from the urban road samples, we calculate the *iF* for primary PM_{10}, ignoring SO_2, which is less important in this sector. We repeat the same exercise as for the industrial sources, employing the same population grid system for the given city and the same meteorological inputs for the ISCLT3 model. The results from the seventy-five road segments sampled are given in table 5.5. The emission-weighted average *iF* for primary PM_{10} is estimated to be $(7.7 \pm 0.6) \times 10^{-5}$. This is one order of magnitude higher than our estimates of industrial sources. The mobile source *iF*s range from 1.8×10^{-5} to 1.5×10^{-4}, smaller than the range for the industrial point sources.

We next examine the relationship between total exposures and distance from the source. Figure 5.6 plots the cumulative *iF* against distance (i.e., how the *iF* rises as more areas are included in the calculation). For the mobile sources, the *iF* reaches 50% of the final value at a distance of 3 km (the final value being that calculated for the entire 50-km domain). This is quite different from the industrial point sources, where the 50% mark is reached beyond 10 km. This implies that the damage from mobile sources is highly concentrated.

We examined the partial relationship between *iF* and total population for industrial sources earlier in 5.3.1. Turning to the mobile sources, their *iF*s are more sensitive to the near-source population than industrial sources. This suggests that it is very important to obtain a high resolution of population density when studying pollution from mobile sources.

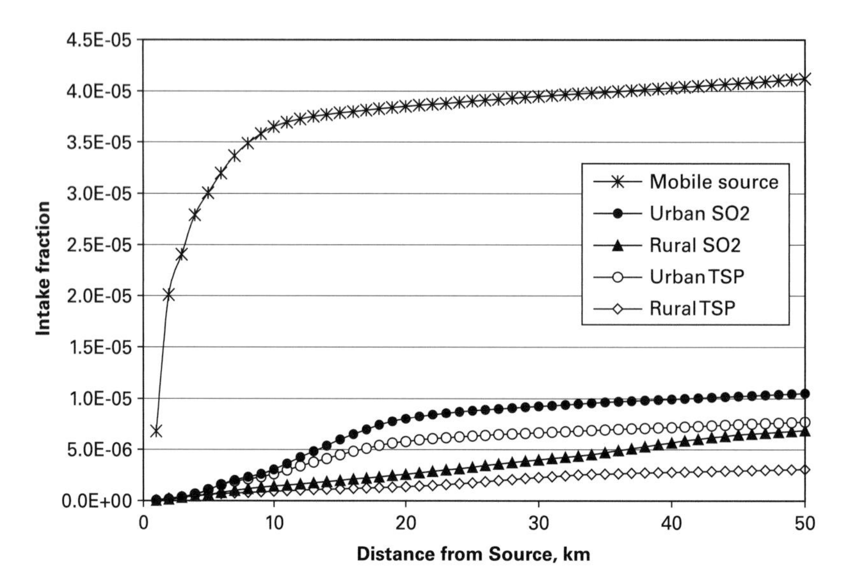

Figure 5.6
Intake fractions versus distance.

5.3.3 Regression Analyses

Levy, Wolff, and Evans (2002) examined whether a limited number of variables representing exposure factors can explain a large portion of the variation in *iF*s derived from air-dispersion models. In this section, we follow that approach by regressing the *iF*s that we estimated for the 541 industrial sources and 75 mobile sources on population distribution and source characteristics. The aim of this is to derive relationships that will allow quick estimates of *iF*s for other sources, based only on summary data.

One main goal of this exercise is to estimate the national *iF*, that is, the average *iF* of all sources in China in the selected sectors. The regressors should ideally include all variables that make up the factors on the right-hand side of equation (5.1): meteorological parameters, source characteristics, and population distribution. As described in section 5.4 below, the summary data for national sources is rather sparse, and this dictates the types and number of regressors we can use. On the basis of the results described in section 5.3.1 above, we examine various combinations of independent variables and finally determine that the critical ones are the at-risk population and the stack height.[5] For the industrial point sources we use the population within 10 km and the population between 10 and 50 km, while for the mobile sources (with their low emission heights) we use population within 5 km and between 5 and 50 km.

Another limitation on the regressors is that we have only five cities and hence only five sets of meteorological values. These factors have to be excluded from the regressions. As we will show, however, they are not as important as one might expect.

The choice of the functional forms of the regressions also involves some tradeoffs. For the population term, the form of equation (5.1) would suggest a linear relationship between the levels or between the logarithms. That the ISCLT3 model is a Gaussian plume model is reflected in the shape of the intake fraction versus stack height graphs in figure 5.5. This suggests something like a logarithmic relationship with the inverse of the stack height. Looking ahead to the purpose of these regressions, however, we have to take into account that not all sources for which we wish to make quick *iF* estimates have all the right-hand side (RHS) values. That is, if we had all the right-hand side variables for the new sources to which we want to apply this reduced-form relationship, we should aim to have the best possible fit; however, when we estimate the national average *iF* of the chemical sector, for example, and we do not have the individual stack heights but only the average height, then we need to worry about the averaging effect. An expression in logarithmic form would not give the correct estimate of the average (i.e., the mean of the logarithm instead of the logarithm of the mean).

The calculations of pollutant concentration and *iF*s are done equally for all three industrial sectors. There should not be any differences between emissions from cement, chemical, or steel plant stacks, in light of the simplicity of the ISCLT3 model and the fact that we must ignore chemical interactions. This argues for pooling the estimates from all three sectors. There is, however, a systematic difference in the location and range of stack heights, and thus we should allow for separate effects.

To suit these various purposes, we estimate different functional forms. These models are reported in tables 5.6 and 5.7, for both TSP and SO_2. In table 5.6, a log-linear model is built up by regressing ln(*iF*) on ln(population) and the inverse of the stack height. In table 5.7, *iF* is related to population in levels and the inverse of the stack height. We report regressions done sector by sector and those where they are pooled and common slopes are imposed.

Turning first to table 5.6a, we see that the fit between ln(*iF*) and just the three aforementioned independent terms is quite good. All the R^2 values are more than 0.7, and the coefficients on $\ln(POP_1)$ (population within 10 km, or 5 km for mobile sources) and 1/SH (inverse stack height) are highly significant and have the expected

Table 5.6
Log-linear regression analyses, form: $\ln(iF) = \lambda + \alpha \ln(POP_1) + \beta \ln(POP_2) + \gamma(1/SH)$

a. Individual Sector Regressions

Pollutant and Sector	Number of Observations	Constant	$\ln(POP_1)$	$\ln(POP_2)$	1/SH	R^2
SO_2						
Cement	169	−20.81 (0.50)	0.55 (0.04)	0.116 (0.04)	5.30 (1.98)	0.75
Chemical	185	−18.41 (0.48)	0.56 (0.04)	−0.063 (0.04)	18.97 (1.67)	0.73
Iron and Steel	187	−18.44 (0.33)	0.42 (0.03)	0.068 (0.04)	17.34 (1.66)	0.83
TSP						
Cement	164	−24.80 (0.53)	0.59 (0.023)	0.34 (0.033)	6.89 (0.853)	0.87
Chemical	126	−28.42 (0.78)	0.80 (0.03)	0.37 (0.04)	13.66 (2.24)	0.87
Iron and Steel	122	−27.186 (0.92)	0.942 (0.06)	0.176 (0.02)	9.929 (1.03)	0.78
PM_{10}						
Transport	75	−10.54 (0.42)	0.89 (0.07)	0.16 (0.06)	n.a.	0.72

b. Pooled Regression over All Industrial Samples

Pollutant	Number of Observations	Constant	$\ln(POP_1)$	$\ln(POP_2)$	1/SH	Dummy, Chemicals	Dummy, Iron	R^2
SO_2	541	−18.90 (0.26)	0.50 (0.02)	0.0008 (0.02)	15.77 (1.05)	0.47 (0.06)	0.41 (0.06)	0.77
TSP	412	−24.57 (0.34)	0.69 (0.02)	0.24 (0.02)	8.91 (0.69)	−0.31 (0.04)	−0.08 (0.05)	0.86

Notes: Standard errors in parentheses. POP_1, population within 10 km (5 km for mobile sources) from source (million); POP_2, population between 10 and 50 km (between 5 and 50 km for mobile sources; million); SH, stack height (meters).

signs. The coefficients for ln(POP_2) (population within 10–50 km, or 5–50 km for mobile sources) are smaller and in some cases insignificant. There are no unusual outliers; over the whole sample the largest regression error is less than 50% of actual value.

For TSP, the ln(POP_1) coefficient is 0.9 for steel mills but much smaller for cement and chemical plants. This may reflect the heterogeneity of the population distribution within that zone for cement and chemical plants; recall that the steel mills are always in the city fringes and not out in the rural areas, as can be the case with chemical and cement plants. Doubling the population everywhere would double the *iF*; since we divide it into POP_1 and POP_2, however, the relation is more complex. In light of the rapid decline in concentration with distance, the coefficients on ln(POP_2) are much smaller than on ln(POP_1).

The coefficients for inverse stack height also differ significantly by sector: 6.9, 14, and 9.9 for cement, chemicals, and iron and steel, respectively. This may be due to either misspecification (that is, the postulated inverse relationship is not applicable to the whole range of heights) or a systematic difference that is not captured in the locations and population distributions across the sectors.

Turning to the pooled regression in table 5.6b, where the sectors are allowed to have different intercepts but have common slopes, the TSP coefficient for 1/SH is 8.9. This means that raising the stack height from 34 m (mean height for cement stacks) to 44 m, holding population at the means, would reduce the *iF* from 7.2×10^{-6} to 6.2×10^{-6}. The coefficient on ln(POP_1) is 0.7 and on ln(POP_2) is 0.2. These are in between the values for the separately fitted regressions and represent the effects of exposed population size that ignore the subtle unobserved differences in location and population distribution captured by the separate regressions. The dummy coefficients indicate that the metallurgical industry has the same intercept as cement industry but that for chemical process industry is significantly lower.

The results for SO_2 are somewhat different, but the coefficients are also highly significant. The R^2 values are somewhat smaller than those for TSP for cement, chemicals, and the pooled regression. In the pooled regression, the R^2 for the SO_2 relationship is 0.8 versus 0.9 for TSP, the coefficient for the less than 10-km population is 0.5 versus 0.7, and the coefficient for 1/SH is 16 instead of 8.9. These differences are due to the less rapid fall in concentration with distance compared to TSP. This means that higher stacks would allow more SO_2 to be blown outside the 50-km domain and that the dispersed populations in the outer suburbs are exposed to higher concentrations. This gives rise to the bigger variance in estimates and the higher coefficient on the stack height term.

Table 5.7
Level regression analyses, form: $iF = \alpha(POP_1) + \beta(POP_2) + \gamma(1/SH) + \lambda$

a. Individual Sector Regressions

	Number of Observations	Constant	POP_1	POP_2	1/SH	R^2
SO_2						
Cement	169	-1.5×10^{-6}	1.52×10^{-5}	4.26×10^{-7}	5.82×10^{-5}	0.72
		(6.65×10^{-7})	(8.26×10^{-7})	(9.34×10^{-8})	(1.37×10^{-5})	
Chemical	185	2.38×10^{-7}	8.36×10^{-6}	-8.20×10^{-8}	1.90×10^{-5}	0.51
		(1.32×10^{-6})	(7.20×10^{-7})	(1.60×10^{-7})	(2.65×10^{-5})	
Iron and Steel	187	1.13×10^{-7}	1.20×10^{-6}	4.12×10^{-7}	2.49×10^{-5}	0.71
		(5.29×10^{-7})	(6.61×10^{-7})	(7.68×10^{-8})	(1.12×10^{-5})	
TSP						
Cement	185	-2.25×10^{-6}	1.31×10^{-5}	7.87×10^{-7}	2.98×10^{-5}	0.80
		(5.0×10^{-7})	(6.35×10^{-7})	(7.12×10^{-8})	(8.17×10^{-6})	
Chemical	164	-8.7×10^{-6}	1.01×10^{-5}	7.93×10^{-7}	1.95×10^{-4}	0.76
		(1.8×10^{-6})	(5.62×10^{-8})	(1.48×10^{-8})	(3.68×10^{-5})	
Iron and Steel	122	-7.6×10^{-6}	1.10×10^{-5}	8.06×10^{-7}	2.19×10^{-4}	0.75
		(1.28×10^{-6})	(1.07×10^{-6})	(7.62×10^{-8})	(1.69×10^{-5})	
PM_{10}						
Transport	75	-7.6×10^{-6}	5.96×10^{-11}	3.06×10^{-6}	n.a.	0.79
		(4.28×10^{-6})	(5.63×10^{-6})	(6.92×10^{-7})		

b. Pooled Regression over All Industrial Sample

	Number of Observations	Constant	POP_1	POP_2	1/SH	Dummy, Chemicals	Dummy, Iron	R^2
SO_2	541	2.33×10^{-6}	8.59×10^{-6}	2.51×10^{-7}	13.15×10^{-5}	3.50×10^{-6}	1.13×10^{-6}	0.59
		(6.65×10^{-7})	(4.17×10^{-7})	(6.62×10^{-8})	(1.16×10^{-5})	(6.23×10^{-7})	(6.03×10^{-7})	
TSP	412	3.59×10^{-6}	1.07×10^{-5}	6.60×10^{7}	9.59×10^{-5}	2.11×10^{-6}	3.72×10^{-7}	0.74
		(6.15×10^{-7})	(3.95×10^{-7})	(5.86×10^{-7})	(9.76×10^{-6})	(6.23×10^{-7})	(6.04×10^{-7})	

Notes: Standard errors in parentheses. POP_1, population within 10 km (5 km for mobile sources) from source (million); POP_2, population between 10 and 50 km (between 5 and 50 km for mobile sources; million); SH, stack height (meters).

In light of the averaging issue discussed above, we next regress the level of *iF* on the levels of POP_1 and POP_2 and inverse stack height. This is reported in table 5.7. The results for TSP are similar to the log-log form discussed above, the R^2 is somewhat lower, and the size of the coefficient for POP_2 is much smaller than that for POP_1. There is some heteroscedasticity in the level form. For SO_2, the level regression gave a lower R^2 than the log-linear model, 0.59 compared with 0.77, which indicates a greater heteroscedasticity problem.

5.4 Sensitivity Analysis

As we described in section 5.2.2, in running the ISCLT3 air-dispersion model, various parameters and options are chosen and held fixed for all simulations, which would cause certain errors in intake fraction results. It should be noted, however, that this study is not aiming at the greatest possible precision in the health exposure assessment, but rather to make rough estimates of the main sources of pollution. In other words, we try to get a good estimate with currently available data and research resources, which is not only a necessary step toward greater precision in future research, but also informative to immediate policy applications. Therefore, the errors up to 30% are regarded as moderate in the following sensitivity analysis on parameter uncertainty.

5.4.1 Scale of Domain

Domain size is obviously important because a larger domain means more population will be included in the exposure calculation. We have already illustrated in figure 5.6 the effect of including more area on the *iF* for two cement stacks, one of which is located in an urban area and the other in a rural zone. For sources in urban areas, the *iF* for SO_2 reaches 50% of its final, 50-km domain value at approximately 14 kilometers from the source, and 95% at 39 kilometers. For rural sources, however, these percentages are reached at 28 and 46 km, respectively. Comparing the two pollutants, TSP is modeled to undergo dry deposition due to gravitational settling and affects fewer people than SO_2, which stays in the atmosphere longer. TSP therefore has a lower intake fraction value than SO_2 at any distance.

We choose a radius of 50 km as our base-case domain in light of the recommendations for ISC models, including ISCLT3. We should note that ISC models have been applied at higher ranges (e.g., up to 100 km from power plants in Thailand by Thanh and Lefevre [2000]). To get an idea of the amount of exposures that we miss by limiting the domain to 50 km, we simulate some cases using a bigger domain, keeping in mind that these longer-range estimates are subject to greater uncertainty and possible bias. We repeat *iF* calculations for all 150 stacks selected in Jinan using a domain of 100-km radius.

For urban sources of SO_2, the $iF_{100\ km}$ is higher than $iF_{50\ km}$ by 2.4% to 24%. The average increase is only 12%. This suggests that our base-case values are reasonable, conservative estimates for evaluating local exposures given the other uncertainties affecting *iF*. For rural sources, however, the bigger domain raises the intake fractions for SO_2 by much more, 36% on average, ranging from 17% to 87%. For the whole sample in Jinan City, the average iF_{SO2} is higher by 24%, and the average iF_{TSP} is higher by 29%.

A more complete analysis would include any possible human exposure (i.e., a domain up to where pollutant concentration falls to zero). Gaussian dispersion models such as ISC are not suitable for ranges in the hundreds of kilometers. We consult the literature to see how iF_{∞} might compare to $iF_{50\ km}$, with a comparison of four studies shown in table 5.8. Thanh and Lefevre (2000) used two different dispersion models to study exposure to SO_2 from power plants in Thailand and estimate that local exposure (less than 50 km) is only 38% of total exposure. The Levy et al. 2002 study of U.S. power plant emissions estimated that approximately 40% of total primary $PM_{2.5}$ exposure occurs within 50 km, and another 30% between 50 and 200 km. Zhou et al. 2002 (in extensions to research of chapter 7) calculates the intake fraction for a power plant in Beijing by using the CALPUFF model and estimates that only 21% of total SO_2 exposure takes place within 50 km and 25% between 50 and 200 km. We run the ISCLT3 model for the same Beijing plant, and our *iF* with the 50-km domain is 30% of the value calculated by Zhou et al. using a

Table 5.8
Influence of modeling scale on intake fraction estimates

Study	Present Study (chapter 5)		Thanh and Lefevre (2000)	Levy, Spengler et al. (2002)	Zhou (2002) and Present Study (chapter 7)
Area	China		Thailand	United States	China
Dispersion models	ISCLT3		SLIM3/ISCST2	CALPUFF	CALPUFF
Emission source	Industry		Power plants	Power plants	Power plants
Item	$iF_{50\ km}/iF_{100\ km}$		$iF_{50\ km}/iF_{\infty}$	$iF_{50\ km}/iF_{\infty}$	$iF_{50\ km}/iF_{\infty}$
Pollutant	TSP	SO_2	SO_2	$PM_{2.5}$	SO_2
Percentage	74	89	38	40	21

1680-km domain. On the basis of these results, we believe that it is reasonable to estimate that iF_{∞} exceeds $iF_{50\,\mathrm{km}}$ by a factor of two or three.

5.4.2 Sensitivity to Population Resolution

The resolution of the grid size depends on the availability of population data. For Jinan City, we have block-level data and are able to use a fine 1 km × 1 km grid. For some rural areas of Jinan, however, we only have county-level data and hence a much cruder grid. We next examine the effects of using county-level population resolution, that is, the impact of assuming that the population is evenly distributed over a large area when in fact it could be concentrated.

To do this, we redraw the population map for Jinan ignoring the detailed block-level information and use only county data for all areas, including the city center. We then repeat the *iF* calculations for the entire sample of Jinan and show the results in figure 5.7. On the horizontal axis is the original iF_{SO2}, and on the vertical axis is the *iF* based on cruder grid cells. Sources in rural areas are marked with

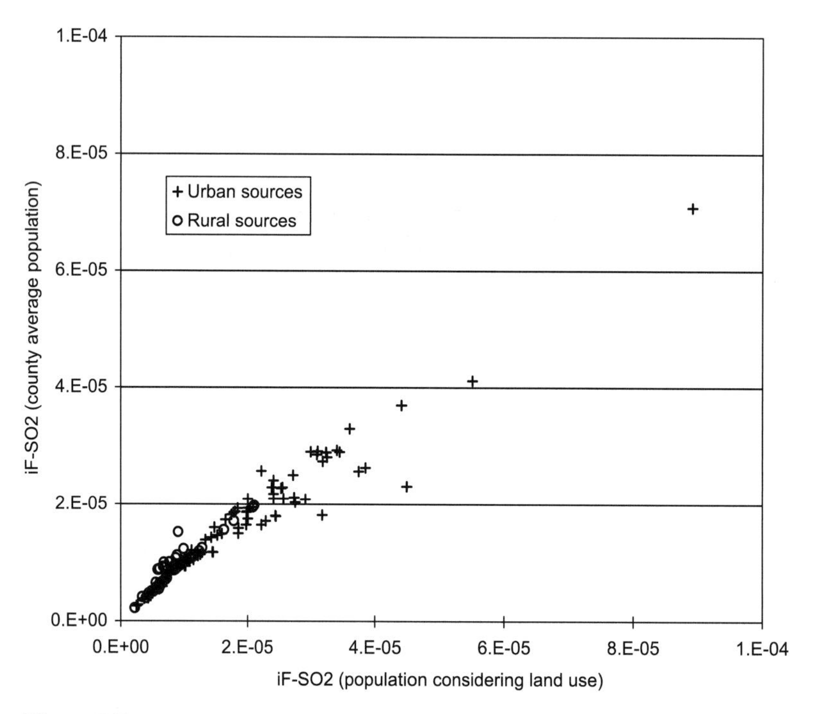

Figure 5.7
Influence of population resolution on intake fraction.

circles, and we see that the two estimates are very close. Urban sources, marked with crosses, on the other hand, have crude *iF*s that are quite a bit lower than those based on more accurate population distributions. The maximum difference for one source is as high as 70%. This indicates that our estimates for the sources in Chongqing might be understated because of the cruder population resolution. For the other three cities—Beijing, Dalian, and Liuzhou—the *iF* estimates would be better than those in Chongqing, though not as good as in Jinan.

5.4.3 Sensitivity to Topographic Factors

In our simulations we set the terrain option to "flat" because it was difficult to obtain terrain data; however, the terrain in some areas of the modeled cities is rugged, with highlands, gorges, and so forth. For example, there is a mountainous region in the southeast of Jinan that includes two hills with elevations above 0.8 km (see appendix C for detailed descriptions). The elevated terrain would affect the dispersion of pollutants.

We repeat our calculations for a cement stack located in the urban area of Jinan, with more complex terrain included in the ISCLT3 run. This change affects the dispersion behavior in many grid cells, as expected. The concentrations are as high as 90% higher in some cells—and as much as 60% lower in others—compared to those assuming flat terrain. At least in this particular case, however, the deviations in different areas roughly offset each other, and the total difference between intake fractions is minimal, about 5%. This is given in figure 5.8a.

We should note that ISCLT3 implements simple algorithms for complex terrain, which differentiates elevations only below stack height. The terrain above stack height is truncated when calculating impacts at receptors above the release height. Accounting for this truncation might have produced differences larger than those estimated. The net difference would still depend on the population location, and thus the error is city-specific.

The ISC short-term model, ISCST3, employs the COMPLEX1 model for applications in complex terrain, which is better than the treatment in the ISC long-term model that we have used so far. We run simulations using ISCST3 for TSP by use of both elevated and flat options. The results are plotted in figure 5.8b. In this case, the difference between intake fractions using elevated versus flat terrain is about 15%. We should also note, however, that even this model is not very accurate in complex terrain, because it assumes that the plume trajectory is unaffected by the terrain. In reality, the trajectory typically deviates when a plume flows over obstacles.

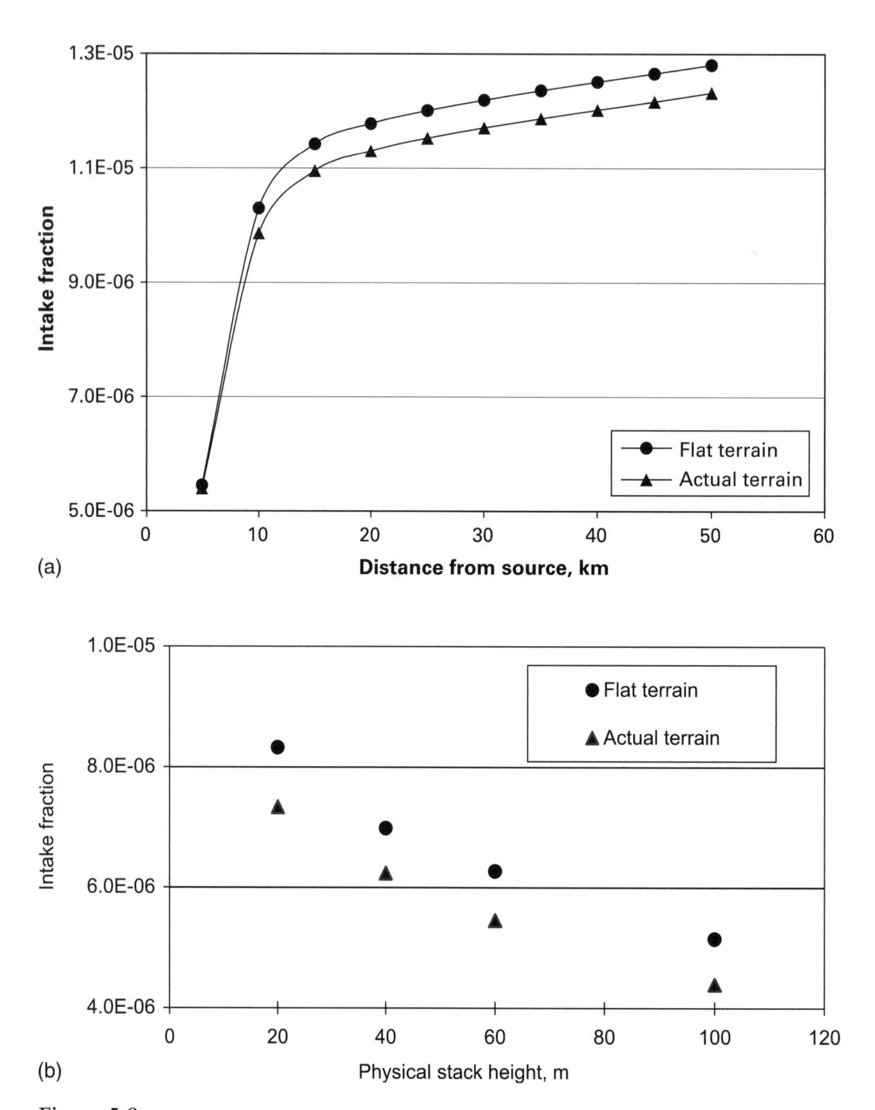

Figure 5.8
Impact of topography on intake fraction of one Jinan cement stack. (a) Impact of elevated terrain on intake fraction by ISCLT3. (b) Impact of elevated terrain on intake fraction by ISCST3.

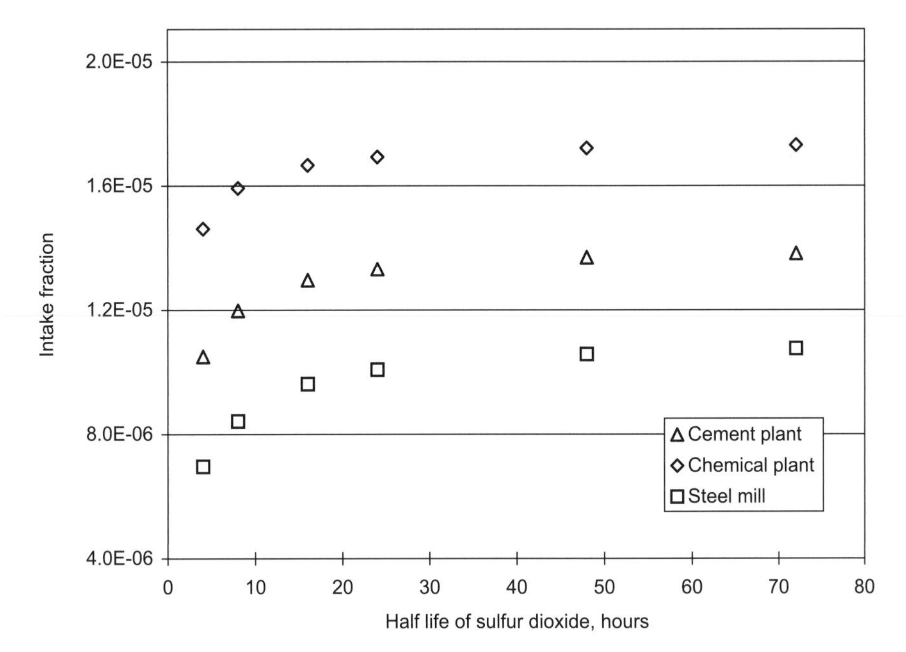

Figure 5.9
Sulfur dioxide intake fractions versus half-life.

We conclude that the topography introduces an error that, though not necessarily large, is difficult to characterize accurately. Analysis in regions with complex topography should ideally take explicit account of it.

5.4.4 Sensitivity to Half-Life of Sulfur Dioxide

The half-life, or decay factor, reflects the mean atmospheric residence time of pollutants and has significant impacts on the total amount of pollutant inhaled. For gases, the mean residence time depends on their reactivity with other air pollutants, water solubility, and other physicochemical properties. We use four hours as the half-life of SO_2, the default value in ISCLT3. The mean residence time of SO_2, however, is believed to be as long as four days (Schwartz 1989), corresponding to a half-life of nearly 70 hours. We run the ISCLT3 model again, changing the SO_2 half-life from in steps from 4 hours to 72 hours. The changes in the *iF*s for the various industrial sectors are given in figure 5.9. For the cases we tested, the iF_{SO2} increases by 8.9% to 21% when the half-life is doubled from 4 to 8 hours and about 16% to 45% when the half-life reaches 24 hours, at which point the impact of increasing half-life has largely leveled. Given this result, one might characterize the influence of half-life on intake fractions as moderate (i.e., less than an order of magnitude).

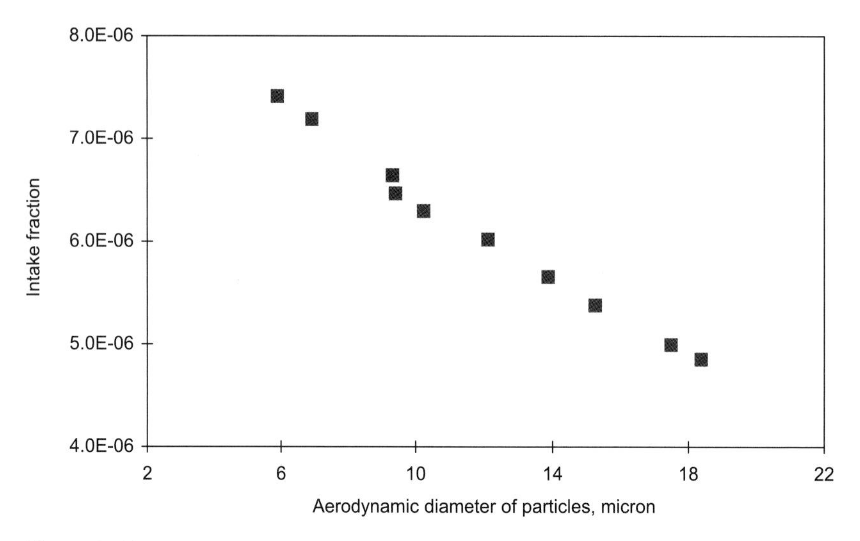

Figure 5.10
Intake fraction versus particle size.

5.4.5 Sensitivity to Particle Size Distribution

Information on particle-size distribution is unavailable for most emission stacks in China. Because the size distribution of particles emitted from stacks differs with fuel quality, combustion processes, and dust-removal equipment, the particle size-distribution, estimated from measurements on boilers in Beijing (Tsinghua University et al. 2002) and the AP-42 report (U.S. EPA 1996; Florig and Song 2000), might vary substantially from the actual distribution. To estimate the impact of this, we ran the ISCLT3 model again for one stack with different values for particle sizes.

Figure 5.10 plots the relationship; the *x* axis is the aerodynamic diameter of particles. This is an important index of particle-size distribution. When the aerodynamic diameter changes from 6 microns (μm) to 18 μm, the *iF* is reduced by as much as 35%. We would classify the impact of particle-size distribution on *iF* as, again, moderate, in light of the estimated uncertainty in other elements of the analysis.

5.4.6 Sensitivity to Dispersion Model Used

Different air-dispersion models have different strengths. The ISC short-term model (ISCST) is considered more accurate than ISCLT (long term) because of the more detailed arithmetic and meteorological inputs. It is more time-consuming to implement and run, however, and thus harder to apply to as many sources as this study

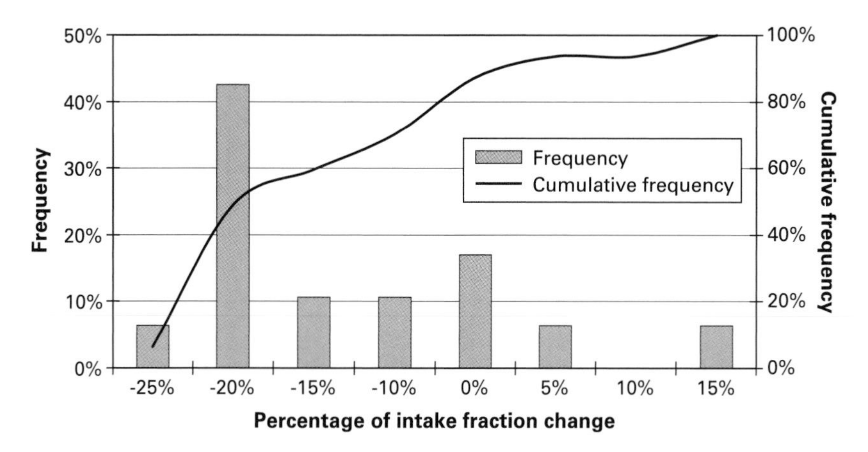

Figure 5.11
Distribution of intake fraction changes. Note: Indexed by $(iF_{ISCST} - iF_{ISCLT})/iF_{ISCLT}$.

requires. To estimate how much is lost because of the averaging done by ISCLT, we use the ISCST3 model to calculate the *iF*s for forty-seven randomly selected stacks in Beijing. In these new runs, the receptor system and population distribution are the same as those used in the base case (ISCLT3) estimates. The main difference is that hourly meteorological data are used instead of annual averages. Beijing is chosen from the five cities because it has the best meteorological data.

The differences in the *iF*s range from −26% to +13%. In figure 5.11, we plot the frequency of changes of various magnitudes for this forty-seven-stack sample. Almost half of the *iF*s derived from ISCST3 are smaller by 20% or more compared to those derived from the long-term model. On average, the *iF*s are 17% lower. In light of the nonlinear way in which wind and temperatures affect concentrations, it is difficult to generalize the results; however, it seems reasonable to conclude that, although it is possible that our chosen model systematically overstates the exposures, the error is not large, at least not large in relation to the other types of errors that we have also examined. Our results are consistent with those from Mestl et al. (2005), which showed that the difference is about 10% under different meteorological conditions.

5.4.7 Sensitivity to Wet Deposition

The ISCLT3 model only allows for dry deposition. To examine the role of wet deposition we turn again to the ISC short-term model. We run ISCST3 twice, with and without the wet depletion option for TSP, for nine sources in Beijing. The differences are as small as 2.44%, perhaps reflecting the low annual precipitation in Beijing of 650 mm. In this case, it is reasonable not to consider wet deposition. This conclusion is consistent with a U.S. EPA study (1998), which indicated that the differences in the maximum concentrations, with and without wet depletion, are only 0.03% and 0.37% for 205 mm and 1,265 mm of annual precipitation, respectively.

We do note that annual precipitation in China varies greatly. It is as high as 2,500 mm annually along the southeastern coast and less than 50 mm in the northwest. A comprehensive estimate of the national average should take this into account.

5.4.8 Summary of Sensitivity Analyses

All the results of sensitivity analyses are summarized in table 5.9. The seven columns give the seven sources of uncertainty that we consider, the last row in each column giving the estimate of the possible effect on the *iF*s. It is clear that the size of the domain analyzed is the most significant influence on intake fraction values.

We believe that our estimates are appropriate to evaluate the local impacts on human health caused by primary SO_2 and TSP emissions, but one should be very cautious when expanding its application to a regional scale. Put another way, our base case provides conservative estimates of total human exposures and, ultimately, health damages due to emissions from these sectors.

5.5 Sector-Average Intake Fractions

In this section we estimate the national average *iF* for emissions from the four selected sectors. These averages are required for the national damage assessment done later, in chapters 9 and 10. As described there (equations [9.20] and [9.21]), we will estimate the total intake of pollutant x by multiplying the national average intake fraction, iF_{xj}^{N}, by the total emissions of x from sector j. We will estimate these national averages by making use of the reduced-form results from the regressions above. To do so we need the values of the variables on the right hand side: population exposed within 50 km, and the stack height. The procedures to obtain national averages of these are described in the next two subsections.

5.5.1 Sector Average Stack Height

The average industry stack height is calculated using different methods for each industry, depending on data availability. We briefly describe these in turn; the details are given in appendix E.

Table 5.9
Summary of uncertainties for intake fraction estimates

Item	Modeling Area (radius)	Dispersion Model	Population Resolution	Elevated Terrain	Half-Life	Aerodynamic Diameter	Wet Depletion
Pollutants	TSP, SO_2	TSP, SO_2	TSP, SO_2	TSP	SO_2	TSP	TSP
Standard assumption	50 km	ISCLT3	1 km × 1 km	Flat	4 hours	6 μm	No
Extreme condition	Infinity (>1,500 km)	ISCST3	County	Actual terrain	72 hours	18 μm	650-mm precipitation
Maximum percentage of *iF* change	+300	−26 +13	−49 +67	−15	+55	−35	−2.5

Cement industry Liu (2001) lists thirteen kiln types and associated output for each type (appendix E, table E.1). Because there are national regulations governing minimum stack heights for the various types of kilns (appendix E, table E.3), we use the output distribution of kiln types to estimate the average stack height, weighted by output:

$$\overline{SH} = \sum_{i} P_i \times SH_i^{std} \tag{5.2}$$

where SH_i^{std} is the minimum permissible stack height for kilns under the national standards, i is an index for kiln types, and P_i is the percentage of kiln output accounted by type i.

By use of data for 1999, this average is estimated to be 35.9 m. This stack height is appropriate for SO_2 emissions. Cement plants produce great quantities of TSP from both combustion and industrial processes, and this should be applied only for combustion emissions.

Chemical process industry There is no convenient national database for the chemical industry similar to that for cement, in part because of the huge variety of products and the large number of plants (52,000). Our approach is to note that much of the emissions are from coal-fired industrial boilers and that there are regulations governing stack heights of boilers. We do not have data on the stock of boilers in the chemical sector and approximate it with the national stock of industrial boilers. The capacity mix of the industrial boilers in 1999 is estimated from the production of boilers in 1990. The national standard for minimum stack heights of industrial boilers is applied in a way identical to that for cement described above (see appendix E, tables E.7 and E.8). With this, we estimate that the average stack height is 35.6 meters.

We note again that this is the average of all boilers, not just those used in the chemical sector. We do not have any evidence that there is a systematic difference but this is obviously an item to check in future research.

Metallurgical industry This is the most difficult sector to estimate because iron and steel works are complex with many processes such as sintering, smelting, coke making, electricity generation, steam supply, and so on. There is a wide variance of stack heights even within a given plant. There is certainly no convenient national database of plant characteristics. We are thus reduced to using information from our five-city sample. The average stack height of the 187 stacks in our sample, weighting by output of products, is 82.0 m.

Although the five cities might seem to represent a very small sample, we should note the concentrated structure of the industry, the biggest four companies producing 23% of the country's steel. One of the big four is the Capital Iron and Steel Company in Beijing. We should also note that there are a total of 187 stacks in our five-city sample. We believe these give a reasonable starting point for our analysis.

5.5.2 Sector-Average Exposed Population

The next step is to estimate the exposed population corresponding to the POP_1 and POP_2 variables in regression equations. There are a total of 7,222 cement plants, 52,000 chemical plants, and 851 large-sized iron and steel works in China. Even if one had the data it would be impossible to estimate populations for all of these pollution sources. Instead, from national lists of enterprises (described in appendix E), we randomly sample fifty cement plants, fifty chemical plants, and eighty iron and steel works, by use of output levels as probability weights. We then locate these selected sources on a national population map that we construct for our study of the electricity sector, as will be described in section 6.2.3 of chapter 6 and appendix B. The source locations are shown in figure 5.12. For each of the 180 selected sources we estimated the population within 10 km and between 10 km and 50 km.

5.5.3 Sector-Average Intake Fraction Estimates

After considering the various functional forms of the regressions presented in tables 5.6 and 5.7, we decide to employ the log-linear form—that is, the regression equation in table 5.6a—in the following manner. For each of the three sectors, we have the average stack height and the population distribution of a sample of national sources. If we also have the individual stack heights of that sample, it would have been a clear matter of applying the regression coefficients to each of the 180 selected sources, calculating the *iF*s, and then calculating the average industry *iF*. Because we have only the average stack height, we apply the SH coefficient to this average and apply the population coefficients to each of the 180 sets of POP_1 and POP_2. These give us 180 *iF*s for TSP and 180 *iF*s for SO_2 reported in appendix E, tables E.5, E.9, and E.10.

The average *iF*s for cement, chemicals, and iron and steel are derived from these 180 pairs of *iF*s using capacity weights. The results are given in table 5.10. All six

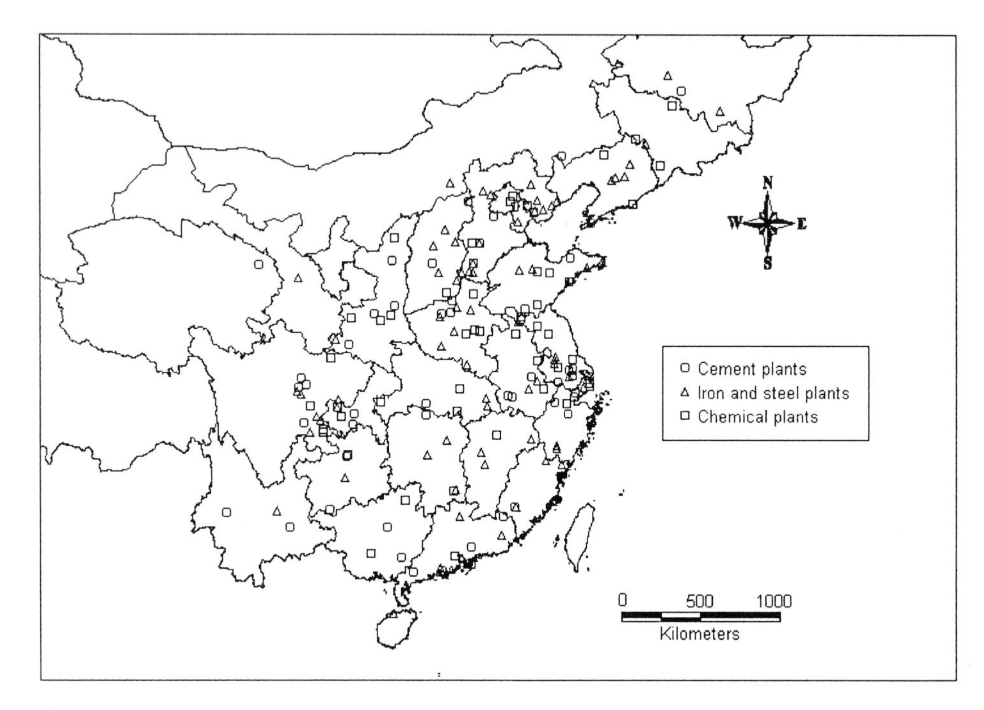

Figure 5.12
Geographic distributions of samples selected to estimate population.

national average *iF*s are on the order of 10^{-6}. Table 5.10 also gives the average of the RHS variables, and for comparison we also calculate the average iF_{xj}^{N} by applying the regression coefficients directly to these national averages of exposed population and stack height. We can see that the two methods are only a bit different, for example, in that iF_{TSP} for cement is 3.46 using the individual source data and 5.52 using the regression with national averages.

The sector with the lowest average *iF*s for both TSP and SO_2 is cement, in light of the comparatively low exposed population around cement plants. This was also the case for the five-city sample. Although iron and steel works have the highest average stack height, the sector still has large average *iF*s because of the high exposed populations resulting from their urban locations. The population within 10 km, averaged over the national sample, is 0.28 million for iron and steel sources, more than twice the 0.14 million figure for cement. As a result, the average *iF* for TSP emitted from steel stacks is 3.8×10^{-6}. This means that the inhaled dose for the population within 50 km is 3.8 grams per metric ton of TSP from steelworks, while it is only 3.4 grams for cement and 3.3 for chemicals.

Table 5.10
Sector-average intake fraction estimates

Sector	Cement		Chemical		Iron and Steel	
Capacity-weighted POP_{10} (million)	0.14		0.22		0.28	
Capacity-weighted POP_{10-50} (million)	2.90		3.94		3.79	
Capacity-weighted Stack Height (m)	35.9		35.6		82.0	
Pollutants	SO_2	TSP	SO_2	TSP	SO_2	TSP
Average *iF* from average coefficients	4.15×10^{-6}	3.52×10^{-6}	6.89×10^{-6}	3.25×10^{-6}	6.36×10^{-6}	3.44×10^{-6}
Average *iF* from individual sources	3.99×10^{-6}	3.46×10^{-6}	6.61×10^{-6}	3.28×10^{-6}	6.03×10^{-6}	3.75×10^{-6}

Note: The average *iF* from average coefficients are calculated by applying the regression coefficients directly to the national averages of exposed population and stack height in this table; the average *iF* from individual sources are derived from 180 pairs of *iF*s using capacity weights.

These national mean *iF*s should be compared to the averages of the five-city sample given in table 5.3. The national average estimates are lower because the five cities represent areas with more dense populations than the typical city, both urban and suburban areas. This is particularly true for chemical plants, where about 60% of SO_2 in the five cities was emitted from stacks located in areas with more than 2,000 persons per km^2. By contrast, in the national sample only 39% of chemical output was from plants near such a high population density. The average population within 10 km is 0.56 million for the five-cities chemical sector sample, compared to 0.22 million for the national sample.

Although these differences between the national and city samples are striking, they do not necessarily mean that our estimates are biased. The regression explicitly takes population into account. Problems may occur if the distribution within the 10-km zone is very different (e.g., if the population within, say, 2 km is a much higher percentage of the POP_1 total in the five-city sample than the one for the national sample). We should point out that we have minimized this issue by using two population variables in our regressions, within 10 km and between 10 km and 50 km, instead of just one total figure. Future research may try to check this by having a bigger sample of cities.

5.5.4 National Mobile Sources

To estimate the national exposures to vehicle emissions precisely, one would need extensive data on vehicle flows and their locations (i.e., how many vehicles of what types use which roads for how many hours). We are unable to find such data for many cities and thus have to rely on the available information for Beijing, Jinan, and Dalian only. That is, the average iF_{PM10} for mobile sources from these three cities given in table 5.5 is the value we use for the national average. This is certainly an insufficient sample and much more work needs to be done, as we are doing in new research. For now, this provides a best guess.

With these limitations, we concentrate on the local-scale effects and hence do not explicitly consider secondary particle formation using more sophisticated dispersion models. In a study on Beijing, He et al. (2002) found that $PM_{2.5}$ constituted about 55% of PM_{10} emissions from vehicles. These fine particles have atmospheric residence times as long as five days (Schwartz 1989), ample time for substantial secondary chemical reactions.

To take this into account we make a crude estimate of the conversion of NO_X into nitrates. We approximate the intake fraction for NO_X by using the results from two studies, Wolff 2000 and Zhou 2002. Wolff estimated iF_{NOx} for both power plants and mobile sources using the CALPUFF model on U.S. data, whereas Zhou estimated it for Chinese power plants in extensions of research presented in chapter 7 (see appendix D for details). Combining these estimates we put the average intake fraction for NO_X (fraction converted from NO_X to nitrates) from vehicles to be 3.0×10^{-6}, which is an order of magnitude lower than that of primary PM_{10} (table 5.5).

According to the number of motor vehicles in China and the estimated emission factors, we calculate the PM_{10} and NO_X emissions from motor vehicles in China were 138.9 kilotons (kt) and 3.2 million tons (Mt), respectively, in the year 2000 (see description in appendix D). By use of our estimated *iF*s, the amount of pollutants from mobile sources inhaled was 11 tons of primary PM_{10} and 10 tons of secondary nitrates. Carmichael, Streets, and Woo 2002 gives a more conservative NO_X emissions estimate, 2.6 Mt, and in this case the dosage would be 8.16 tons of secondary nitrates. That is, regardless of the estimates of emissions, the amount of secondary particles breathed in is of the same order of magnitude as that of primary particles, a high emission offsetting a low *iF*. In light of this, we believe we should not ignore the health impacts of secondary particles even though our estimates are very crude.

5.6 Discussion

5.6.1 Comparisons with Other Studies

The closest study to this is Zhou 2002, also summarized in chapter 7 here. It uses CALPUFF, a model suitable for estimating air dispersion up to hundreds of kilometers. As chapter 7 will summarize, Zhou estimates *iF*s for power plants in different locations in China. Our average *iF*s for industrial sectors are of the same order of magnitude; however, the maximum *iF*s for both SO_2 and TSP emitted by industrial sectors are an order of magnitude higher than those for power plants. In addition, both the average and maximum *iF*s for mobile sources are one order higher than those for industrial sectors (see table 5.11).

The *iF*s for airborne particulate matter from power plants were estimated by Smith (1993) to be roughly on the order of 10^{-5} for developing countries. Phonboon (1996) computed the *iF* of TSP from an oil refinery in Thailand and reported that it varies from 7.0×10^{-6} to 1.5×10^{-5} as the half-life of TSP changed from 1 to 72 hours. The SO_2 *iF*s for France estimated by Spadaro and Rabl (1999) are

Table 5.11
Comparison of intake fraction estimates with other studies

	Country	Source Type	Pollutants	Average *iF*	Minimal *iF*	Maximum *iF*
Present study (chapter 5)	China	Industrial Source	TSP	7.15×10^{-6}	7.10×10^{-7}	5.7×10^{-5}
			SO_2	6.65×10^{-6}	2.38×10^{-7}	4.75×10^{-5}
		Mobile Source	PM_{10}	7.72×10^{-5}	1.84×10^{-6}	1.54×10^{-4}
Zhou (2002) and present study (chapter 7)	China	Power Plants	PM_{13}	1.80×10^{-6}	6.70×10^{-7}	5.20×10^{-6}
			SO_2	4.80×10^{-6}	1.80×10^{-6}	8.90×10^{-6}
Nigge (2001)	Germany	Mobile Source	PM_{10}	4.95×10^{-6}	3.10×10^{-6}	1.26×10^{-5}
Spadaro and Rabl (1999)	France	Stationary	SO_2	n.a.	1.70×10^{-6}	1.57×10^{-5}
Phonboon (1996)	Thailand	Refinery	TSP	n.a.	7.00×10^{-6}	1.50×10^{-5}

between 1.7×10^{-6} and 1.6×10^{-5}. Nigge (2001) gives the range of intake fractions for PM_{10} emissions from vehicles from different locations in Germany to be 3.1×10^{-6} to 1.3×10^{-5}.

Compared with these other results, the range of the *iF*s that we estimate for industrial sectors is larger, due mainly to the wide range of population densities along with the variation in stack heights. For example, in Jinan the population density in the urban area is more than 5,000 persons per square kilometer, whereas in rural areas it is only hundreds of persons per square kilometer. The population concentration in urban areas means a high exposure rate to emissions from urban traffic. The PM_{10} *iF* for vehicle emissions is 7.7×10^{-5} on average, and ranges from 1.8×10^{-5} to 1.5×10^{-4}. These are one order of magnitude higher than the Nigge figures for Germany, where the maximum population density is only about 2,000 people per square kilometer.

5.6.2 Secondary Particles

We have described how secondary particles are likely to be of the same magnitude in effect as primary particles in the transportation sector. Our estimates are rough, relying on a small sample. This is, however, a rapidly growing source of pollution and further research should be done to clarify the emission-exposure relationship of secondary pollutants by using advanced, long-distance air-dispersion models and to do so for a bigger sample. These, and other issues related to human exposures to mobile-source pollutants in China, are the subject of new research by members of the current team.

5.7 Conclusions

We estimate the exposures to primary air pollutants emitted from three highly polluting industrial sectors and from mobile sources using the *iF* methodology. We calculate the fraction of emissions breathed in by the population close to the emission sources using data from a sample of five cities in China, and a simple air-dispersion model. The average *iF*s estimated for industrial sectors and mobile sources are much higher than that those for power plants because of their low emission heights and high population densities. This is relevant because much research has focused on the power sector, perhaps because it is the largest user of coal. This study highlights the importance of considering these other sectors.

Compared to estimates derived from other countries, the *iF*s for industrial sectors in China vary across a much larger range. This is due to the range of population densities that include unusually high values and a large range of stack heights. The average estimates, however, are comparable to countries with similar population densities.

By use of the data from the five-city sample, we estimate a very simple reduced-form expression for *iF*s as a function of population sizes around the source and stack heights. These simple regressions fit remarkably well and may be used to provide a quick, rough estimate of the *iF*s of other sources when running full air-dispersion models is too costly or time-consuming. We use them to estimate the national average *iF*s for the cement, chemical, and iron and steel sectors by combining results with national industry data.

While further investigation is needed to determine the magnitude of uncertainty, these national average *iF*s can be used to estimate the local human health damage due to TSP and SO_2 emissions from industrial sectors and urban traffic in China. An example of this is chapter 9 in this volume. These damage estimates by sector will be an important indicator for determining pollution-control priorities across different emission categories and can serve as a basis for benefit-cost analysis of control strategies.

In light of the simple nature of our methodology, there are limits to the application of our approach for the estimation of health damages. The Gaussian plume model used here is only suitable for the short-range dispersion of primary air pollutants. For many combustion pollutants, atmospheric dispersion is significant over hundreds, or even thousands, of kilometers (Seinfeld and Pandis 1998; Curtiss and Rabl 1996). This means both local and regional effects are important. The estimation of such regional effects would need to use more advanced models to simulate long-distance transportation. Furthermore, exposures to secondary pollutants such as ozone and secondary fine particles are substantial and should be included in future refinements of these methodologies and applications.

Acknowledgments

This research was funded by generous grants to the China Project of the Harvard University Center for the Environment (HUCE) and Tsinghua University from the China Sustainable Energy Program of the Energy Foundation, and to the HUCE China Project from the Henry Luce Foundation. This chapter reports analyses and detail expanding on results presented previously in an article in *Science of the Total Environment* (Wang et al. 2006).

Notes

1. The total industrial emissions are from SEPA (2000a) and sector emissions are from SEPA (2000b), which both give statistical data for 1999. In the Chinese statistical system "industry" refers to mining, manufacturing, and utilities. Industry accounts for more than 80% of total TSP emissions in 1999.

2. For Beijing and Jinan, the hourly meteorological data are supplied by local EPBs and then processed to the data required by ISCLT by the authors. For the other three cities, the annual meteorological data are taken from Environmental Impact Assessment (EIA) Reports from local EPBs (e.g., the meteorological data of Dalian are from the "EIA report on the #4 boiler construction in Dalian Beihaitou power plant, 1995").

3. Shougang Group is one of the largest steel manufacturers in China and has a steel output of about 8 Mt per year in 2000. Its SO_2 emissions account for nearly 50% of total emissions from the five steel companies sampled. Its location determines its high *iF* and the large share of total SO_2 emissions gives the steel sector the high average *iF*.

4. We group the sources by stack height: 10–15, 15–25, ... 120+ meters.

5. Other regression results not reported here are available from the authors, which include emission temperature, gas exit velocity, and mixing height.

References

Bennett, D. H., T. McKone, J. Evans, W. Nazaroff, M. Margni, and O. Jolliet. 2002. Defining intake fraction. *Environmental Science and Technology* 36 (9):207A–211A.

Carmichael, G., D. G. Streets, and J. H. Woo. 2002. Gridded TRACE-P emission inventories. Available at www.cgrer.uiowa.edu/EMISSION_DATA/index_16.htm.

China Environment Yearbook. 2000. Beijing: China Environmental Science Press. In Chinese.

Curtiss, P. S., and A. Rabl. 1996. Impacts of air pollution: general relationships and site dependence. *Atmospheric Environment* 30 (19):3331–3347.

Florig, H. K., and G. Song. 2000. Industrial reporting of air pollution emissions in China: a case study of one medium-sized industrial city. March. Available at http://www.andrew.cmu.edu/user/kf0f/Emissions_Reporting_slides_3-00.pdf.

He, Kebin, Qiang Zhang, Yongliang Ma, Fumo Yang, Steven H. Cadle, Tai Chan, Mulawa A. Pat, Chak K. Chan, Xiaohong Yao. 2002. The source apportionment of $PM_{2.5}$ in Beijing. Conference paper presented at 224th ACS National Meeting, August 22, Boston.

Künzli, N., R. Kaiser, S. Medina, M. Studnicka, O. Chanel, P. Filliger, M. Herry, F. Horak, V. Puybonnieux-Texier, P. Quénel, J. Schneider, R. Seethaler, J.-C. Vergnaud, and H. Sommer. 2000. Public-health impact of outdoor and traffic-related air pollution: a European assessment. *Lancet* 356:795–801.

Levy, J. I., J. D. Spengler, D. Hlinka, D. Sullivan, and D. Moon. 2002. Using CALPUFF to evaluate the impacts of power plant emissions in Illinois: model sensitivity and implications. *Atmospheric Environment* 36:1063–1075.

Levy, J. I., S. K. Wolff, and J. S. Evans. 2002. A regression-based approach for estimating primary and secondary particulate matter intake fractions. *Risk Analysis* 22 (5):893–901.

Liu, Z. J. 2001. Today's tendency of cement industry of China. *Cement Technology* 1:5–8. In Chinese.

Mestl, Heidi Elizabeth Staff, Kristin Aunan, Jianghua Fang, Hans Martin Seip, John Magne Skjelvik and Haakon Vennemo. 2005. Cleaner production as climate investment: Integrated assessment in Taiyuan City, China. *Journal of Cleaner Production* 13 (1):57–70.

Nigge, K. M. 2001. Generic spatial classes for human health impacts. Part I. Methodology. *International Journal of Life Cycle Assessment* 6 (5):257–264.

Phonboon, K. 1996. Risk assessment of environmental effects in developing countries. Ph.D. diss., Harvard School of Public Health.

Schwartz, S. E. 1989. Acid deposition: Unraveling a regional phenomenon. *Science* 243:753–763.

Seinfeld, J. H., and S. N. Pandis. 1998. *Atmospheric chemistry and physics: from air pollution to climate change*. New York: John Wiley and Sons.

Smith, K. R. 1993. Fuel combustion, air pollution, and health: The situation in developing countries. *Annual Review of Energy and Environment* 18:529–566.

Spadaro, J. V., and A. Rabl. 1999. Estimates of real damage from air pollution: Site dependence and simple impact indices for LCA. *International Journal of Life Cycle Assessment* 4 (4):229–243.

State Environmental Protection Administration (SEPA). 2000a. Report on the State of the Environment in China 1999. Beijing: SEPA. Available at http://www.zhb.gov.cn/english/SOE/soechina1999/air/air.htm.

State Environmental Protection Administration (SEPA). 2000b. Environmental Statistical Data for 2000 and 1999. Beijing: SEPA. Available at http://www.sepa.gov.cn/english/SOE/soechina2001/english/statistics/sector-waste-gas-e.pdf.

Thanh, B. D., and T. Lefevre. 2000. Assessing health impacts of air pollution from electricity generation: the case of Thailand. *Environmental Impact Assessment Review* 20 (2):137–158.

Tsinghua University, Beijing Urban Planning and Design Research Institute, Beijing Academy of Environmental Protection, Beijing Institute of Labour Protection Science, China Research Academy of Environmental Science. 2002. Study on the strategy and countermeasures of air pollution control in Beijing. Technical report of project funded by the Ministry of Science and Technology, People's Republic of China. Beijing. In Chinese.

U.S. Environmental Protection Agency (U.S. EPA). 1992. Development and evaluation of a revised area source algorithm for the Industrial Source Complex Long Term Model. Report no. EPA-454/R-92-016. U.S. Environmental Protection Agency, Research Triangle Park, N.C.

U.S. Environmental Protection Agency (U.S. EPA). 1994. Development and testing of a dry deposition algorithm, revised. Report no. EPA-454/R-94-015. U.S. Environmental Protection Agency, Research Triangle Park, N.C.

U.S. Environmental Protection Agency (U.S. EPA). 1995. User's guide for the Industrial Source Complex (ISC3) dispersion models. Volume I: User instructions. Report no. EPA-454/B-95-003a. U.S. Environmental Protection Agency, Research Triangle Park, N.C.

U.S. Environmental Protection Agency (U.S. EPA). 1996. Compilation of air pollutant emission factors, AP-42, 5th ed., Volume I: Stationary point and area source, Appendix B.2, generalized particle size distributions. September. Available at www.epa.gov/ttn/chief/ap42/appendix/appb-2.pdf.

U.S. Environmental Protection Agency (U.S. EPA). 1998. Industrial waste air model technical background document. Report no. EPA 530-R-99-004. U.S. Environmental Protection Agency, Office of Solid Waste, Washington, D.C. December.

Wang, Shuxiao, Jiming Hao, Mun S. Ho, Ji Li, Yongqi Lu. 2006. Intake fractions of industrial air pollutants in China: Estimation and application. *Science of the Total Environment* 354:127–141.

Wolff, S. K. 2000. Evaluation of fine particle exposures, health risks and control options. Ph.D. diss., Harvard School of Public Health.

Zhou, Y. 2002. Evaluating power plant emissions in China: Human exposure and valuation. Ph.D. diss., Harvard School of Public Health.

6

Local Population Exposure to Pollutants from the Electric Power Sector

Bingjiang Liu and Jiming Hao

When people think of energy and air pollution, a classic image that often comes to mind is of a plume flowing from a tall smokestack at an electric power plant. It is in some respects a fair association, because electric power is the greatest fossil fuel–consuming and air-polluting sector in many nations. This is true in China, with power generation consuming nearly half of the country's coal in 2002, as chapter 1 mentioned. Thermal power plants can be very large and are thus concentrated sources of pollution. The emissions from one poorly located power plant can affect the air quality of an entire city.

The popular image of power plants as top polluters, symbolized by their belching tall stacks, however, may be rather simplistic. Concentrating combustion of coal in large facilities has at least some benefits to air quality over other alternatives that are familiar in China. For instance, larger plants tend to have higher combustion efficiencies than smaller plants, so less coal is needed to generate the same quantity of power. This reduces total emissions. Electricity is a cleaner form of energy to heat a home than a small boiler in a building basement, at least from the perspective of the residents of the surrounding neighborhood. As will be shown later in chapter 9, the damage per ton of emissions from small stacks at manufacturing factories seems to be far worse than the damage per ton from power plants.

In fact the smokestack does help to symbolize a core research concept of this book. Smokestacks locate the heaviest concentrations of waste gases and particles far above the ground, away from people near the sources. Of course, meteorological conditions and characteristics such as flue gas temperature will then affect each plume, sending it downwind to neighboring regions. As it spreads, it may become less concentrated, which would reduce the health risk, but its geographical extent also may grow, which could raise the risk by putting it in contact with more people. If an ocean or uninhabited desert is downwind, on the other hand, it might affect fewer people, not more.

These are the conflicting factors that the intake fraction (*iF*) method, introduced in preceding chapters, helps to explore for different sectors.[1] In this chapter, we calculate the *iF*s of primary particulates and sulfur dioxide (SO_2) from the electric power sector, thus demonstrating the methodology and examining the influence of various parameters on those relationships.

A national database of the electric power industry especially constructed for this study includes approximately 75% of total capacity of coal-fired power generation. To represent the entire power sector, we randomly select a sample of plants that have the requisite meteorological data. We use an "impact pathway" approach widely applied in air pollution health-damage estimation in developed nations to estimate intake fractions of primary particles and SO_2 from each stack in the sample.

In light of the limited data and research resources, we estimate the air pollution damage for the entire electric power sector by using summary relationships derived from this sample data. To implement this approach, we develop a set of regression equations linking *iF*s of primary particles and SO_2 with population distribution and source characteristics. These equations provide a quick estimate of *iF*s, and hence of health damage, with reasonable accuracy based on easily observed source characteristics and without running data-intensive and time-consuming air-dispersion models for every power plant.

These results are applied in chapter 9 to estimate health damages for the entire power sector. These applications demonstrate how these *iF* calculations and regression equations can serve as tools for policy makers who need aggregate information to analyze national-level power-sector policies.

6.1 Brief Overview of China's Electric Power Industry

China's current power capacity and generation are second only to that of the United States. According to *CEPY* (2004), total installed generating capacity in 2003 was approximately 391 gigawatts (GW), including a thermal power share of more than 290 GW. In terms of output, thermal power accounted for 83% of the 1905 terawatt-hour (TWh) total. This thermal power is fueled predominately by coal, which on a fuel-share basis constitutes about 90% of the energy used. Electricity-generating capacity in China expanded 3.2 times from 1990 to 2003 (SPC 2003). Capacity additions exceeded 10 GW annually after 1990, whereas electricity generation grew by about 10% per year during the same period (*CEPY* 2004). China's

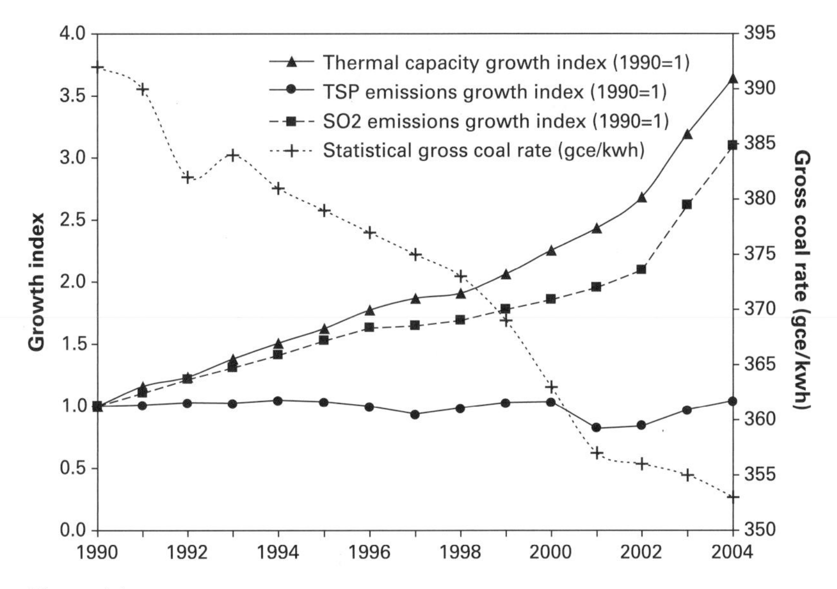

Figure 6.1
Pollutant emissions and energy efficiency in China's electric power industry, 1990–2004.
Sources: NBS 2004; CCEY 2004; MOP 2004.

electric power industry has aimed to upgrade generating technologies and narrow the gap between China and the West with regard to energy efficiency and environmental loadings. Figure 6.1 shows the progress made by the thermal power industry since 1990. During this period, thermal generation increased by 3.6 times, but the gross coal rate decreased from 392 gce/kWh (grams of coal equivalent per kilowatt-hour) in 1990 to 355 gce/kWh in 2003 and emissions of primary particulates barely changed. SO_2 emissions, however, grew only modestly slower than power output, rising by 3.0 times.

Before 2002, China had an integrated approach to the control of SO_2 emissions, including designation of SO_2 and acid rain control zones, closure of small coal-fired power plants, and limitations on the mining of high-sulfur coal (Finamore and Syzymanski 2002). As a result, many power plants switched from high-sulfur to low-sulfur coal, and more than 15 GW of small coal-fired plants were shut down. SO_2 emissions increased at a slower rate compared to electricity output during 1990–2002, as shown in figure 6.1. In 2003, these policies were insufficient to keep power-sector emissions under control, because of skyrocketing national electricity demand and capacity expansions exceeding 50 GW in total. In 2004 alone, China brought new generating capacity online equal to the entire capacity of California. In response, China's government enacted tough programs against emissions from the power sector, including implementation of a decree of the state council to raise fees levied on sulfur emissions, to enact strict emission standards for pollutant emissions, as well as to advance ongoing design and trial applications of a cap-and-trade system. Under full implementation of these regulations, almost all newly built coal-fired generation units will install flue gas desulfurization (or FGD), whereas existing units will install it gradually and many small coal-fired generation units will be closed. These measures are expected to reduce SO_2 emissions very substantially by 2010.

Power generation is the largest coal consumer in China—as in the world—and will continue to be so far into the future. China's large coal reserves suggest that most added generating capacity will be coal fired, with 75% of coal production increases used to meet the capacity expansion (UNEP and NEPA 1996; Zhou and Zhou 1999). The country expects to add about 200 GW coal-fired power capacity by 2010 and to more than double current capacity by 2020. Although the power industry has undergone major shifts in the recent past and is likely to see further transformation in terms of efficiency improvements and emission reductions, human exposure to pollutants raises challenging questions.

Under the pre-2002 policy framework where the primary focus was to keep air pollution from exceeding thresholds in areas adjacent to given sources, the effect was to encourage high stacks that in practice dispersed concentration impacts to an entire region. The aggregate effect of many such sources was overlooked but may outweigh the health benefits viewed only on a local scale. With implementation of the new regulations to reduce SO_2 emissions, policy makers are required to assess their costs and benefits retrospectively. It is therefore imperative to look forward from the current policy to a future risk-based policy framework that can fully assess the true social cost, notably health damages, from power generation. To date, such assessments, particularly for the whole power sector, have not been successfully conducted because of both the unavailability of data and methodological limitations. The *iF* method was specifically developed with this in mind, to serve as a step toward estimating such total population risk despite the data limitations that exist. We note also that our modeling was done on the best data set that was available at the time of research. As described above, China's power sector has substantially expanded capacity and generation in recent years, but the fundamental applicability of the methods described here remains the same.

6.2 Methodology to Calculate Intake Fraction

As explained in chapter 4, *iF* is defined as a dimensionless ratio between the amount of a pollutant inhaled and the amount of the pollutant emitted (Bennett et al. 2002). As in the previous chapter, we adapt the formulaic definition from chapter 4 to specify how we apply *iF*, and the parameters we use, in the research of this chapter. For a given pollutant, influencing health through the inhalation pathway, from a given source:

$$iF = \sum_{d=1}^{n} (POP_d \times C_d \times BR)/EM, \tag{6.1}$$

where POP_d is the population at location d; C_d is the incremental annual pollutant concentration at location d in micrograms per cubic meter (μg/m^3); BR is the breathing rate, assuming in our analysis a nominal $BR = 20$ m^3/day for the entire population; EM is the emission rate of the pollutant from the source in micrograms per day (μg/day); and n is the number of receptors in the domain, in our case $n = 2{,}400$.

We apply equation (6.1) to each stack of the sampled power plants. First, for each stack, we use an air-dispersion model to calculate C_d, the incremental annual concentration of the target pollutant, at each of the 2,400 grid cells. We then employ a geographical information system (GIS, in ArcView) to digitize POP_d, the population in each grid cell.

To emphasize the joint distribution of concentration and population in the intake fraction definition, we can rewrite equation (6.1):

$$iF = \left(\sum_{d=1}^{n} C_d \times POP_d \middle/ \sum_{d=1}^{n} POP_d \right) \times \sum_{d=1}^{n} POP_d \times BR/EM, \quad \text{or}$$
$$iF = C_{pw} \times POP \times BR/EM, \tag{6.2}$$

where C_{pw} is the incremental population-weighted concentration and POP is total population in the given domain. C_{pw} is an aggregate concentration measure, reflecting the distribution of incremental concentration and population over the domain. The numerator $C_{pw} \times POP \times BR$ represents the rate of pollutant inhaled from the source released at rate EM.

Our procedures are described in more detail below in section 6.3. To summarize, our first aim is to estimate the power-sector *iF* at the source level, that is, at the level of the individual stacks in our sample. Because most power plants have multiple

1 2	A	B	C	D	E	F	G	H	I	J	K	L	M	N	O	P	Q	R	S	T	U	V	W
1	Plan	Power plant	Geographic	Year	Unit	Stack	Gross			Coal	SO_2		Particular matter				Nitride oxide		Stack				
2	No.	name	Location	in use	Capacity	Capacity	coal rate	LHV	Annual	burned	S	Emiss	Ash	Emis	Conc	PM10	Emis	Conc	H	D		V	olum
3			Lo/La	yr.	MV	MV	gce/kwh	KJ/KG	hours	10 kt/yr.	%	t/h	%	t/h	mg/m^3	t/h	t/h	mg/m^3	(m)	(m)	T	m/s	m^3/s
4			[1]	[2]	[3]	[4]	[5]	[6]	[7]	[8]	[9]	[10]	[11]	[12]	[13]	[14]	[15]	[10]	[16]	[16]	[17]		[18]
5	1	Liaoning Power	123°58'/41°51'	1966	5x50	250	367	20640	5278	85.2	0.50	1.180	27.53	1.840	1304	1.104	1.61	505	100	4.5	90	32.8	521
28	9	HLJ MDJ No.2	129°31'/44°46'	1981	2x100	200	362	17330	4912	81.2	0.30	0.554	40.32	1.131	974	0.519	1.64	1096	150	4.0	80	33.2	417
53	18	Jilin Changshan	124°50'/45°7'	1972	1x12+1x6	18	469	14040	5135	8.0	0.52	0.138	41.51	0.372	3300	0.108	0.16	1500	120	2.0	155	14.3	45
73	27	Beijing Guohua	116°30'/39°55'	1968	1x100+2x50	200	263	23970	5000	21.0	0.58	0.940	17.42	0.250	98	0.173	0.39	524	100	4.2	140	22.8	316
85	34	Tianjing	117°19'/39°5'	1982	2x50+1x100	200	420	22000	5935	65.5	0.57	1.072	13.00	0.307	300	0.141	1.10	1074	95	5.0	120	20.8	409
91	36	Shanxi Taiyuan	112°33'/37°47'	1978	200	200	375	25160	5800	52.3	0.94	1.421	20.62	3.578	1700	2.147	0.90	1037	210	5.2	70	14.2	302
124	47	Hebei Handan	114°28'12"/	1960	6x12	72	431	20850	6200	40.0	0.39	0.446	30.10	1.700	2400	1.020	0.64	1107	60	3.0	110	32.0	226
171	63	Mongolia	113°1'/41°0'	1990	2x200	400	341	18392	6500	145.1	0.60	1.860	18.86	1.249	700	0.987	2.22	1244	210	6.0	125	25.6	723
203	74	SH Shidongkou	120°24'/31°28'	1988	2x300	600	340	22604	5000	140.0	1.88	7.972	26.70	0.374	142	0.295	1.71	650	240	6.5	130	32.5	1079
215	77	Jiangsu Wangting	121°08'10"/	1989	1x300	300	354	21566	5110	84.0	0.80	0.968	31.26	0.378	325	0.299	0.97	650	120	5.5	125	25.4	602
258	94	Zhejiang Jiaxin	30°36'10"	1995	2x300	600	335	22797	4998	136.5	0.41	1.585	11.00	0.517	200	0.408	1.68	650	240	6.5	120	31.1	1033
292	109	Fujian Fuzhou	119°28'30"	1989	2x350	700	304	22441	6445	139.6	0.63	2.960	19.77	0.405	200	0.320	1.32	650	210	7.0	125	21.3	819
309	118	Anhui Huaibei	116°45'/33°59'	1978	2x50+2x125	350	354	18260	6640	128.6	0.37	1.190	31.61	3.000	1490	1.800	1.93	1251	150	5.5	73	22.8	542
338	126	Jiangxi Jiujiang	116°2'/29°44'	1988	2x125	250	351	20695	4774	68.3	0.81	1.940	26.00	0.990	1120	0.594	1.42	1136	180	5.0	75	22.6	444
359	136	SD Jinan	116°45'47"	1983	2x100+1x125	325	381	23421	5469	69.8	1.48	3.022	23.82	0.530	378	0.419	1.27	1029	210	6.0	116	17.3	488
414	162	Henan	113°36'/34°45'	1971	3x6	18	639	20348	7267	11.7	0.40	0.109	26.30	0.084	600	0.050	0.16	1151	100	2.0	80	15.9	50
456	179	Hubei Huangshi	115°4'/30°13'	1960	4x50+5x6	230	512	19870	5982	96.1	2.35	6.418	28.81	0.818	600	0.375	1.60	1172	115	5.0	115	27.4	538
475	187	Hunan Changsha	112°/28°	1984	1x3	3	778	23700	5600	1.5	2.00	0.094	21.35	0.046	1700	0.027	0.03	1020	50	1.0	71	12.0	9
497	197	Guangdong	113°33'58"	1993	2x300	600	332	23844	4500	158.0	0.91	4.280	22.39	0.818	374	0.646	2.24	650	210	7.0	110	34.9	1342
522	208	Guangxi Liuzhou	109°17'/24°24'	1960	3x12	36	701	18755	6043	22.8	5.00	2.910	47.00	1.012	3300	0.607	0.38	1226	60	3.0	65	14.9	105
533	215	Hainan Haikou	110°10'/20°00'	1986	2x50	100	472	21658	4477	27.3	0.80	0.828	29.00	1.161	2100	0.697	0.61	1096	150	4.5	70	12.1	193
537	216	Sichuan Baima	104°55'/29°30'	1982	94.5	94.5	556	18840	3344	25.5	3.00	3.889	32.90	0.435	700	0.199	0.76	1222	120	3.5	70	22.5	217
560	227	Chongqing	106°12'/29°24'	1985	4x50	200	501	20959	5000	76.3	3.33	6.267	25.64	0.600	500	0.275	1.52	1125	80	4.0	85	39.2	492
567	230	Yunnan Kaiyuan P		1969	4x6	24	661	12370	5258	14.8	1.20	0.421	12.20	0.203	1200	0.059	0.28	1660	60	2.5	130	14.1	69
584	238	Guizhou Kaili	108°6'/26°28'	1970	2x3	6	651	20934	4876	2.7	3.07	0.172	21.54	0.100	1546	0.060	0.05	1126	35	1.5	108	10.6	19
603	248	Shaanxi Weihe	108°43'/34°21'	1971	2x50	100	401	21000	5121	33.0	3.00	2.734	36.00	1.884	3300	0.547	0.64	1123	100	4.3	135	16.3	237
626	258	Gansu Qingyuan		1986	2x200	400	351	26290	5585	108.0	0.53	1.742	18.98	0.616	300	0.487	1.92	937	180	6.0	110	28.3	800
641	267	Qinghai	101°41'/36°54'	1977	86	86	471	19030	4776	34.3	0.35	0.400	28.72	2.418	3939	1.451	0.71	1212	80	4.0	71	16.4	206
647	269	Ningxia	106°46'/39°14'	1960	2x12	24	882	17900	4819	16.0	2.25	1.267	34.78	0.726	2800	0.436	0.33	1270	80	2.5	80	19.0	93
656	273	Xinjiang Urumqi	87°35'/43°50'	1974	1x50	50	512	24750	4208	16.1	0.76	0.336	20.06	0.657	1700	0.191	0.38	984	60	3.0	101	20.8	147
672	278	Kalamayi PP	84°49'/45°46'	1982	150	150	447	25000	5249	16.4	1.45	0.770	22.00	0.541	1700	0.157	0.31	976	120	3.0	135	18.7	132
673		Total/average				160888		20926	4848	39518.0	0.89	1.159	25.40	0.471		0.287	1.02		153				

Figure 6.2
Picture of database for China's power industry.

stacks, we next aggregate these results using emission weights into *iF*s at the plant level. Finally, we estimate the *iF* of the entire electric power sector by aggregating these plant level *iF*s from our sample using a regression-based method.

6.2.1 Pollutants Concerned and Emission Sources Sampled

We identify two classic pollutants: SO_2 and particulate matter, represented here as either total suspended particulates (TSP) or particulate matter less than 10 microns in aerodynamic diameter (PM_{10}). We concentrate on these principally because they have been proven to pose major threats to human health. Emission data for SO_2 and TSP are available and relatively reliable and corresponding dose-response coefficients from domestic and international epidemiological studies are among the most widely studied compared to other pollutants. Here we do not consider secondary sulfates, fine particulates chemically formed from SO_2 in the atmosphere, but this is studied in chapter 7. The *iF*s for both sets of pollutants are used later to estimate health damages of the power sector.

In 2000, there were at least 1,037 thermal power plants in China, according to incomplete statistics of CEY (2002). Ideally we would apply equation (6.1) to each plant and aggregate them to get the *iF* for the entire power sector. This is infeasible,

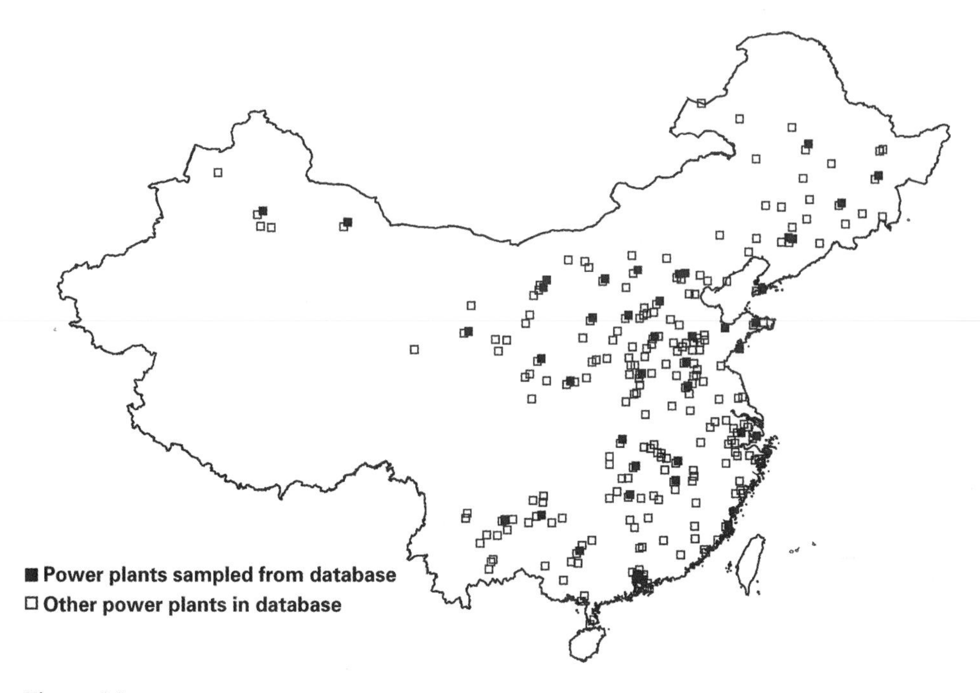

Figure 6.3
Locations of coal-fired power plants in database and sample.

in light of the time-consuming nature of calculating *iF*s. As we show below, however, our sample of plants provides a good representation of the whole sector and thus gives a good estimate of the aggregate *iF*.

To select representative power plants (and stacks), we compile information on as many power plants as possible and construct a national database of the industry. It includes approximately 75% of total capacity of coal-fired power generation, and we assume that their characteristics are representative of the entire sector. The database consists of 686 stacks distributed across 278 coal-fired power plants. An image of the database is shown in figure 6.2, and locations of the plants are shown in figure 6.3. (Details on the references and formulas for all data in the database are given in a lengthy note.[2])

We randomly select a sample of 160 stacks from 52 power plants of those in the national database, excluding those where the meteorological data necessary for modeling is unavailable. The total capacity of the 52 power plants amounts to 48 GW, roughly 22% of total national thermal capacity at the time. Detailed

Table 6.1
Comparison of characteristics of power plants in the entire country, the national database, and the selected sample

Unit Size	Sector		Database		Sample	
	Capacity (GW)	Percent	Capacity (GW)	Percent	Capacity (GW)	Percent (%)
Generator capacity:						
>300 MW	90.9	42.7	76.3	48.9	31.2	59.0
200–300 MW	38.8	18.2	35.2	22.5	10.2	19.4
100–200 MW	36.6	17.2	28.6	18.3	7.40	14.0
50–100 MW	18.3	8.6	10.1	6.5	2.60	4.9
25–50 MW	10.8	5.1	3.28	2.1	0.846	1.6
12–25 MW	10.2	4.8	2.17	1.4	0.477	0.9
6–12 MW	7.2	3.4	0.558	0.4	0.102	0.2
<6 MW	0	0.0	0.012	0.0	0.012	0.0
Total	213	100	156	100	52.9	100
	Number	Percent	Number	Percent	Number	Percent
Generator numbers:						
>300 MW	262	7.4	223	17.7	83	24.0
200–300 MW	193	5.4	175	13.9	51	14.7
100–200 MW	321	9.1	258	20.4	67	19.4
50–100 MW	355	10.0	202	16.0	52	15.0
25–50 MW	418	11.8	130	10.3	33	9.5
12–25 MW	817	23.0	178	14.1	39	11.3
6–12 MW	1180	33.3	92	7.3	17	4.9
<6 MW	0	0.0	4	0.3	4	1.2
Total	3,546	100	1,262	100	346	100

Notes: The sector figures include all thermal generators with unit capacity larger than 6 MW (CEPY 2002), and the figures in the database and sample are based on statistics of the study team. Our database capacity roughly accounts for three-quarters of national thermal capacity, whereas the sample accounts for one-quarter. In terms of generator numbers, the database and sample account for approximately 35% and 10% of the thermal power sector. Percentage columns may not add up to 100% because of rounding.

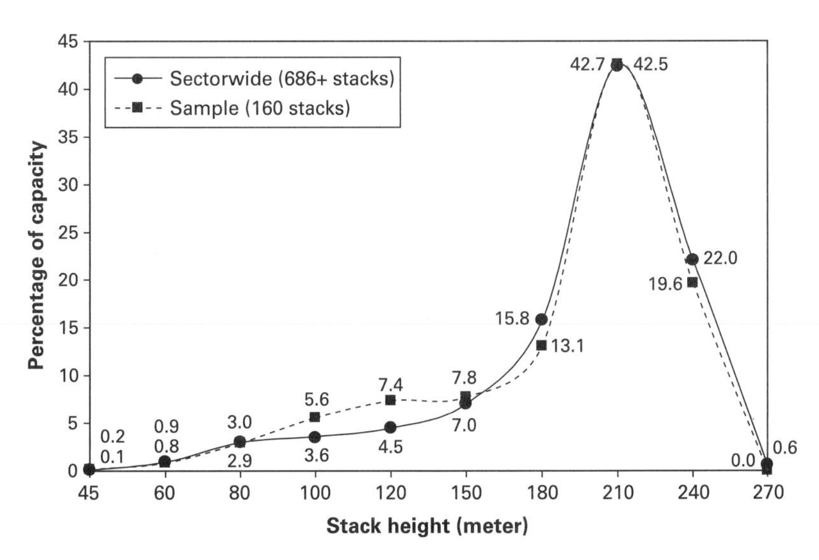

Figure 6.4
Comparison of stack height distribution in the database and sample.

information for the 160 stacks is listed in table A.1 of appendix A. To show how this sample and the national database compare to the entire sector, we tabulate the size distribution of all three in table 6.1. We can see that the sample distribution is quite close to that of the whole sector. We next compare the distribution of stack heights because it is known to be a key determinant of intake fractions in past studies (Levy et al. 2002; Levy, Wolff, and Evans 2002a, 2002b). The distributions of both database and sample are plotted in figure 6.4, and we see that our sample is fairly representative.

6.2.2 Incremental Annual Concentration

We apply the ISCLT (Industrial Source Complex—Long Term) air-dispersion model to calculate the incremental annual concentration of SO_2 and primary particulates generated by each stack, around each of which is a modeling domain of 100 by 96 kilometers (km). The model assumes source emissions are carried in a straight line by the wind, mixing with the surrounding air both horizontally and vertically to produce pollutant concentrations with a normal (or Gaussian) spatial distribution. The dispersing pollutants are presumed to be effectively chemically stable in the analyzed domain. The model is typically applied only up to a distance of 50 km from the source (Schnelle and Dey 1999). ISCLT requires such input data as source characteristics, meteorology, and receptor location. Appendix A provides a detailed introduction and description of the model and data inputs.

The characteristics and conditions of sources include stack height, stack inner diameter, stack gas temperature and exit velocity, and pollutant emissions. We must note that particle size distribution (PSD, also called mass fraction) in the chimney is not available, because few measurements of dust-removal equipment are reliable in China. Moreover, it is common that a single chimney might vent flue gases from several boilers equipped with different particle-removal devices. To estimate particle size distribution, we assume that there are no chemical reactions or aggregations when particle-carrying flue gases merge into one chimney and use an average of the PSDs of contributing sources (see appendix A).

The joint frequency distribution of wind speed class, wind direction sector, and stability category are critical meteorological inputs to run ISCLT. These data are taken directly from environmental impact assessments of the selected power plants.

Each stack has a domain of 100 km in the east-west direction by 96 km in the north-south direction, resulting in 2,400 Cartesian receptor cells with resolution 2 km × 2 km. One of the inputs into ISCLT is the terrain, and we assume a flat terrain for both the source and the surrounding domain for two reasons. First, ISCLT models do not incorporate complex terrain well (Schnelle and Dey 1999). Second, complex terrain simulations in surrounding areas result in air concentrations that are highly dependent on site-specific topography. Simulations from flat terrain produce values that are generalizable, which serves the objectives of our study.

6.2.3 Gridded Population

We developed and coded a county-level population dataset within a China map digitized in ArcView GIS format, the details of which are given in appendix B. There are 2,569 county-level administrative units, each represented by a distinct polygon in the GIS, that cover a total population of 1.24 billion in 1999. This includes eighteen county-level units for Hongkong and one for Macao but excludes data for Taiwan. With this database, a user can transform the spatial resolution of population from administrative units into a grid that precisely matches the size and resolution of the air-dispersion model output. We assume that population density in each administrative unit is uniform (i.e. population in the unit is evenly distributed regardless of land-use patterns). Our 100 km × 96 km domains often cover multiple cities and counties, and an assumption of uniform density allows us to calculate the domain population. We count a proportion of a county's total population based on the fraction of its area falling within the domain.

6.3 Results

6.3.1 Intake Fractions of Sulfur Dioxide and Particulates: Individual Sources

Using equation (6.1), we calculate *iF*s of SO_2 and particulates (TSP) from the 160 stacks distributed across the 52 coal-fired power plants. Figure 6.5 illustrates the intake fraction calculations for SO_2 from a 120-meter chimney at the Wangting power plant, where the domain contains a population of about nine million. Figure 6.5a represents the gridded incremental concentration distribution of SO_2, determined by application of the ISCLT dispersion model to the source data.[3] The darker shades represent higher SO_2 concentrations dispersed from the source at the center of the domain. The higher concentrations reflect prevailing wind directions and are naturally higher when they are closer to the source. Figure 6.5b represents the gridded population distribution, with higher population represented by darker cells. Figure 6.5c depicts the patterns of dosage of SO_2 at each grid cell, representing the total amount of SO_2 inhaled by people living in the cell. These values are equivalent to the SO_2 concentrations of figure 6.5a multiplied by the population of figure 6.5b and a constant breathing rate of 20 m^3/day. The darker cells represent higher dosages. As anticipated, the cells with high dosages stretch discontinuously northwestward and southeastward because those areas have both dense populations and high concentrations along prevailing winds. The discrete jumps in values in figure 6.5c reflect the quantum geographical variations of our population data; as figure 6.5a shows, in contrast, the incremental concentration only changes gradually. The figure also shows the strong impact of population density on inhaled dosage. The city of Changzhou can be seen as the high-population region in the upper-left corner of figure 6.5b. The gridded dosages of SO_2 there are very high, close to those nearer the source, despite much lower SO_2 concentrations.

The sum of all 2,400 gridded doses amounts to 30 kilograms (kg) of SO_2. That is the total from this source inhaled annually by the population of nine million in the domain. The *iF* of SO_2 of this stack is this sum divided by total emissions from it, 3.57×10^{-6}, a dimensionless number. Adding *iF*s of SO_2 from all five stacks at the plant, we estimate that the people in the domain inhale 240 kg of SO_2 from the plant annually, an average of 27 mg per capita.[4] People living downwind of the plant inhale proportionally more as indicated by the different shades in figure 6.5a.

Intake fractions of SO_2 calculated for the 160 sampled stacks are shown in figure 6.6. The analogous figure for TSP, which has a very similar form, is presented as figure A.1 in appendix A. The geometric mean *iF*s of SO_2 and TSP are 2.77×10^{-6} and 2.61×10^{-6}, respectively. *iF*s of TSP are somewhat lower than those of SO_2,

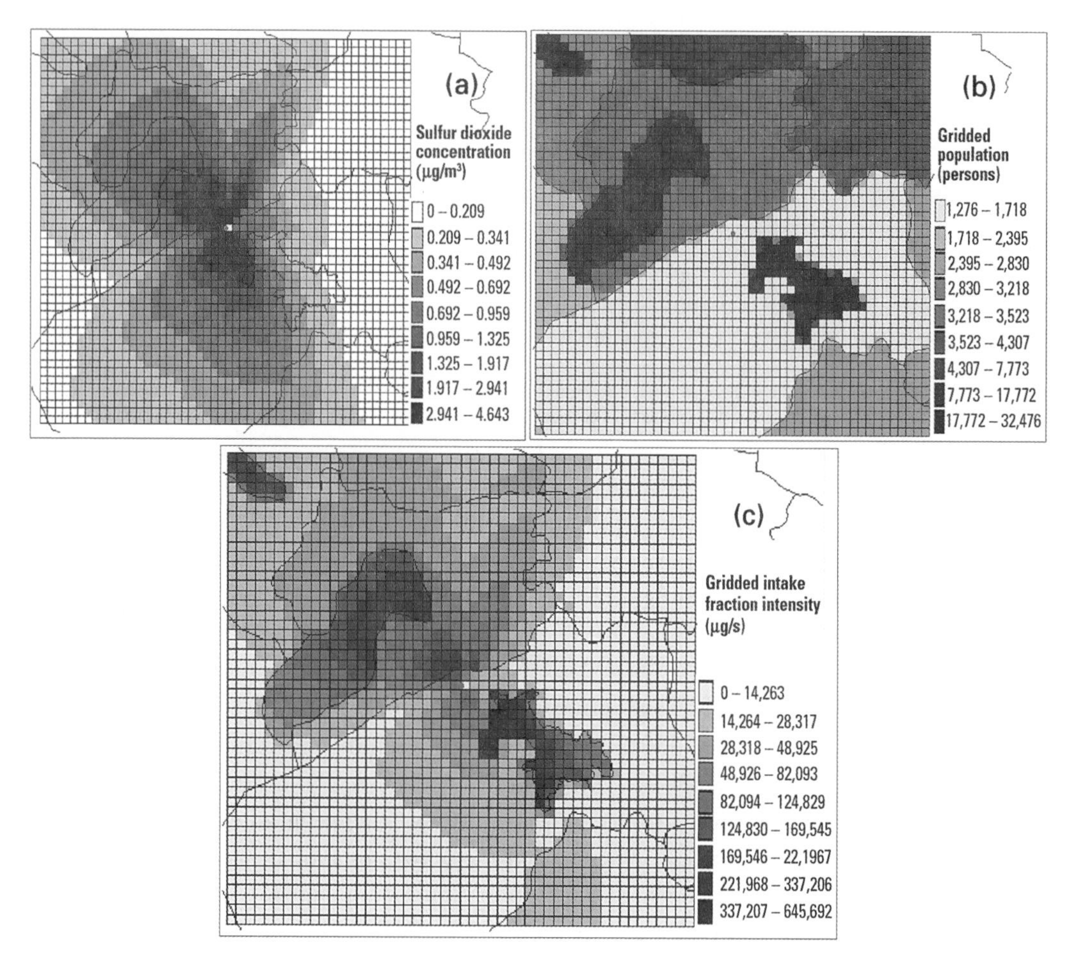

Figure 6.5
Intake fraction intensity of sulfur dioxide at grid cell from one stack of Wangting plant. Note: The plant is located at longitude 120°25′ and latitude 31°25′. (a) Concentration distribution of sulfur dioxide at grid cell through ISCLT. (b) Population distribution for grid cells via GIS ArcView. (c) Gridded intake fraction intensity (cell population dose).

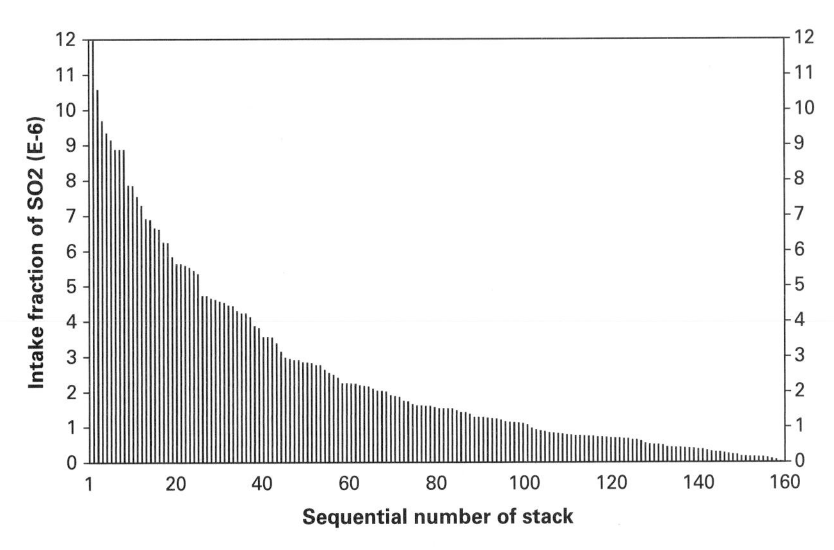

Figure 6.6
Intake fractions of sulfur dioxide from 160 stacks in 52 power plants.

largely because of the deposition of coarse particles within the 50-km range. *iF*s of SO_2 range from 1.80×10^{-8} to 1.28×10^{-5}, a span of three orders of magnitude. Particulate *iF*s also span three orders of magnitude, from 1.6×10^{-8} to 1.23×10^{-5}.

The highest *iF*s of both SO_2 and TSP occur at the Beijing Guanghua cogeneration power plant, and the lowest occur at the Hami power plant in Xinjiang. The primary reason for the enormous difference is the exposed population, with the domain of the Beijing Guanghua cogeneration power plant including 14.2 million people and that of Hami including only 38,000 people. Second, in the case of Hami, because the higher concentrations and denser population do not coincide, the pollutants from the plant tend to fall into less populated areas. For the whole sample, the variation of *iF*s can largely be explained by differences among population distributions and source characteristics rather than differences in meteorological conditions.

Figure 6.7 shows how TSP *iF*s change with the domain size, for five stacks with different source characteristics at the Zhengzhou power plant.[5] *iF*s for SO_2 have similarly shaped curves, though the values are different. For this analysis, note that *iF*s come from different stacks at the same plant (i.e. the population distribution and meteorological data are the same). The differences in stack height, flue release conditions, and particle size distribution generate the differences in *iF*s.

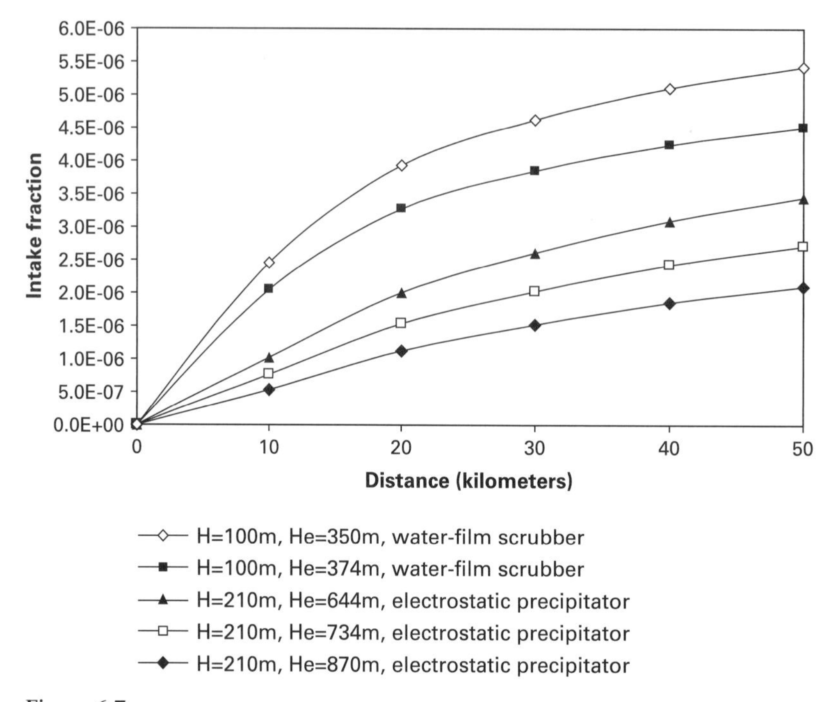

Figure 6.7
Correlation between intake fraction of total suspended particulates and source characteristics. Note: H is stack height; He is effective stack height.

First, one would expect that the higher the physical stack height, the lower the *iF*, though one can imagine unusual circumstances where this might not be true. We can see this by comparing the 100-m stack graphs in figure 6.7 with the 210-m graphs. As figure 6.7 also shows, however, other factors also influence *iF*, accounting for the slight differences between the two 100-m stacks, or between the three 210-m stacks. In particular, the effective stack height—which includes not just physical stack height, but also the plume rise that depends on the flue gas release conditions—plays a key role. Second, *iF*s from lower stacks are higher and increase faster close to the source. Concentrations in the vicinity of the lower stacks are higher, and, because the lower stacks often use inferior particle collectors, the TSP emissions are coarser. Coarse particles have a faster settling velocity, bringing more of the pollutant into contact with people near the source. The finer particles (smaller than 10 μm) emitted by better particle collectors act more like gaseous pollutants and tend to travel farther from the source before coming into contact with people. (Details on the particle size distribution resulting from different particle collectors—the grade removal efficiency—are in table A.2 of appendix A.) Third, the *iF*s of TSP from the sources do not reflect the full coverage of emissions inhaled by humans because all of the *iF* curves are still rising at the domain limit of 50 km—that is, the concentrations have not fallen to zero yet.

An important factor in *iF*s of pollutants from power plants is the definition of the area affected by the sources. It is well documented that the range of analysis should extend to thousands of kilometers from the emission source to fully assess the level of potential damages (Krewitt et al. 1999; Levy et al. 2002; Evans et al. 2002). This is because, far from the emission source, the incremental pollutant concentrations can be decreasingly small but the affected population can be increasingly large, resulting in considerable dosages. The mechanisms affecting pollutant concentrations change with distance, and there are local and regional processes. As Thanh and Lefevre (2000, p. 140) describe, at the local range, "chemical reactions in the atmosphere are assumed to be negligible and therefore have little influence on the concentration of primary pollutants.... At the regional range, by contrast, it is generally assumed that pollutants have been vertically mixed throughout the height of the mixing layer, but chemical transformation of pollutants and their wet and dry deposition play decisive roles in determining the spatial distribution of pollutant concentrations. For example, SO_2 will be chemically transformed into SO_4, which also contributes to the concentration of airborne particulates and causes adverse effects at the regional range."

6.3.2 Intake Fractions of Sulfur Dioxide and Particulates: Sample Plants

We can calculate the *iF*s of both SO_2 and TSP of the fifty-two sample power plants through emission-weighted averages of *iF*s of the various stacks at each of the plants. This effectively aggregates source characteristics of various stacks into one fictitious stack, which is reasonable because meteorological conditions and the population distribution pattern will be the same for every stack at a plant. In practice, assessing the contribution to local ambient concentrations from one plant with multiple stacks also effectively aggregates all stacks into one (Schnelle and Dey 1999).

The main purpose for calculating the average *iF*s of SO_2 and TSP from the fifty-two sample power plants is to give a sense of the intertwined effects on health damages of such factors as population, meteorology, technology, and other parameters. Figures A.2 and A.3 in appendix A illustrate *iF*s of SO_2 and TSP of the sample power plants, respectively. These figures resemble the distribution of *iF*s for stacks shown in figure 6.6, but over a narrower range of values.

For SO_2, the highest and lowest average power plant *iF*s are Beijing Shijingshan and Xinjiang Hami, respectively. Plants with higher *iF*s of SO_2 are clearly located in densely populated areas and those with lower ones in sparsely populated areas. This is true, even though plants located in unpopulated areas generally have lower stack heights—which would logically contribute to higher intake—than those of plants in more populated areas. Thus, the effect of population appears to outweigh the effect of stack height.

Among the averaged plant *iF*s for TSP, Hami still ranks lowest, whereas the Weihe power plant in Shaanxi has replaced Beijing Shijingshan as the highest. We can use this to illustrate the factors that may affect *iF* results. For instance, we can explore how differences in two pollutants affect *iF*s while other factors (meteorology, stack height, total populations, and their pattern of distribution) remain the same. First, the main distinction of the two pollutants is in the nature of their dispersion. SO_2 is a gas that disperses comparatively widely with an assumed Gaussian distribution both vertically and horizontally. TSP consists of particles subject to gravitational settling and yields a plume with a vertical distribution that is no longer Gaussian—its centerline loses height with distance from the source—in what is termed a "tilted plume" (U.S. EPA 1995).

This means that the TSP *iF* may favor nearer-source population concentrations (i.e., it will be higher if more people are near the source to inhale the pollutant). If there are highly effective electrostatic precipitators (ESPs) on flue gas streams, however, as there are in Shijingshan but not on the one stack of Weihe, TSP consists primarily of finer particles that behave much like a gas. The *iF*s of both TSP and SO_2 for Beijing Shijingshan are nearly identical as a result. Given the same meteorology and population, once the ESPs have removed the influence of coarser particles on settling rates, the calculation for TSP and SO_2 yield similar values.

Now we might explore how these effects relate to other factors. The total population in the domains of Weihe and Shijingshan are 16.03 million and 12.98 million, respectively. Taken alone, it is hard to say how Weihe's higher population might favor its *iF* for TSP over SO_2, relative to Shijingshan's; however, Weihe is located in an area with roughly equal distance to three large- or medium-size cities—Xi'an, Xianyang, and Weinan. That distance, moreover, happens to be roughly equal to the distance at which ground-level concentrations of TSP from Weihe peaks. Therefore, the distribution of Weihe's population and its TSP may be coincidentally concentrated in the same places.

In contrast, Shijingshan is located at the western boundary of a large city, Beijing, and both populated and unpopulated areas exist in close proximity to the source. It seems to lack a population distribution that makes it so vulnerable to TSP exposure, taking into account that TSP from this plant is also more gaslike in behavior. With regard to SO_2, on the other hand, for similarly complex reasons, Shijingshan must differ from Weihe in the match of its population distribution to the more widely dispersed ground-level concentrations of gaseous pollutants.

6.3.3 Intake Fraction of Sulfur Dioxide and Particulates: Sector Regressions

There is a growing need for aggregated environmental information about the entire electric-power sector (and other sectors) for use in national-level policy analysis. As described previously, the *iF* will eventually (in chapters 9 and 10) be coupled with dose-response functions for pollutants and the monetary value of various effects—a version of the so-called impact pathway approach—to yield an estimated marginal cost per ton of pollutant. We focus in this volume on health effects, which dominate the costs of pollution.

Applying the impact pathway approach to each individual power plant to calculate marginal cost and then summing them through a bottom-up aggregation would be an ideal approach to determine sector-level damages. In 2000, however, there were more than 1,000 coal-fired power plants in the entire Chinese power industry. Such analysis and aggregation would be extraordinarily time-consuming. Although this is widely recognized as the most scientifically refined method to calculate marginal cost per ton of pollutants from a sector (Krewitt et al. 1999; Levy et al. 2002b), it is often impractical. For our purposes and in light of the limitations of Chinese data, such an approach is impossible, yet the policy need for estimates of sector incremental damage remains.

Alternatively, we could simply take the geometric mean of the intake fractions from our 160 sample stacks described previously and assume they are representative of the sector. An important limitation of this "multiple-source" approach is that it ties our *iF* to the particular conditions currently existing in the sector and cannot be used to suggest pollution intake accurately in the future, after those conditions have evolved. In part because the power sector is growing and modernizing so swiftly, its average attributes—such as stack heights—are also changing. It will be advantageous to have an approach for determining sector *iF*s that can take account of these transformations.

Our next use of our results is therefore to construct *iF* regression equations for the entire power sector. These equations are based on our sample data on 160 stacks, for each including *iF*s of both SO_2 and TSP, source characteristics, meteorological parameters, and at-risk population within domains of 100 km $\times$ 96 km. We attempt to construct models in a way that maximize the generalizability to different locations and sources while including as many potential variables as possible. This can serve as a policy analysis tool so that decision makers can quickly estimate human exposures to emissions across the sector with reasonable accuracy. Therefore, we avoid not only parameters that are especially plant-specific (such as atmospheric stabilities and wind-profile characteristics), but also variables that could decrease interpretability and difficult-to-obtain sector-level conditions (such as ground roughness near sources). Avoiding such parameters will result in only testing variables that have both a strong theoretical basis for predicting *iF* and explainable meaning in light of limits of data availability for sector analysis. The logical way to construct regression equations is to test all the potential variables available from the plant level and then reduce those that are difficult to explain or have infeasible data requirements at the sector level.

One term that has strong influence on intake fraction is the number of people within the 100 km × 96 km domain. We take total population in the domain as one variable because it can easily be estimated in most settings and because it also has a strong theoretical relationship to *iF*. Although distinguishing near-source and far-source population might increase accuracy, it is difficult to interpret the implication (Levy et al. 2002b).

A variable differentiating population in reference to prevailing directions of wind would surely increase both accuracy and interpretation. At the extreme, assume a wind rose (a representation of wind patterns) indicates a single wind direction at all times. The *iF* would be zero if no population were distributed downwind of the source, even if millions lived in the other districts of the domain. We conducted some preliminary analysis to test this variable but ultimately found it too difficult and time-consuming at this stage of our research to calculate the population in the prevailing directions of wind for each plant. Thus, we use total population as one variable, the data of which are described in appendix B.

The second included term is the effective stack height, which includes both physical stack height and plume rise. The Gaussian air-dispersion equations built into the ISCLT model clearly show that effective stack height would affect the distribution of concentrations as a function of distance from the source (see appendix A). The plume centerline begins at the effective height rather than the physical stack height. Because the effective stack height usually is several times higher than the physical stack height in coal-fired power generation, it strongly effects ground-pollutant concentration. If the plume is emitted free of turbulent zones caused by stack downwash and any building or terrain effects, both source characteristics and meteorological factors will influence the rise. The source characteristics include the velocity and temperature of the effluent at the top of the stack and the diameter of the top. Meteorological factors include an inverse relationship with wind speed, temperature of the air into which the flue gas is emitted, atmospheric stability, and wind-shear variation with elevation above ground (Schnelle and Dey 1999). It has been the usual practice to use the wind speed adjusted from its measured height to the release height (i.e., typically the stack height). We can see that, although many factors are associated with ground concentration, a correlation between concentration and effective stack height is clear: the higher the plume goes, the lower will be the resultant ground-level concentration. Because plume rise is a function of the physical stack height, the inner diameter of the stack, gas exit velocity, exit temperature, ambient temperature, wind speed at exit, and stability, we use physical stack height and plume rise as two distinct variables in regression equations. It is noteworthy that plume rise changes not only between plants, but also within a plant. So we assume that plume rise only refers to conditions of stability class C and D because the statistical results from meteorological data in the fifty-two power plants show that frequency of wind represented by these categories account for more than 50% of the samples. We use the default Briggs formula of the ISCLT model to calculate the plume rise of each stack by using various parameters tabulated in appendix A.

The third term is wind speed, measured at anemometer height. It is both a proxy of atmospheric stability and a critical factor affecting plume rise. When a plume leaves the chimney, it keeps rising by initial momentum because of the stack gas velocity and its temperature. Eventually the plume is bent horizontally by the wind. Thus, mean wind speed is a critical factor in determining plume rise. The faster the wind blows, the larger the area that the source affects. Therefore, we take the mean wind speed at anemometer height of 10 m above ground as a variable in the regression equations.

The fourth term is mixing height. This is defined as the atmospheric height beneath which a pollutant can fully mix vertically. In the case where a stable layer of air exists above mixing height, the line of demarcation between the layers serves as an upper limit in the vertical motion (U.S. EPA 1995). The lower layer is usually well mixed. A pollutant released in the lower layer will spread horizontally and vertically, but the vertical spread will cease when the top of mixed layer is encountered. We should note that, if the effective stack height exceeds the mixing height, the plume is assumed to fully penetrate the elevated inversion and the ground-level concentration is set equal to zero (U.S. EPA 1995). Because effective stack height for

Table 6.2
Log-linear regression equations for sulfur dioxide (SO_2)

Variable	Only POP	POP and H_s	POP, H_s and U	POP, H_s, U and H_{plume}	POP, H_s, U, H_{plume} and H_{mix}
Constant	−29.15	−28.83	−28.60	−28.27	−28.48
Population, POP	1.03 (17.0)	1.10 (25.4)	1.10 (25.6)	1.10 (26.1)	1.10 (26.3)
Stack height (10^3 m), H_s		−8.4 (12.6)	−8.5 (12.8)	−5.6 (4.7)	−5.3 (4.4)
Mean wind speed (m/s), U			−0.11 (1.8)	−0.27 (3.3)	−0.26 (3.6)
Plume rise (10^{-4} m), H_{plume}				−10.5 (3.0)	−11.8 (3.3)
Mixing height (10^{-4} m), H_{mix}					3.3 (1.7)
R^2	0.646	0.823	0.828	0.837	0.840

Note: *t* values listed in parentheses.

some sources under certain stabilities can reach more than 1,000 m, penetrating the mixing height does occur. Because mixing height is directly related to atmospheric stability and class D stability is most common, we use the mixing height under this class as a variable of the regression equations. The mixing height for the sample plants is listed in table A.1.

Other variables such as population distribution differentiated by land-use patterns, surface roughness, and others might affect the accuracy of regression equations. Including these variables would require major data and processing efforts, however, and we expect them to have a minor affect on the accuracy compared to the major parameters already described.

The combination of total population within a domain of 100 km × 96 km, stack height, plume rise, annual mixing height, and wind speed is able to estimate the logarithm of the modeled *iF* with R^2 of 0.840 and 0.838 for SO_2 and TSP, respectively (see the right-most columns in tables 6.2 and 6.3). The *iF*s of both SO_2 and TSP are greater with more surrounding population, with lower stack heights, lower plume rises, higher mixing height in the domain, and lower wind speed. The greater the absolute *t* value, the more significant is the variable. We can clearly see that population is by far the most significant variable, while others have lesser significance. According to the significance of the variables, they can be ranked as follows: population,

Table 6.3
Log-linear regression equations for total suspended particles (TSP)

Variable	Only POP	POP and H_s	POP, H_s and U	POP, H_s, U and H_{plume}	POP, H_s, U, H_{plume} and H_{mix}
Constant	−29.16	−28.88	−28.49	−28.24	−28.52
Population, POP	1.03 (17.6)	1.09 (25.0)	1.10 (26.0)	1.09 (26.1)	1.10 (26.6)
Stack height (10^{-3} m), H_s		−7.7 (11.5)	−7.9 (12.0)	−5.8 (4.9)	−5.4 (4.6)
Mean wind speed (m/s), U			−0.19 (3.3)	−0.28 (3.9)	−0.31 (4.4)
Plume rise (10^{-4} m), H_{plume}				−7.4 (2.1)	−9.2 (2.6)
Mixing height (10^{-4} m), H_{mix}					4.5 (2.3)
R^2	0.662	0.816	0.828	0.833	0.838

Note: *t* values listed in parentheses.

physical stack height, wind speed, plume rise, and mixing height. It is noteworthy that as the coefficient of population approaches 1, the *iF*s of both SO_2 and TSP are proportional to the population. Meanwhile, other variables have log relationships with *iF*. We plot fully modeled versus regression-based *iF*s of SO_2 and TSP for the sample data in figures A.4 and A.5, both illustrating that the regression results are fairly good (each has $R^2 = 0.84$).

Tables 6.2 and 6.3 list coefficients for various functional forms. The right-most column represents the full regression that can serve all purposes, both where one has only partial information and where one has full information. Other columns represent the reduced forms by dropping the variable with minimum *t* value. It clearly shows that population within the domain and physical stack are significant, and excluding wind speed, plume rise, and mixing height either alone, combined, or altogether only slightly decreases predictive power. Regressions with only two variables, population and stack height, predict *iF* nearly as well as the full regression: $R^2 = 0.823$ for SO_2 and 0.816 for TSP. But if only population within the domain is known, the accuracy to predict *iF* will be reduced considerably, as reflected in an $R^2 = 0.646$ for SO_2 and 0.662 for TSP. In addition, the coefficients of the two full regressions (i.e. for SO_2 and TSP) are similar, as are those of the corresponding two reduced-form regressions, though the final *iF*s are slightly lower for SO_2 than for

TSP.[6] This correspondence would be unlikely if we could take into account secondary particle formation, which occurs at longer range than the 50-km limits of our modeled domains. Chapter 7 reports on the project's parallel analysis that includes secondary particles in its assessment of *iF* from power sector emissions.

The full regression is not generalizable and applicable to sectorwide analyses, in light of the difficulty of gaining and interpreting sectorwide averages of wind speed, plume rise, and mixing height. The full regression is applicable to individual power plants when issues of new construction arise or when a portfolio of existing power plants is considered for control measures. In particular, the full regression can help estimate the health damage of new and innovative power projects, a critical requirement of Chinese regulations relating to environmental impact assessments but not yet fully implemented because analytical tools are lacking.

We should note, however, that applying the full regression of TSP to some individual sources where coarse particles form a large proportion of emissions will introduce inaccuracies. This is because the full regression excludes particle-size distribution, which could be a theoretically significant variable. For instance, applying the full regression to estimation of health damage from the same source but with different particle collectors will give the same result. In reality, whether a multicyclone collector or electrostatic precipitator is installed, for example, will strongly affect the nature of TSP emissions and thus the true *iF*.

The reduced-form regressions can serve different purposes, largely depending on the availability of data. For national policy questions that might affect large groups of sources—such as the ongoing closure of small power plants across the entire country—it may be infeasible to collect all the data necessary to estimate health benefits of such a closure using the full-form regression. The reduced regression including just population and stack height, however, can provide quite good accuracy for such a policy analysis, as reflected in the R^2 values for both SO_2 and TSP. We plot fully modeled versus regression-based *iF*s of SO_2 and TSP of our sample plants in figures 6.8 and 6.9. As anticipated, both figures show that the regression results are fairly good. Hence, we believe that this reduced regression provides a reasonable prediction of the *iF*s of SO_2 and TSP at a sector level for electric power.

To calculate the average *iF* of SO_2 and TSP of the power sector, one must know the mean population and average physical stack height in the industry (see the reduced regression with inclusion of population and stack height in tables 6.2 and 6.3). As explained previously, our national power database consists of 686 stacks distributed across 278 coal-fired power plants. From this database, we can calculate

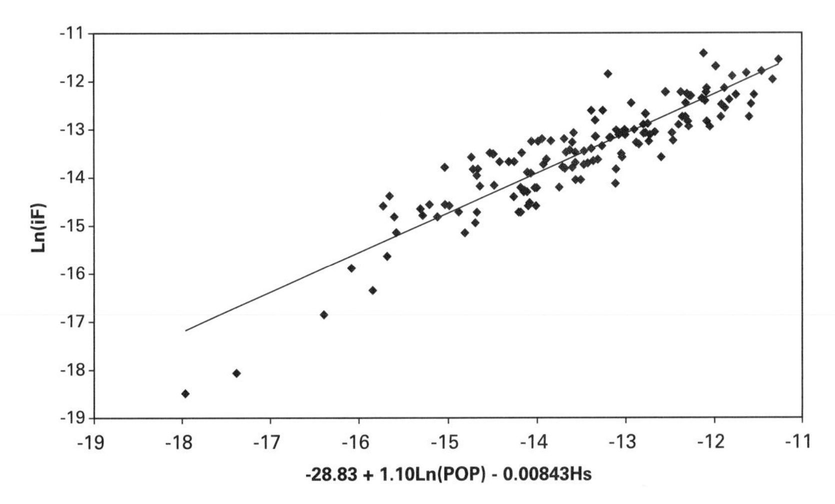

Figure 6.8
Modeled intake fraction of sulfur dioxide versus regression value. Note: $R^2 = 0.82$.

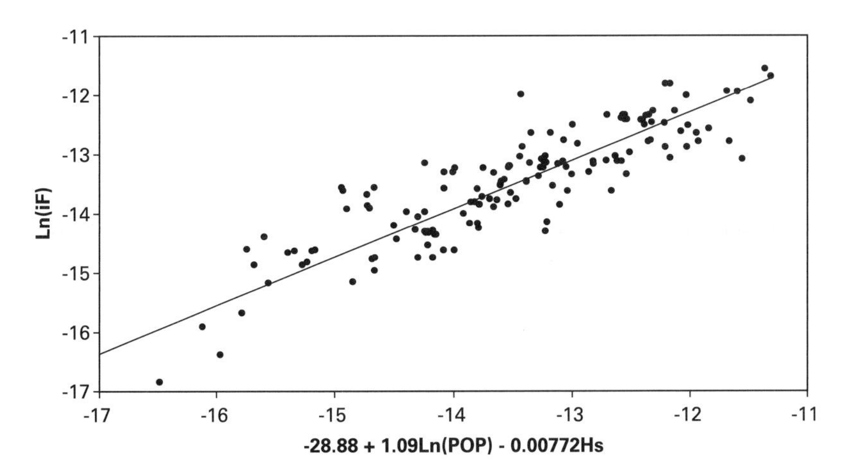

Figure 6.9
Modeled intake fraction of total suspended particulates versus regression value. Note: $R^2 = 0.81$.

the geometric mean stack height of 686 stacks in 2000 to be 153 m. We take this as the mean stack height in the power industry in the year 2000. We also locate all 278 power plants in the digitized Chinese national map and apply the same approach as we did for the 52 sample plants to get an average domain population. That average was 6.02 million in 1999.

With these two figures, we can directly estimate the *iF*s of SO_2 and TSP in China's electric power industry through the reduced regression equations in tables 6.2 and 6.3. The regression-based power-sector *iF*s of SO_2 and TSP are 2.38×10^{-6} and 2.16×10^{-6}, accordingly. To help interpret these estimates, we have averaged sector characteristics and estimated that, for each ton emitted, 2.38 g of SO_2 and 2.16 g of TSP are inhaled within the domain of each power plant.

By contrast, the geometric mean *iF*s of SO_2 and TSP on the basis of our 160 sample sources are 2.77×10^{-6} and 2.61×10^{-6}. The two results are pretty close, which implies that our sample appears reasonably representative of the entire power sector. Also these results suggest that regression-based and multiple-sources-based aggregations are comparable for our immediate purposes. It is important to note that the 160 sample sources represent only current conditions. If we want to know the average *iF* at some time in the future, it would be ideal to redo the entire procedure, including somehow reselecting a sample that represents average source characteristics at that time. Because this sampling would be assumption-laden and the repeated dispersion modeling time-consuming, we can alternatively apply the current reduced regression equations to predicted average attributes (population and stack height) to yield a reasonable projected *iF*.

Because of the technical limitations of the ISCLT model such as a recommended domain of only 50 km, our results can be more precisely described as "*iF*s of primary pollutants in the local range." To have a sense of *iF* of pollutants over a larger range, the parallel study in chapter 7 uses the CALPUFF air-dispersion model instead of ISCLT and shows the *iF* of primary fine particles from Beijing Shijingshan power plant is 1.5×10^{-5} within a domain of 3,360 km × 3,360 km, 8.4×10^{-6} for primary SO_2, 6.0×10^{-6} or sulfate, 6.5×10^{-6} for nitrate (see chapter 7 and Zhou et al. 2002). Hence, if one accepts the assumptions of that model, our estimation using ISCLT does not fully assess the impact of the electric power industry, by underestimating the total exposures and impacts. CALPUFF is a much more data-intensive and time-consuming model, thus limiting the number of sources that can be modeled. The ISCLT-based results of this chapter, though limited in respect to domain size and secondary pollution effects, have the advantage of yielding equations based on far more source data. This may improve the accuracy of the results for this study's purposes. The ISCLT-based results may also prove particularly useful for local-level health risk assessment and policy implementation.

6.4 Conclusion: Summary, Future Research, and Applications to Policy

We apply a new and useful concept—*iF* of pollutants—to translate emissions directly into human exposures in the electric-power sector of China. We employ the ISCLT air-dispersion model and a population database in GIS ArcView to calculate *iF*s of SO_2 and TSP of 160 stacks at 52 sample power plants. These were selected randomly from those for which necessary data were available. From this, we conclude the following:

- This study demonstrates the feasibility and applicability of calculating *iF*s for a sample of Chinese power plants.
- The *iF*s are strongly dependent on site-specific conditions. The geometric mean *iF* of SO_2 is 2.38×10^{-6}, while it is 2.16×10^{-6} for TSP. The *iF*s of SO_2 for individual stacks of our sample range from 1.80×10^{-8} to 1.28×10^{-5}, a span of three orders of magnitude. Similarly, *iF*s of TSP range from 1.6×10^{-8} to 1.23×10^{-5}, also a difference of three orders of magnitude. These figures are best defined as the *iF*s of primary pollutants from the power industry in the local range. Secondary pollutants are not taken into account at this stage of research.
- Regression-based equations to estimate *iF*s of SO_2 and TSP provide a tool to predict exposures to emissions from plants that are not directly modeled. Both full- and reduced-regression equations require limited and easy-to-obtain data to estimate *iF*s quickly. Even the reduced-regression equation that includes just population and stack height can estimate intake fraction of SO_2 and TSP for the entire power sector with reasonable accuracy. Application of such sectorwide regression equations may be the most practical quick means to estimate human exposures given current data constraints.
- Population, source characteristics, stack age distribution, and meteorological parameters are key determining factors of the level of *iF*. *iF*s are significantly proportional to population exposed to the source. Physical stack height, wind speed, plume rise, and mixing height are secondary variables to estimate *iF*.

For future research and policy applications, we note that the Chinese central government has a number of policies in place and under development designed to limit emissions of the electric power sector. A major immediate policy of the government is to shut down small generating units, to create an electric power source structure more compatible with the long-run developmental needs of the national economy, in terms of both efficiency and environmental protection. At the same time, local governments have an interest in keeping these small units operational to the ends of their technical lifespans, to protect prior investments, generate local revenue, and provide electricity and jobs to local residents. Fortunately, there is a common objective for both central and local governments: to minimize the negative impacts of power plants on the health of citizens. Thus, quantitative analysis of the benefits and costs of small generation units proposed for closure would provide helpful information to further action.

In principle, one could apply air-dispersion models to each of thousands of small units to estimate the health damage caused by the pollutants emitted and add them up. At the same time, a parallel assessment would need to be taken of the health damage from emissions of large units that would generate the electricity substituting for that lost from the closed small units. Obviously, this procedure would require enormous efforts and would be essentially infeasible within a reasonable timeline of practical policy action. The studies of *iF*s described above and elsewhere in this volume indicate that useful quantitative assessments of the health implications of the policy of small plant closures may be technically feasible, simple, and fast.

China's central government is also implementing an ambitious program to significantly reduce sulfur emissions from the power sector. In 2004, there were about 20 GW of generating capacity with flue-gas desulfurization (FGD) systems installed, reducing emissions of SO_2 emissions by about 1 million tons. By 2005, about 60 GW with FGD were to be in use, and by 2010, the government plans the total capacity with FGD to be about 300 GW, at a total investment cost of more than 100 billion yuan. Applying *iF* methods to the program would give policy makers a quantitative sense of avoided health damages.

At the same time, we must recognize that the government faces swiftly escalating demand for electricity because of economic growth, with more than 50-GW power capacity installed nationwide in recent years. This in turn is affecting the development of energy resources for the power-generating system, including the quality of coal provided to energy markets. In 2004, the coal quality is known to have significantly deteriorated, leading to a dramatically increase of SO_2 emissions. It is valuable to know the health damage of such deterioration.

These cross-cutting factors of new control policies, expanding demand, and the nature of energy supply will continue to shape air pollution from the power sector and the associated health damages. Quantitative assessment tools are thus critically needed, both to inform the public of the impacts of these developments and aid the government in the design of efficient and effective policy responses.

Acknowledgments

The research of this chapter was funded by a generous grant to the China Project of the Harvard University Center for the Environment and Tsinghua University Institute of Environmental Science and Engineering, from the China Sustainable Energy Program of the Energy Foundation. It was supplemented by grants to the Harvard China Project from the V. Kann Rasmussen Foundation and from the Henry Luce Foundation for Harvard-China research exchange.

Notes

1. To avoid the confusion between *iF* and health damage, we reiterate here that health damage is a function of the *iF* (see equation [9.11] in chapter 9). The total health damages also depend on the dose-response coefficient and the level of total emissions.

2. The data of more than 120 power plants come from environmental impact assessment (EIA) reports conducted on individual power plants. Each EIA records information on not only preexisting generation units and stacks, but also ones planned at the time of the EIA and subsequently constructed. Data for some preexisting stacks, such as pollutant emissions, removal efficiency of particle collectors, flue gas temperature and volume, are physical measurements; all data for newly planned stacks are based on well-accepted design formulas described in notes below. We should note that these EIA reports were completed in different years ranging from the mid-1980s to 2002. We extrapolate certain data—such as fuel consumption and pollutant emissions as well as flue gas volumes—from reporting years to 2000 according to heat rates. Other data are assumed to be unchanged, such as gross coal rate, stack height and inner diameter, as well as type of particle collectors. In a few plants, information on new stacks added or old stacks removed after the EIA report was completed is estimated on the basis of known unit additions or retirements in the plants. For example, the EIA report for Wuhu Power Plant (phase III) was completed in 1987. This report recorded that there was one 80-meter stack linking six units (2×0.6 MW $+ 4 \times 1.2$ MW) in 1960, and that a second stack 150 m high linking two units (2×125 MW) would be erected in 1991. We input all information for the two stacks in the database; however, there were two additional units each of 125 MW added in the period 1996–2000 (*CEPY* 1999). In this case we assume that all information for the units linked to the third stack is identical with the second capacity expansion. These EIA reports are available at individual plants, institutes that have been licensed to undertake such EIAs, and the China Evaluation Center on Environmental Engineering Projects, under the State Environmental Protection Administration (SEPA) of China.

Information for the other roughly 150 power plants comes from mixed reference sources. Ideally, the EIA reports are the best source of information. Collecting them, however, requires agreement of multiple authorities, a time-consuming process that was practically infeasible for every plant in this study. In the case of these latter 150 plants, different reference sources provide specific information for corresponding columns of the database. Certain information for six power plants in Beijing, twenty plants in Shandong province, and seventeen plants in

Hunan province directly derive from project-based sources (Beijing University et al. 2003; Eliasson and Lee 2003; Li and Hao, 2003).

We report the sources of data column-by-column below:

Column 1: *Geographical locations* of power plants in longitude and latitude come from three sources. The first is the more than 120 EIA reports for individual power plants mentioned above. The second is the large-point-source (145 power plants) database of China for RAINS-Asia (IIASA 2002). The third is a power plant database (about 200 power plants) specifically designed for the National Eighth Five-Year-Plan Key Project (MOST 1997). Where there are multiple stacks erected in various years at one plant, we assume that the geographical location of these stacks is identical.

Column 2: *Year in use* refers to the date of initial full-scale, commercial operation. In general, if multiple flues from various generators merge into one stack, we consider the year when the last linked unit came into operation as the year in use for the stack.

Column 3: *Unit capacity* linked to one individual stack comes from the 120 EIA reports and three additional sources (*CEPY* 1998; MOST 1997; MOP 1996). The EIA reports provide unit capacity of each stack. For the other plants, the three sources, respectively, provide plant-based parameters (MOP 1996), unit-based information (*CEPY* 1998), and incomplete stack-based coefficients (MOST 1997). The combination of three sources generates unit capacity for given stacks. Because the three sources provide information prior to 1996, we extrapolate such information to the year of 2000 by counting new additions within the period 1996–2000 at each plant. The new unit additions are available in yearbooks (*CEPY* 1997–2002). For example, the total capacity of Tangshan Power Plant reached 1700 MW in 1995 (MOP 1996) and a new 300-MW unit was added in 2000, whereas six 25-MW units were retired (*CEPY* 2001). This left a total of 1,850 MW at the end of 2000. To attribute stacks to generating capacity, we assume various units shared the various stacks as follows: the first stack served six units built in the 1960s (6 × 25 MW), the second served two built in 1977 (2 × 125 MW), the third served two built in 1978 (2 × 250 MW), the fourth served two built in 1984 (2 × 200 MW), the fifth served two built in 1987 (2 × 200 MW), and the sixth stack served the last unit built in 2000 (1 × 300 MW). Both the unit capacity (*CEPY* 1998) and stack information (MOST 1997) contribute to the breakdown of the plant capacity in those cases where an EIA report is not available.

Column 4: *Stack capacity* is the sum of unit capacity linked to an individual stack (3).

Column 5: *Gross coal rate*, also called the heat rate, is the fuel consumed in gce necessary to generate 1 net kWh of electric energy under full-load conditions. We assume this parameter would not vary annually because of the dominance of most new units coming into operation nationwide since the 1980s. We are aware that generating units used beyond their twentieth year of operation suffer drops in efficiency, reliability, and capacity (Tavroulareas and Charpentier 1995).

Column 6: *The latent heat value (LHV)*, is heat content reported as the net or lower heating value in kilojoules (kj)/kg of coal burned. We derive it directly from the EIA reports and a national coal database (BRICC 1994). We assume that it is constant for each plant and the same for various units unless specified in EIA reports.

Column 7: *The annual operation hours* (or hours in service during 2000) of various generating units are 1996-recorded data (MOP 1996; *CEPY* 1998), adjusted to 2000 according to the ratio of annual hours of thermal units in 1996 and 2000 at provincial level (*CEPY* 2000). See the example in (8).

Column 8: *Coal burned* is the summed amount of fuel consumed by various generation units linked to one individual stack in the year 2000. It varies with the operation load. Coal burned in each generation unit is equal to the product of unit capacity, gross coal rate, annual generation hours (or capacity factor), and a conversion factor from standardized to physical dimensions (based on the fuel heat value). We adjust annual generation hours of various units in different years to 2000 on the basis of provincial ratios. For example, how much coal was burned by the two units linked to the second stack in Tanshan Power Plant General in 2000, we take a three-step procedure. First, the 1996 data (*CEPY* 1998) for annual generation hours for units 1 and 2 in 1996 were 6400 and 5594 hours, and the gross coal rates were 347.9 and 241.9 gce/kWh (*CEPY* 1998), respectively. The standardized coal burned in 1996 was 125 MW $\times$ 6400 hours $\times$ 347.9 gce/kWh = 278.3 ktce for unit 1 and, analogously, 241.9 ktce for unit 2. The total standardized coal consumption for the stack was thus 520.2 ktce in 1996. Second, because the annual generation hours of thermal power in Hebei province in 1996 and 2000 were 6,586 and 5,562 hours, respectively (MOP 1996; *CEPY* 2000), we extrapolate standardized coal consumption for the two units in 1996–2000 as 520.2 ktce $\times$ 5562 hours/6586 hours = 439.3 ktce. Third, because the heat value of coal burned in the plant was 19,534 kj/kg, we convert the standardized coal consumption to physical amounts as 439.3 ktce $\times$ (29,260 kj/kg)/(19,534 kj/kg) = 658 kilotons. This gives us that total quantity of coal burned in 2000 for this stack. The actual type of coal burned—anthracite, bituminous, lignite, sub-bituminous, and so forth—determines the actual emissions. In light of the lack of detailed data, however, we have to ignore this.

Column 9: *Sulfur content* in coal, percent by weight, has the same assumptions as (6). The sulfur content in coal consumed in plants with EIA reports is taken on an "as burned" basis, whereas others are obtained from the assumed fuel supplier on an "as received" basis (BRICC 1994).

Column 10: *Sulfur dioxide emission*, ton emitted per hour, is calculated through a well-accepted formula, $EM(SO_2) = 2 \times 85\% \times S\% \times FUEL \times (1 - \eta/100)$. S, sulfur content, is the numeric value in (9); *FUEL*, the amount burned hourly, is the quotient of coal burned annually in (8) divided by the annual operation hours in (7); and η, percent sulfur removal efficiency by weight, is set to zero unless specified. 85% is the statistical conversion ratio of sulfur in coal to gaseous sulfur dioxide (i.e., 85% of sulfur in coal chemically converts to SO_2). The remainder collects either in stack ash or at the bottom of the boiler.

Column 11: *Ash content in coal* by weight has similar assumptions with LHV in (6) and sulfur content in (9).

Column 12: *Total particulate matter emissions*, tons emitted from the stack per hour, is the sum of emissions from various units linked to that stack. For each preexisting stack in the plants with EIA reports, we directly derive particle emissions for corresponding stacks from the reports regardless of load conditions. For planned stacks in plants with EIA reports and all stacks in those plants without EIAs, we estimate the particle emissions of each generating unit and aggregate emission of various units to the stack emissions on the basis of emitted particle concentration and flue volume calculated in (18). We assume that emitted particle concentrations of each unit are equal to the required emission standard based on the expectation that they were in compliance by end of 2000 (Zhang, Zhan, and Chen 2002). As such a standard is a function of coal ash, plant location, unit size, operation date of unit, particle collector type, and firing type (SEPA 1997), we assume that there is no switch of coal variety, upgrade of particle collector, or innovation of firing technology after the unit is put into operation.

Column 13: *Concentration of total particles*, milligrams per cubic meter of stack gas, is from the EIA reports or calculated through national emission standards of air pollutants for thermal power plants.

Column 14: *PM_{10} emissions* are calculated by using equations and coefficients described in appendix A. As all coal-fired units added during the 1990s were required to be equipped with electrostatic precipitators (Zhang, Zhan, and Chen 2002), we assign the same particle-size distribution for these units. In reality, small-sized, coal-fired units equipped with Venturi scrubber and cyclone remover in far areas to city, and many units built before 1990 were retrofitted to ESP.

Column 15: *Nitrogen oxide emissions*, tons emitted from the stack per hour, is the product of coal burned hourly [8] multiplied by the NO_X emission factor—0.00995 ton/ton of coal burned (Hao, Tian, and Lu 2002). Because almost all of the units with generating capacity of 300 MW and above have been equipped with one of several types of low-NO_X burners (Zhang, Zhan and Chen 2002), we assume these generating units comply with national emission standards—650 mg/m^3 for dry bottom boilers and 1,000 mg/m^3 for wet bottom boilers—to estimate the NO_X emission for these units by multiplying with the flue gas volume (SEPA 1997).

Column 16: *Stack height (H)* and *inner diameter (D) at top of the stack* come from the EIA reports or the MOST power plant database. If neither is available, we assume the two parameters are equivalent to stacks of a similar unit size and operation year.

Column 17: *Flue temperature (T) at exit of stack* is calculated by using the data on gas temperature at boiler exit (*CEPY* 1998), and assuming a 5°C drop per 100 meters along the stack (SEPA 1997).

Column 18: *Flue gas volume (V)*, m^3/s, is calculated by the following equation except for those stacks measured and reported in EIA reports: $V = FUEL \times (1 - q_4/100) \times [(Q_{LHV}/4704 + 1.65) + 1.0161(\alpha - 1)V^0]$, in which $V^0 = Q_{LHV}/4145 + 0.5$; Q_{LHV} is the latent heat value of coal in kj/kg coal; q_4 is the heat loss rate by coal leakage and $= 0.9\%$; $FUEL$ is the coal amount burned in kg/s; α is the air excess coefficient $= 1.5$; V^0 is the theoretically calculated required volume of gas flue to burn 1 kg of coal in m^3/kg.

3. Air-dispersion models use both Cartesian and polar grid systems. ISCLT, for example, can generate a domain of 96 km × 100 km with a resolution of 2 km × 2 km, or a polar domain of concentric gridded circles within a radius of 50 km. ISCLT results have been verified to be unbiased within a radius of 50 km, our 96 km × 100 km rectangular domain includes some areas beyond the recommended 50-km range. For ease of calculation we have followed other studies is using a rectangular domain because this is a trivial bias compared to other sources of uncertainty.

4. There are five stacks, each 120 m high, linked to a 300-MW power generator equipped with an electrostatic precipitator. For the five stacks, SO_2 emissions are 0.968, 2.023, 1.358, 1.358 and 1.694 ton/hour, respectively. Intake fractions of SO_2 from the stacks are 3.4×10^{-6}, 3.05×10^{-6}, 4.35×10^{-6}, 4.35×10^{-6} and 3.57×10^{-6} for the same domain, respectively.

5. The Zhengzhou cogeneration plant is located at 113°45′ longitude and 34°45′ latitude. There are five stacks with physical height ranging from 100 to 210 meters. The plume rise height for each stack is calculated assuming an atmospheric stability class D (neutral) because D is the dominant one in Zhengzhou.

6. ISCLT results are only unbiased within 50 km, hence the size of our domains. Within this distance, particles emitted from typical particle collectors are dominated by those of aerodynamic diameter of less than 10 μm, which behave like gaseous SO_2. This is why the coefficients of the regressions are very similar. The coefficients are not exactly the same, however, and *iF* of TSP is somewhat higher most likely because of the coarse fraction of TSP from low-efficiency collectors. These incur dry deposition with 50 km. We should also point out that the model assumes that chemical formation of secondary particles from gaseous SO_2—which would increase *iF* for SO_2—does not occur substantially within this distance.

References

Abt Associates Inc., ICF Consulting, and E. H. Pechan Associates, Inc. 2000. The particulate related health benefits of reducing power plant emissions. Report for the Clean Air Task Force. Available at http://www.cleartheair.org/fact/mortality/mortalityabt.pdf.

Beijing Research Institute of Coal Chemistry (BRICC), China Institute of Coal Science. 1994. National coal database: new functions of data storage by computer. Project no. 91–203. In Chinese.

Beijing University, Tsinghua University, Beijing Research Institute on Environment, and Beijing Research Institute on Labor Protection. 2003. Study report on comprehensive strategies to control Beijing air pollution. In Chinese.

Bennett, D. H., T. E. McKone, J. S. Evans, W. M. Nazaroff, M. D. Margni, O. Jolliet, and K. R. Smith. 2002. Defining intake fraction. *Environmental Science & Technology* 36 (9):206A–211A.

China electric power yearbook (CEPY). 1997–2004. Beijing: China Statistics Press. In Chinese.

China Environment Yearbook (CEY). 2001–2002. Beijing: China Environment Press. In Chinese.

Crapanzano, G., L. D. Furia, M. Pavan, S. Ascari, M. Fontana, A. Lorenzoni, and F. Maugliani. 1997. ExternE National Implementation: Italy. Available at http://externe.jrc.es/it.pdf.

Dorland, C., H. M. A. Jansen, R. S. J. Tol, and D. Dodd. 1997. ExternE National Implementation: The Netherlands. Available at http://externe.jrc.es/nl.pdf.

Eliasson, B., and Y. Y. Lee, eds. 2003. *Integrated assessment of sustainable energy systems in China: A framework for decision support in the electric sector of Shandong province*. Dordecht: Kluwer Academic Publishers.

European Commission (EC). 1997. EC DGXII, Science, Research and Development, JOULE. Aggregation of ExternE Results. ExternE Core Project report. EC, Brussels.

Evans J. S., S. K. Wolff, K. Phonboon, J. I. Levy, and K. R. Smith. 2002. Exposure efficiency: An idea whose time has come? *Chemosphere* 49 (9):1075–1091.

Finamore, B. A., and T. M. Szymanski. 2002. Taming the dragon heads: Controlling air emissions from power plants in China: An analysis of China's air pollution policy and regulatory framework. *Environmental Law Reporter* 32 ELR:11439–11458.

Hao, J. M., H. Z. Tian, and Y. Q. Lu. 2002. Emission inventories of NO_X from commercial energy consumption in China, 1995–1998. *Environmental Science and Technology* 36 (4):552–560.

International Institute for Applied System Analysis (IIASA). 2002. The "Large Point Source" database of RAINS-ASIA model. Available at http://www.iiasa.ac.at.

Krewitt, W., T. Heck, A. Trukenmuller, and R. Friedrich. 1999. Environmental damage costs from fossil electricity generation in Germany and Europe. *Energy Policy* 27:173–183.

Levy, J. I, J. K. Hammitt, Y. Yanagisawa, and J. D. Spengler. 1999. Development of a new damage function model for power plants: Methodology and applications. *Environmental Science and Technology* 33:4364–4372.

Levy, J. I, J. D. Spengler, D. Hlinka, D. Sullivan, and D. Moon. 2002. Using CALPUFF to evaluate the impacts of power plant emissions in Illinois: model sensitivity and implications. *Atmospheric Environment* 36:1063–1075.

Levy, J. I., S. K. Wolff, and J. S. Evans. 2002a. Modeling the benefits of power plant emission controls in Massachusetts. *Journal of the Air and Waste Management Association* 52:5–18.

Levy, J. I., S. K. Wolff, and J. S. Evans. 2002b. A regression-based approach for estimating primary and secondary particulate matter intake fraction. *Risk Analysis* 22(5):893–902.

Li, J., and J. M. Hao. 2003. Application of intake fraction to population exposure estimates in Hunan province of China. *Journal of Environmental Science and Health, Part A* 38 (6):1041–1054.

Ministry of Power (MOP). 1996. *Statistics of power industry in 1995 (No. 60).* Beijing: China Power Industry Press. In Chinese.

Ministry of Power (MOP). 1998. *Development and environmental protection of China's electric power industry.* Beijing: China Power Industry Press. In Chinese.

Ministry of Science and Technology (MOST). 1997. Study report on China's acidic deposits and its impact on ecological environment. Project no. 85–912–01. Beijing. In Chinese.

National Bureau of Statistics of China (NBS). 2001. *China statistical yearbook 2001.* Beijing: China Statistics Press. In Chinese.

Schnell, K. B., and P. R. Dey. 1999. *Atmospheric dispersion modeling compliance guide.* New York: McGraw-Hill.

State Environmental Protection Administration (SEPA). 1997. Emission standard of air pollutants for thermal power plants. Project no. GB 13223–1996. Beijing: China Environmental Science Press. In Chinese.

State Power Corporation of China (SPC). 1996. *Statistics of power industry in 1995 (No. 60).* Beijing: China Power Industry Press. In Chinese.

State Power Corporation of China (SPC). 2003. Historical capacity and generation mix. Available at www.sp.com.cn/zgdl/dltj/d0104.htm. In Chinese.

Tavoulareas, E. S., and J. P. Charpentier. 1995. Clean coal technologies for developing countries. World Bank technical paper no. 286, energy series. Washington, D.C.: World Bank.

Thanh, B. D., and T. Lefevre. 2000. Assessing health impacts of air pollution from electricity generation: The case of Thailand. *Environmental Impact Assessment Review* 20:137–158.

United Nations Environmental Programme (UNEP) and the National Environmental Protection Agency of China (NEPA). 1996. Incorporation of environmental considerations in the energy planning in the People's Republic of China. Beijing: Chinese Environmental Science Press. Available at http://www.seib.org/leap/Reports/China%20UNEP+NEPA.pdf.

United States Environmental Protection Agency (U.S. EPA). 1995. User's guide for the industrial source complex (ISC3) dispersion models. Volume II. Description of model algorithms. Report no. EPA-454/B-95-003b. Research Triangle Park, N.C.

Zhang, X. L., Z. H. Zhan, and J. L. Chen. 2002. Environmental protection and sustainable development in power industry during the 9th five-year plan. In *China electric power yearbook (CEPY)* 2002. Beijing: China Power Industry Press. In Chinese.

Zhou, Fengqi, and Dadi Zhou. 1999. Study on long term energy development strategies of China. Beijing: China Planning Press. In Chinese.

Zhou, Y., J. I. Levy, J. K. Hammitt, and J. S. Evans. 2002. Estimating population exposure to power plant emissions using CALPUFF: A case study in Beijing, China. *Atmospheric Environment* 37:815–826.

7

Population Exposure to Power Plant Emissions Using CALPUFF

Ying Zhou, Jonathan I. Levy, James K. Hammitt, and John S. Evans

7.1 Introduction

Epidemiological studies have shown a significant association between both short- and long-term ambient particulate matter (PM) exposures and increased mortality and morbidity risk (Dockery et al. 1993; Samet et al. 2000; Pope et al. 2002). Examples of the former include death due to heart and lung disease; examples of the latter include cardiovascular disease, pneumonia, and chronic obstructive pulmonary disease.

Power plants are significant emitters of sulfur dioxide (SO_2) and nitrogen oxides (NO_X), which are harmful at high concentrations and contribute to the formation of atmospheric fine particulates. An estimated 11 million tons of SO_2 were emitted from coal-fired power plants in the United States in 2003, which accounted for about 69% of the total estimated SO_2 emissions in the country (U.S. EPA 2005). In China, an estimated 7 million tons of SO_2 were emitted from coal-fired power plants in 1996, which accounted for 35% of SO_2 emissions from all industrial sectors (MOP 1998).

Because the power sector is such an important source of pollution, we focus on it in this chapter. We use the intake fraction (*iF*) methodology described in chapter 4. Although several studies have estimated the *iF*s for power-plant emissions in the U.S. (Smith 1993; Lai et al. 2000; Wolff 2000; Evans et al. 2002; Levy et al. 2002), they could well differ in developing country settings, because of differences in stack characteristics, meteorology, and proximity of power plants to the centers of population. Furthermore, currently available *iF* estimates from developing countries (Phonboon 1996; Wang and Smith 1999; Kandlikar and Ramachandran 2000) and those derived in chapters 5 and 6 use simplified models (e.g., Gaussian plume models), which may not capture the full impact of a pollution source.

The research reported in this chapter was developed in part to evaluate the feasibility and tradeoffs of using a more sophisticated but time-consuming dispersion model than that employed in chapters 5 and 6, ISCLT (Industrial Source Complex—Long Term). The aims of this study are thus threefold: First, to demonstrate how the *iF* for emissions from one power plant in China can be calculated using a detailed, long-range transport model (CALPUFF) through a case study; second, to test how sensitive the results are to key assumptions within the model and compare the results to the U.S. estimates; third, to explore the plant-to-plant variation in *iF* by selecting power plants throughout China and to estimate typical *iF*s for these plants.

How these results might be used, compared particularly to the more extensive but simpler power sector analysis of chapter 6, is discussed in chapter 9 and appendix D. The work presented in this chapter is drawn from Zhou 2002, much of it has appeared previously in Zhou et al. 2003, and another part of it has been reported in Zhou et al. 2006.

7.2 Methods

7.2.1 Intake Fraction

We refer back to intake fraction (*iF*) as defined and used in chapters 4–6, to specify parameters used in the calculation of *iF* in this chapter:

$$iF = \frac{\sum_{d=1}^{n} POP_d \times C_d \times BR}{EM}.$$

Our modeling domain is divided into 120 × 120 grid cells, *d* indexing these cells. Then POP_d is the population at location *d*, C_d is the concentration at location *d* (grams/cubic meter, g/m^3), *BR* is the breathing rate (m^3/day), and *EM* is the emission rate (grams/second, g/s). The concentration at location d is estimated using CALPUFF. A constant breathing rate of 20 m^3/day is used in the calculations.

7.2.2 Pollutants Modeled

As Jonathan Levy and Susan Greco described in chapter 4, fine particles (those smaller than 2.5 microns, $PM_{2.5}$) pose the greatest risk to health. We remind the reader that atmospheric fine particles can be further divided into primary and secondary ones, a distinction that will be particularly relevant in the analysis of this chapter. Primary fine particles are formed directly during the combustion process, while secondary fine particles are formed in the ambient air by the chemical reactions of gaseous precursors (e.g., SO_2, NO_X) during atmospheric transport.

This study focuses on particulates, as well as their precursor gases, but does not include other power-plant pollutants, such as mercury, carbon dioxide (CO_2), and organics. We calculate *iF*s for primary particles, SO_2, secondary sulfate (defined as ammonium sulfate—$[NH_4]_2SO_4$—inhaled per unit of SO_2 emissions), and secondary nitrate (defined as ammonium nitrate—NH_4NO_3—inhaled per unit of NO_X emissions). In the following discussion, sulfate and nitrate are abbreviated as SO_4 and NO_3, respectively.

7.2.3 Modeling Structure

A detailed long-range transport model is needed to model the atmospheric processes and the resulting pollutant concentrations in the air. CALPUFF is the modeling system recommended by the United States Environmental Protection Agency (U.S. EPA) for simulating long-range transport (Interagency Workgroup on Air Quality Modeling [IWAQM]; U.S. EPA 1998). The CALPUFF modeling system includes three main components: CALMET, CALPUFF, and postprocessing and graphical display programs. CALMET is a meteorological model that develops hourly wind and temperature fields on a three-dimensional gridded modeling domain. Associated two-dimensional fields, such as mixing height, surface characteristics and dispersion properties, are also included in the file produced by CALMET (Scire et al. 2000).

To develop meteorological data for CALMET, we use the PSU/NCAR mesoscale model, as recommended by the IWAQM. The Fifth-Generation NCAR/Penn State Mesoscale Model (MM5) is the latest in a series and we use the current release for the MM5 modeling system, version 3. In addition, Four-Dimensional Data Assimilation (FDDA), a method of running a full-physics model while incorporating observations, is used in our MM5 analysis, as recommended by IWAQM.

CALPUFF is a Lagrangian puff model that simulates continuous puffs of pollutants released from a source into the ambient wind flow. As the wind flow changes from hour to hour and place to place, the path each puff takes changes according to the new wind-flow direction (Scire et al. 2000). Puff diffusion is Gaussian, and concentrations are based on the contributions of each puff as it passes over or near a receptor point. In addition, CALPUFF allows for the estimation of both primary and secondary pollutant concentrations. Figure 7.1 shows the major model components used in this study.

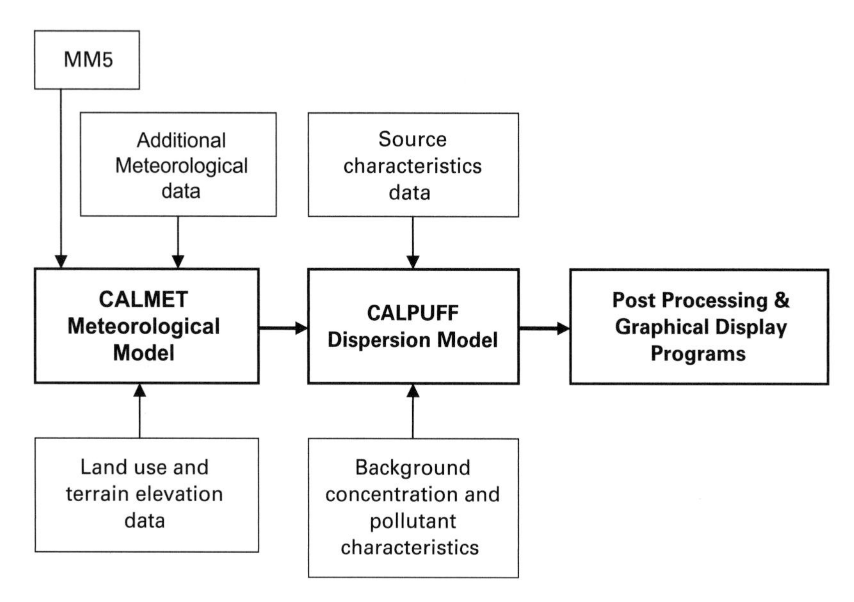

Figure 7.1
Major components of the CALPUFF modeling system.

7.2.4 Domain Coverage

For the case study, we evaluate the aggregate impacts of one power plant in Beijing. The southwest corner of the domain is located at longitude 93°E, latitude 18°N; the northeast corner at longitude 136°E, latitude 45°N. There are 120 receptor grids in both the x and y directions, with a grid spacing of 28 kilometers (km) (figure 7.2). The entire domain, 3360 km × 3360 km, covers most of China's area and all of its heavily populated regions.

7.2.5 Source Characteristics

The power plant chosen for the case study is located in Shijingshan, which is one of Beijing's eight urban districts. The plant has four boilers, each with a capacity of 200 megawatts (MW). There are two stacks at the plant: stack 1 serves boilers 1, 2 and 3; stack 2 serves boiler 4. Table 7.1 summarizes the emission rates, type of coal burned, and stack characteristics. There is no significant difference between the characteristics of the two stacks, except that stack 2 has lower emission rates because it serves only one boiler. Therefore, we focus on stack 1 in this analysis.

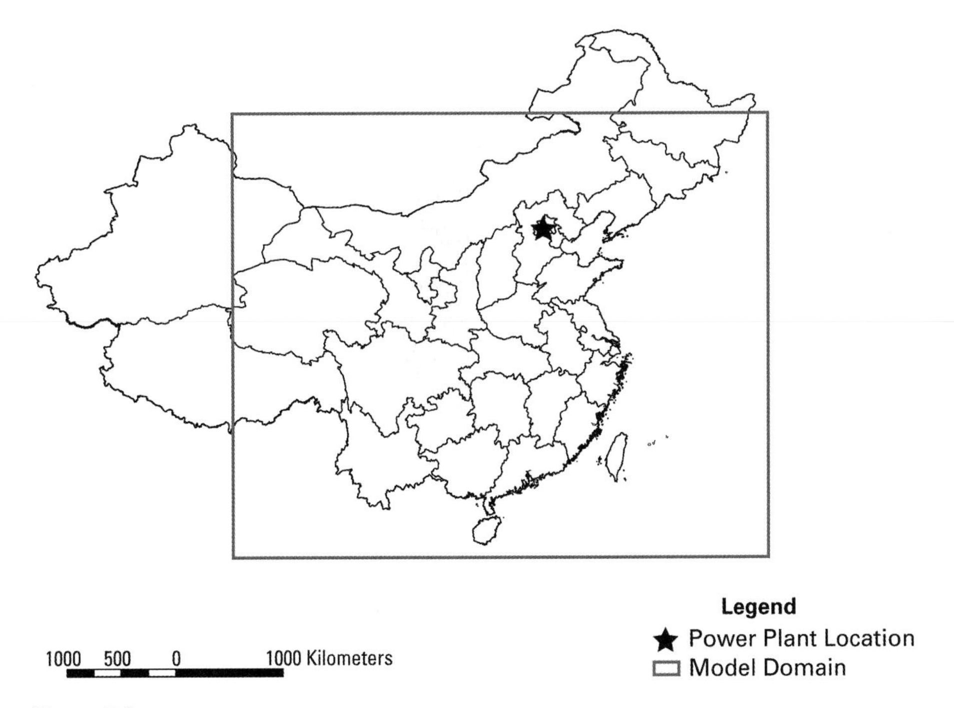

Figure 7.2
Modeling domain and location of the Shijingshan power plant used in the case study.

Table 7.1
Summary information of Shijingshan power plant, Beijing

Variable	Stack 1	Stack 2
Stack height (m)	210	210
Stack diameter (m)	7	4
Exit velocity (m/s)	27.4	28.7
Exit temperature (K)	418	418
SO_2 emission (ton/year)	32,234	10,745
NO_X emission (ton/year)	17,228	5,743
PM_{10} emission (ton/year)	3,520	1,173
$PM_{2.5}$ emission (ton/year)	1,667	556
Annual coal burned (ton)	3 million	
Dust removal equipment	Electrostatic precipitators	
Sulfur content (%)	0.9	
Ash content (%)	22	

7.2.6 Meteorology Model Setting

MM5 To develop a meteorological field for CALPUFF, we use the MM5 model with FDDA as the initial meteorological field to CALMET. The MM5 domain center is at longitude 102°E, latitude 36°N. There are ninety receptor grids in the x direction and seventy-five receptor grids in the y direction, with a grid spacing of 60 km. This domain covers China's entire area. The Lambert conformal map projection with model default settings is used. We use the model default of twenty-three vertical layers.[1]

Input global terrain and land use data have a resolution of 30 minutes, which is approximately 56 km at the equator. The input meteorological data for the MM5 runs are from the National Centers for Environmental Prediction (NCEP) Global Tropospheric Analysis, NCEP ADP Global Surface Observations, and NCEP ADP Global Upper Air Observation Subsets. During the MM5 model run, the program automatically searches for the meteorology observations in the dataset within the defined domain.

CALMET The CALMET domain is the same as the final CALPUFF domain as described in section 7.2.4. The Lambert conformal map projection is used to account for the earth's curvature. The two standard parallel latitudes are set at 20°N and 60°N. The reference longitude and latitude are set at 100°E and 25°N. These quantities are used to adjust observed and prognostic winds to fit the Lambert conformal mapping.

In addition to the meteorological field developed by MM5, twenty-two surface stations, three upper-air stations, and fifty-five precipitation stations are used in the CALMET run. Both the surface and upper-air data are obtained from the National Climatic Data Center (NCDC). The formats of the obtained datasets are in surface airways hourly (TD-3280) and NCDC upper air digital files (TD-6201), respectively.

Because no hourly precipitation data are available for stations in China, daily precipitation data are obtained from the Climate Prediction Center (CPC) global summary of daily/monthly observations. Assuming uniform precipitation intensity within each day, we obtain hourly precipitation values by dividing those for daily precipitation by twenty-four, also used for incorporating wet deposition in CALPUFF runs. CALMET default parameters are used to interpolate wet deposition rates between the observation stations. We also test the sensitivity of the *iF* results if we assume that the daily precipitation all happens within 1 hour of a day, with no precipitation for other hours.

CALMET uses fewer vertical layers than MM5, in part because air pollution modeling does not require detailed information on the upper atmosphere. Ten vertical layers with cell face heights at 0, 20, 40, 80, 160, 320, 640, 1,000, 1,500, 2,200 and 3,000 m are used in the CALMET runs. CALMET automatically interpolates from the MM5 model grid system to the CALMET grid.

7.2.7 CALPUFF Settings

Six species are modeled in CALPUFF: SO_2, SO_4, NO_X, nitric oxide (HNO_3), NO_3, and primary $PM_{2.5}$. The MESOPUFF II chemical transformation method is used in the CALPUFF run. Background ozone (O_3) and ammonia concentrations are set at model default levels of 80 and 10 parts per billion (ppb), respectively.

Both wet and dry deposition are included in the model run. Dry deposition includes gas-phase SO_2, NO_X, and HNO_3, and particle-phase SO_4, NO_3, and primary fine particles. We use the option for full treatment of spatially and temporally varying gas/particle deposition rates predicted by a resistance-deposition model.

7.2.8 Population Information

County-level population data for the year 1999 are used to calculate population exposures, as described in appendix B. An add-in package called "spatial analyst" to Geographical Information System (GIS) software ArcGIS is used to convert the county-level population data to match the concentration data calculated using CALPUFF. The grid resolution of the CALPUFF model domain was described in section 7.2.4.

The dataset contains a total of 2,569 counties, with a total population of 1.2 billion in year 1999. Seventeen counties have missing population data for 1999, and 1990 population data[2] for those counties are used in the calculation.

7.2.9 Dispersion Model Simulation Details

The year 1995 is chosen for the analyses because it was a typical year meteorologically (i.e., not an El Niño or La Niña year) for China (*China Meteorology* 1999). We calculate *iF*s for periods in each of the four seasons, February 14–24, May 14–24, August 15–25, and November 14–24. In each ten-day simulation, the first six days have continuous emissions, whereas the last 4 days have no emissions.

The 10-day run-length for each season is chosen based on consideration of the lifetime of pollutants modeled in the atmosphere, as well as the capacity limit of the computing facilities available. Preliminary calculations show that a small fraction of the pollutants modeled remains in the atmosphere at the end of the 10-day simulation. This means that most of the pollutants have enough time to be removed from the atmosphere in the run period (e.g., because of deposition or chemical transformation). As a result, running CALPUFF for a longer period of time would not alter the estimates substantially.

We also evaluate the dependency of the *iF* estimates on the number of days in this 10-day period when we model emissions, by using November for these sensitivity calculations. The *iF* estimates reflect two tendencies. First, for scenarios in which emissions are limited to only 1 or 2 days, *iF* values are highly variable, because short-term changes in weather conditions are not averaged out to as great an extent as they are in those scenarios in which emissions last for a longer period of time. Second, for scenarios in which emissions occur during most of the 10-day simulation period, the *iF* estimates are lower, because a substantial amount of emitted material does not have a chance to reach people by the end of the simulation run. After considering these two tendencies, we choose to have continuous emissions in the first 6 days in our simulation.

7.2.10 Power Plant Sample

We also conduct a broader analysis than the case study, sampling power plants from across China to ensure geographic heterogeneity and allow for the evaluation of predictors of *iF*s. To select plants for the sample, we use the list of power plants in China provided by the RAINS-Asia model developed by the International Institute for Applied Systems Analysis (IIASA). The RAINS-Asia model has been developed to analyze future trends in emissions, estimate regional impacts of resulting acid deposition levels, and evaluate costs and effectiveness of alternative mitigation options. RAINS-Asia includes a total of 213 large power plants with capacity greater than 500 MW in China (Bertok et al. 2001).

The modeling domain defined in this study covers thirty of the thirty-three administrative units in Mainland China, including twenty-one provinces, three autonomous regions, four municipalities, and two special administrative districts.[3] The power plants listed in RAINS-Asia cover twenty-nine of the thirty administrative units in our domain[4] and there are from one to eighteen large power plants in each of the twenty-nine administrative units. Using this list, we randomly select one plant from each administrative unit to produce a sample of twenty-nine power plants. Figure 7.3 shows the locations of the chosen power plants.

Although RAINS-Asia provides the location and capacity information for 213 power plants, it does not include detailed source characteristics information, such as stack height and emission rates for each plant. To estimate *iF* for the twenty-

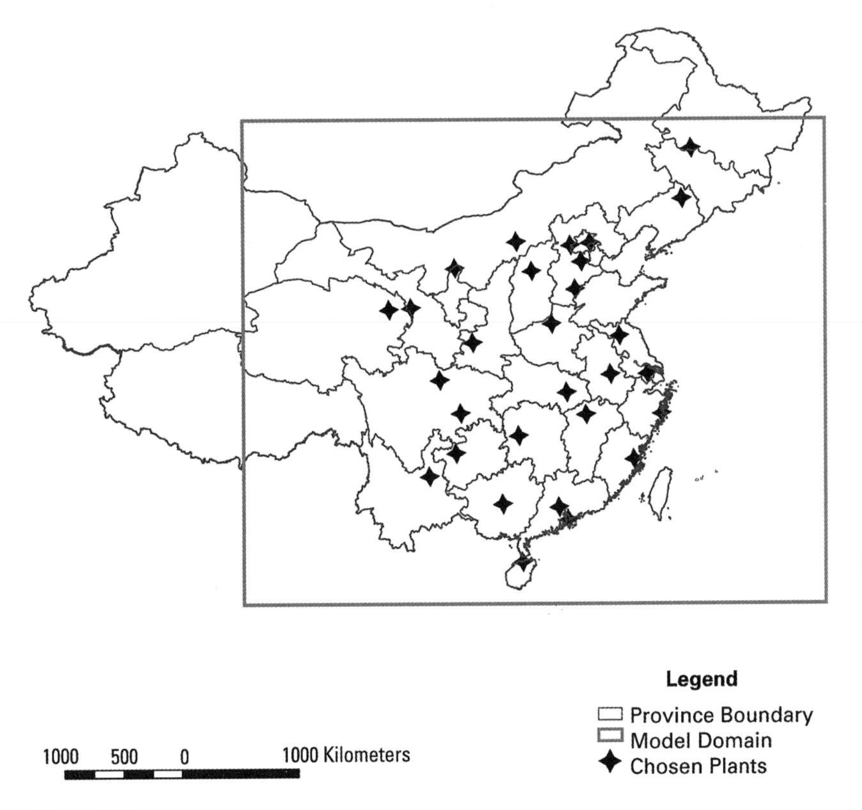

Figure 7.3
Modeling domain and location of the twenty-nine power plants in the national sample.

nine plants in our sample, we used the same source characteristics information as that of the Shijingshan power plant in Beijing (table 7.1), with a capacity of 800 MW.

7.3 Results

7.3.1 Intake Fraction Estimates for Shijingshan Power Plant Case Study

The annual average *iF*s for each pollutant in the case study are calculated by averaging the *iF*s of the four seasons in 1995. The average *iF* is 8.4×10^{-6} for SO_2, 6.0×10^{-6} for SO_4, 6.5×10^{-6} for NO_3, and 1.5×10^{-5} for primary fine particles. This would mean that in the case of primary particles, for example, for every metric ton of particles emitted, the combined effect of dispersion and removal will result in an exposure pattern across the affected population leading to inhalation of 15 g of

Table 7.2
Intake fraction estimates for representative months of different seasons ($\times 10^{-6}$)

	February	May	August	November
SO_2	13	5	8	8
SO_4	11	3	6	4
NO_3	15	2	2	7
$PM_{2.5}$	25	9	13	14

this material during a year. Table 7.2 lists the *iF* values calculated for representative months of each season.

In all four seasons, the *iF* for primary fine particles is the highest, whereas the *iF* for SO_2 is the second highest. Furthermore, the *iF*s for all the pollutants are higher in the colder months (February and November) than in the warmer months (May and August), which is likely because of meteorological conditions. For example, the averages of rainfall observed in the fifty-five precipitation stations, which were used as an input to CALMET, were 27 and 44 mm in the run periods of May and August, respectively, whereas they were 13 and 4 mm in February and November. Moreover, this trend of higher *iF*s in the colder months is particularly strong for NO_3, which is consistent with the finding that higher temperature inhibits the formation of aerosol ammonium nitrate (West et al. 1999).

Figure 7.4 shows how *iF* changes with distance from the power plant in the November run. It shows that it takes about 200 km for SO_2, 300 km for primary fine particles and NO_3, and 500 km for SO_4 to reach half of their overall intake fractions. Furthermore, primary fine particles and SO_2, which are directly emitted, have higher *iF* values than the secondary particles at any distance.

7.3.2 Removal of Pollutants by Different Mechanisms

The pollutants in the air are removed by transformation, wet depletion, dry depletion, or advection (i.e., being blown out of the domain). In figure 7.5, we show the fraction that is removed and the distribution over these removal mechanisms for each pollutant and for representative months of each of the four seasons.

Sulfates are formed through transformation from SO_2 and the removal rates are calculated as a percentage of the mass of total sulfates. In the SO_4 plot in figure 7.5, the entire bar for each season shows the fraction removed. There are three components representing the percentage removed because of advection out of the domain,

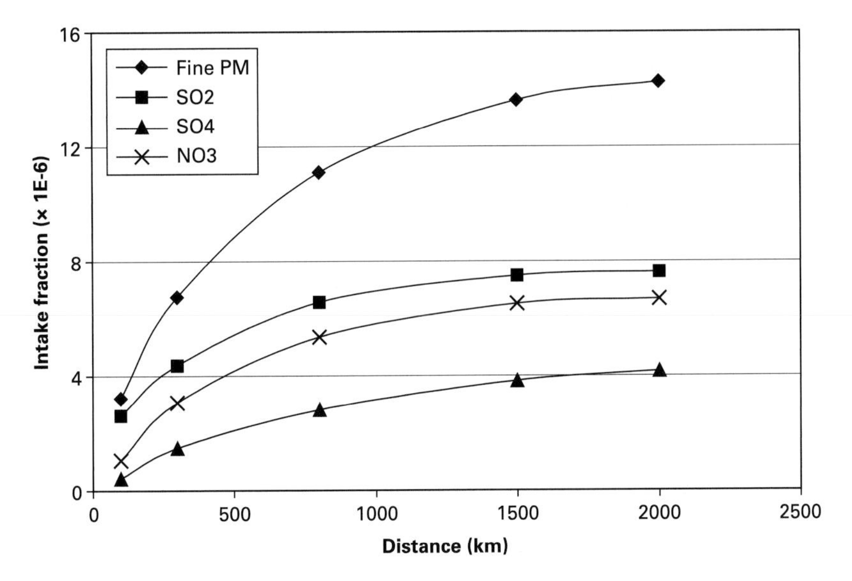

Figure 7.4
Intake fraction versus distance.

wet depletion, or dry depletion. The difference between the bar and 100% is the fraction remaining in the domain.

SO_2 is removed because of advection, wet depletion, dry depletion, or transformation to sulfates. As shown in the plot, only a *iF* remains in the domain. Primary fine particles and nitrates are mostly washed out by the rain (wet depletion), except during the dry winter months.

All four pollutants see a greater percentage of wet depletion in the two wetter and warmer months (May and August) than the two colder months (February and November).

7.3.3 Sensitivity Analysis for Shijingshan Power Plant Case Study

To test the influence of parameter uncertainty and variability on the resulting *iF* estimates, the following sensitivity analyses are performed by using the November meteorology settings.

Sensitivity to particle-size distribution We assume that secondary particles are entirely in the $PM_{2.5}$ range, because SO_4 and NO_3 are predominantly observed in the fine mode (Seinfeld and Pandis 1998). We use the CALPUFF default geometric mass

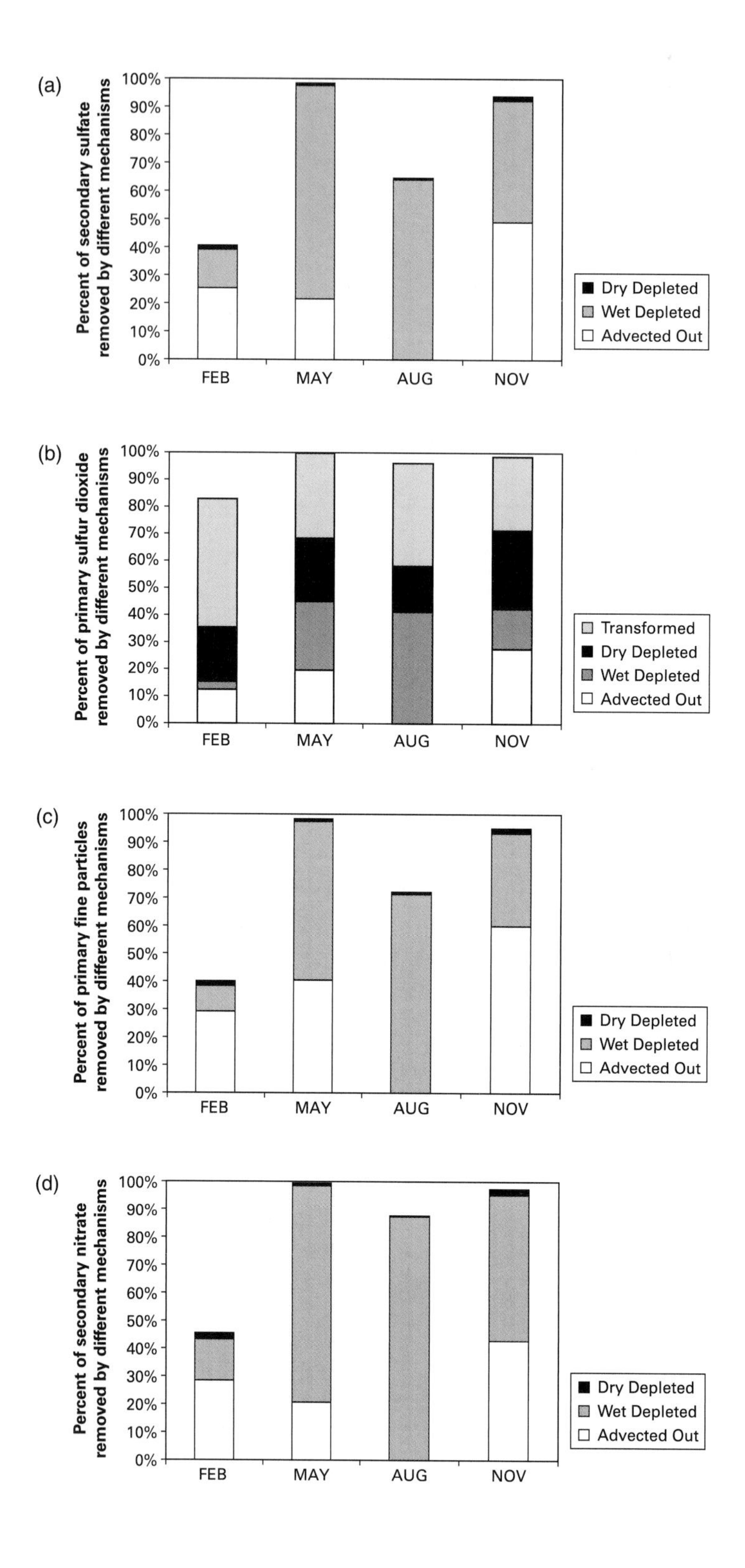
(a)
Percent of secondary sulfate removed by different mechanisms
100%
90%
80%
70%
60%
50%
40%
30%
20%
10%
0%
FEB
MAY
AUG
NOV
Dry Depleted
Wet Depleted
Advected Out
(b)
Percent of primary sulfur dioxide removed by different mechanisms
Transformed
Dry Depleted
Wet Depleted
Advected Out
(c)
Percent of primary fine particles removed by different mechanisms
(d)
Percent of secondary nitrate removed by different mechanisms

Table 7.3
Particle diameters that correspond to 12.5, 37.5, 62.5, and 87.5 percentiles of total mass (based on He and Wu 1999)

	Percentile (%)			
	12.5	37.5	62.5	87.5
Analysis 1	0.9	3.0	6.0	12.2
Analysis 2	0.7	1.9	3.5	6.3
Analysis 3	1.5	4.1	7.4	13.5
Analysis 4	1.3	3.9	7.4	14.0
Analysis 5	0.5	2.9	7.6	20.6
Analysis 6	1.4	3.8	7.0	13.1
Average diameter (μm)	1	3	7	13

mean diameter of 0.48 microns (μm) and geometric standard deviation of 2 in the calculation.

The particle-size distribution of primary particles, however, can vary substantially for different combustion processes and dust-removal equipment, and information on particle-size distribution is not available for most power plants in China. Therefore, the following sensitivity analysis on particle size will focus on the size distribution of the primary particles.

A report in the Chinese literature (He and Wu 1999) reported six Weibull distributions to characterize the particle size distribution of the emissions, based on the testing of several boiler and electrostatic precipitator (ESP) types. Table 7.3 shows the particle sizes that correspond to the 12.5th, 37.5th, 62.5th and 87.5th percentile of the total mass based on the six distributions reported. These are the midpoints of the four quartiles of the distribution, and the corresponding diameters can be used to represent particle sizes of the four quartiles to simplify calculations. The typical diameters averaged across the six distributions that correspond to the above percentiles are 1, 3, 7, and 13 μm. The *iF*s for each of these four diameters are 2.7×10^{-5}, 1.9×10^{-5}, 1.1×10^{-5}, and 7.3×10^{-6} for November, which demonstrates the expected decrease with increased particle size.

Figure 7.5 (*Opposite*)
Percent of pollutants removed by advection, transformation, wet depletion, and dry depletion during different seasons. (a) Secondary sulfate. (b) Sulfur dioxide. (c) Primary fine particles. (d) Secondary nitrate.

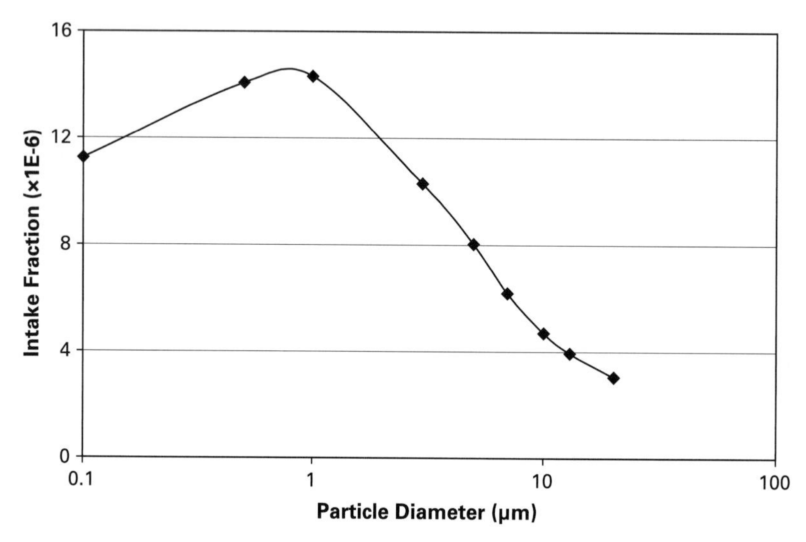

Figure 7.6
Intake fraction versus primary particle size.

To further explore the relationship between *iF* and particle size, we calculate *iF*s for different particles sizes. Figure 7.6 shows the relationship between *iF* and particle size. The *x* axis is particle diameter plotted on a logarithmic scale. It shows that *iF* for particles peaks at the particle size between 0.5 and 1 μm. This is because for very small particles (diameter less than 0.1 μm), Brownian diffusion coefficients increase as particle diameter decreases, whereas for larger particles (diameters greater than 1 μm), eddy diffusivity becomes important, as do gravitational settling and particle inertia in causing deposition (Sehmel 1980).

Sensitivity to the stack height We evaluate how *iF* estimates change with stack heights ranging from 35 to 240 meters (m), the common stack height range for power plants in China. In general, the higher the stack, the lower the *iF*; however, the *iF* estimates of different pollutants modeled are not very sensitive to the change in stack height. For example, there is only a 5–10% difference in the resulting *iF* estimates when stack height changes from 35 m and 240 m.

Sensitivity to the background ammonia concentration In the formation of secondary particles, ammonia in the atmosphere preferentially neutralizes SO_4, while the NO_3 partitioning is shifted toward the gas phase when the SO_4 concentration is high and toward the aerosol phase when SO_4 is low (Seinfeld and Pandis 1998).

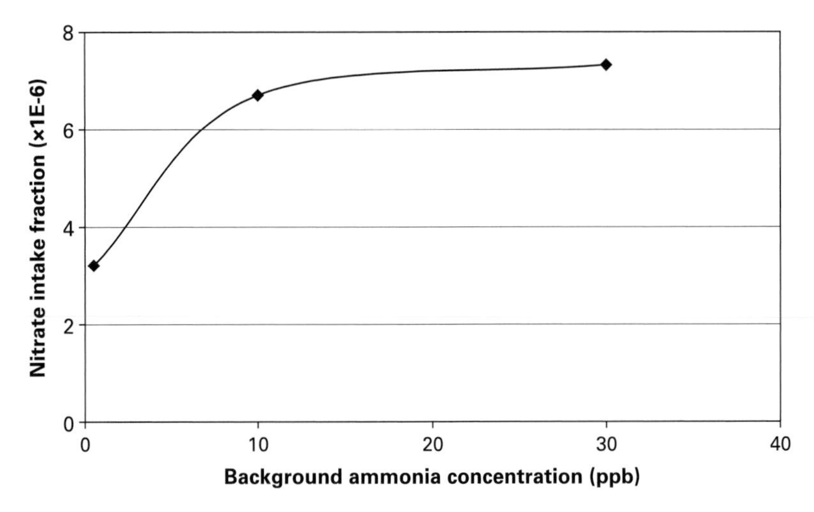

Figure 7.7
Nitrate intake fraction versus background ammonium concentration.

Background ammonia concentration is one of the input variables to CALPUFF; however, information on ammonia concentration throughout the domain defined in China is very limited. We found information on the background ammonia concentration in one report in the Chinese literature. It reported thirty-three sample averages based on samples collected in different regions of China in different seasons (Sun and Wang 1997). The average ammonia concentration based on these samples was 11.5 ppb, with a standard deviation of 9.7 ppb.

The CALPUFF default ammonia concentration is 10 ppb, which is used in the basic runs. We also test the sensitivity of *iF* estimates to the background ammonia concentration by varying the ammonia concentration between 0.5 and 30 ppb (figure 7.7).

The NO_3 *iF* increases with the increase in the background ammonia concentration, whereas the *iF* for primary particles, SO_2, and SO_4 do not depend on the ammonia concentration. This is because the ratio of SO_4 to SO_2 emissions is influenced by the oxidizing power of the atmosphere, which does not depend on the background ammonia concentration. Unlike SO_4 formation, however, the NO_3 formation process is reversible and equilibrium is established between nitric acid, ammonia, and ammonium nitrate (Scire et al. 2000). Therefore, variations in the background ammonia concentration can shift the equilibrium and change the *iF* for NO_3.

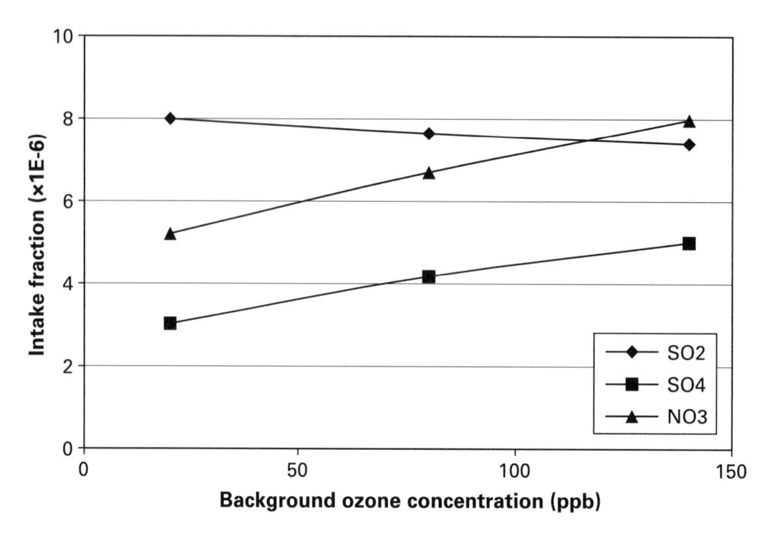

Figure 7.8
Sulfur dioxide, sulfate, and nitrate intake fractions versus background ozone concentration.

Sensitivity to the ozone (O_3) concentration The background O_3 concentration is another input variable required by CALPUFF. During daytime, the MESOPUFF II mechanism for chemical transformation uses O_3 concentration (along with radiation intensity) as surrogates for the hydroxyl radical concentration during the day (Scire et al. 2000). Default nighttime oxidation rates of 0.2% and 2% for SO_2 and NO_X, respectively, are used in the model run. CALPUFF provides two options for the specification of O_3 concentration. The first is to use hourly O_3 data from a network of stations; the second is a single, user-specified background O_3 value. Because there is little O_3 monitoring data publicly available in China, the second option of a single O_3 value is used. The model default O_3 concentration of 80 ppb is used in the calculation.

Figure 7.8 shows how *iF* estimates of SO_2, SO_4, and NO_3 change when the background O_3 concentration changes between 20 and 140 ppb. The *iF* for primary fine particles does not depend on background O_3 concentration. At the background O_3 concentration of 20 ppb, the *iF* for SO_4 and NO_3 decreases by 38% and 29% from the base case, respectively, whereas that of SO_2 increases by 4%. At the background O_3 concentration of 140 ppb, the *iF* for SO_4 and NO_3 increases by 17% and 16% from the base case, respectively, whereas that of SO_2 decreases by 3%.

In general, the higher the background O_3 concentration, the more SO_2 transformed to SO_4, which results in a higher SO_4 *iF* and a lower SO_2 *iF*.

Sensitivity to emission rate Experience in the United States shows that reductions in airborne SO_4 concentration may cause the total secondary fine particles to respond nonlinearly, because nitric acid gas may transfer to the aerosol phase. When this occurs, reductions in SO_2 emissions will be much less effective than expected at reducing fine particles (West et al. 1999).

To test whether there is a nonlinear response to changes in emission rates in this case, we perform sensitivity analyses on emissions rates. When the SO_2 emission rates are doubled or halved from the base-case emission rates, the *iF* for SO_4 remains the same, which shows that the ambient SO_4 concentration increases or decreases proportionally to SO_2 emissions. The *iF* for NO_3 decreases slightly (less than 0.1%) when SO_2 emission rates are doubled and increases slightly (less than 0.1%) when SO_2 emission rates are halved.

When the emission rates of NO_X are doubled and halved, the *iF* for NO_3 increases and decreases, respectively, from the base case by about 5%. This shows that the ambient concentration of NO_3 increases or decreases more than proportionally to the change in emission rates. In other words, when the emission rates of NO_X increase, the equilibrium established between nitric acid, ammonia, and ammonium nitrate shifts toward forming more ammonium nitrate and vice versa.

The change in NO_X emission rates does not influence the *iF* of SO_2, SO_4, and primary fine particles, nor does the change in primary particle emission rates influence the *iF* for any pollutants modeled.

The CALPUFF modeling system currently does not allow the use of actual background ammonia concentrations. In addition, we did not model the impact from all the pollution sources in the domain. Therefore, although the model results do not show a significant nonlinear response, it might have a larger effect in reality if many power plants are controlled simultaneously.

Sensitivity to precipitation In the basic runs, we assume uniform precipitation intensity and divide daily precipitation values by 24 to obtain hourly precipitation estimates. This assumption generates the precipitation scenario of light rain over a relatively long period of time, which is likely to be more efficient at pollutant removal than heavy rain over a short period of time and therefore is likely to result in overestimation of pollutant removal and underestimation of the *iF*s. To test the sensitivity of the *iF* results to this assumption, we also calculate *iF*s assuming daily precipitation occurring in 1 hour (5–6 PM) of the day with no precipitation in the other hours of the day. In the four 10-day periods for which we calculate *iF*s, there is more precipitation in August than the other three seasons, and we test the

Table 7.4
Summary statistics of the annual average intake fraction estimates for the twenty-nine plants modeled

	SO_2	Sulfate	Nitrate	Primary PM_1	Primary PM_3	Primary PM_7	Primary PM_{13}
Mean	4.8E-06	4.4E-06	3.5E-06	1.0E-05	6.1E-06	3.5E-06	1.8E-06
SE	3.5E-07	2.8E-07	3.2E-07	6.9E-07	5.6E-07	3.4E-07	1.9E-07
Minimum	1.8E-06	7.3E-07	8.0E-07	2.8E-06	1.7E-06	1.1E-06	6.7E-07
Maximum	8.9E-06	7.3E-06	7.1E-06	1.9E-05	1.2E-05	8.2E-06	5.2E-06

Note: PM_x = particulate matter with diameter equal to x μm.

sensitivity to precipitation in the August run period. The *iF*s of SO_2, SO_4, NO_3, and primary $PM_{2.5}$ increase by 7%, 25%, 18%, and 12%, respectively, when daily precipitation is assumed to occur in 1 hour of the day. The true precipitation pattern of the run period is likely to be in between these two assumptions.

7.3.4 Intake Fraction Estimates for the National Power-Plant Sample

Table 7.4 lists the summary statistics of the annual average *iF*s for the twenty-nine power plants of our national sample. The smallest primary fine particles, with diameters of 1 micron (PM_1), have the highest *iF*s, the average over the twenty-nine plants being 1×10^{-5}. This means that for every metric ton of PM_1 emitted, 10 g are eventually inhaled by people in the domain. PM_3 has the second-highest *iF* (6×10^{-6}), followed by SO_2 (5×10^{-6}), SO_4 (4×10^{-6}), PM_7 (4×10^{-6}), NO_3 (4×10^{-6}), and PM_{13} (2×10^{-6}). Among the primary particles, the coarser the particles, the lower the *iF*. By averaging the *iF* estimates for PM_1, PM_3, PM_7, and PM_{13}, the *iF* of primary particles including these four sizes is 6×10^{-6}. For all the pollutants modeled, plants with the lowest *iF* estimates are located either close to the border of the domain (e.g., plants 15 and 18) or in the least densely populated areas (e.g., plants 5 and 21).

7.3.5 Sensitivity of Intake Fraction Estimates for the National Power-Plant Sample

In the basic runs on plants of the national power plant sample, we used a stack height of 210 m. To test the sensitivity to stack height, we calculate *iF*s for five power plants assuming a stack height of 35 m (table 7.5). Similar to results for the stack height of 210 m, the bigger the particle size, the smaller the *iF* at the stack height of 35 m.

Table 7.5
Annual average intake fraction estimates for five plants from the national sample at the stack height of 35 m versus 210 m

Plant	PM_1	PM_3	PM_7	PM_{13}	SO_2
Beijing:					
35 m	1.7E-05	1.4E-05	9.5E-06	6.3E-06	9.9E-06
210 m	1.5E-05	1.2E-05	8.2E-06	5.2E-06	8.4E-06
Difference (%)	11	13	16	21	18
Guangxi:					
35 m	8.6E-06	3.5E-06	2.2E-06	1.3E-06	4.0E-06
210 m	7.7E-06	2.7E-06	1.5E-06	8.0E-07	3.3E-06
Difference (%)	11	30	46	63	21
Hubei:					
35 m	1.3E-05	7.8E-06	4.4E-06	2.2E-06	6.1E-06
210 m	1.3E-05	7.4E-06	4.1E-06	2.0E-06	5.7E-06
Difference (%)	5	5	8	10	7
Ningxia:					
35 m	8.8E-06	5.9E-06	3.0E-06	1.4E-06	3.4E-06
210 m	8.7E-06	5.6E-06	2.8E-06	1.3E-06	3.3E-06
Difference (%)	2	5	6	7	3
Sichuan:					
35 m	5.1E-06	2.9E-06	1.1E-06	7.0E-07	2.2E-06
210 m	5.0E-06	2.8E-06	1.1E-06	6.7E-07	2.2E-06
Difference (%)	3	4	1	5	4

Note: PM_x = particulate matter with diameter precisely equal to x μm. Plant names/locations are identified in Zhou 2002 (figures 1 and 6).

Figure 7.9 shows the percentage difference of the annual average *iF* of primary pollutants between the stack heights of 35 and 210 m. We find that the lower the stack height, the higher the *iF*. For primary particles, the percentage difference ranges from 6% to 21%, and the larger the particles, the higher the percentage difference, as expected. This shows that stack height has a greater influence on the *iF* of big particles.

Secondary particles are not emitted from the source directly, but are formed in the atmosphere. Ammonium sulfate and ammonium nitrate are predominantly in the fine particle mode (Seinfeld and Pandis 1998), which has long residence time, can travel long distances in the atmosphere, and is not strongly influenced by the local

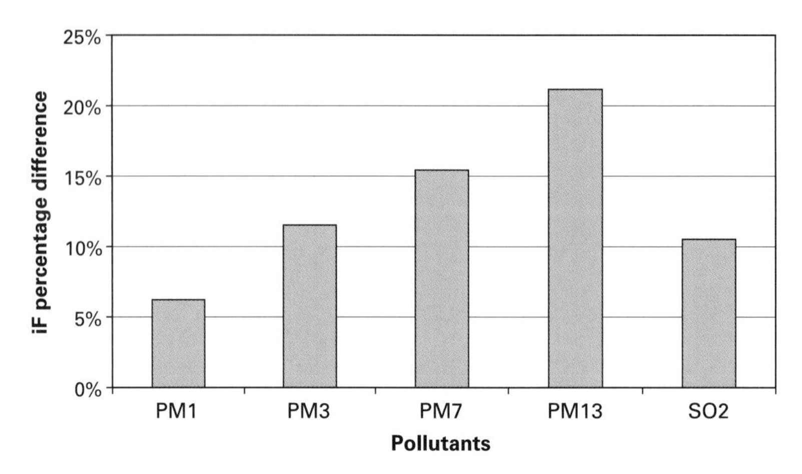

Figure 7.9
Percentage increase of primary pollutant intake fractions from decrease of stack height from 210 m to 35 m at five plants from the national sample.

population density as primary particles. Therefore, the stack height at which pollutants are emitted has a limited impact on the *iF*s of secondary particles. For example, in a previous study using a regression-based approach for estimating primary and secondary particulate matter *iF*, stack height was not a significant predictor for either SO_4 or NO_3 *iF*s (Levy, Wolff et al. 2002). In this analysis, we randomly chose a few plants and calculated the *iF* of NO_3 and SO_4 for the low stack height of 35 m. The differences in the secondary particle *iF* between the stack heights of 35 m and 210 m are less than 1.5%.

7.4 Discussion and Conclusions

7.4.1 Consistency Check

We model the incremental concentration change in the atmosphere caused by emissions from one power plant in Beijing. Ideally, if we could obtain atmospheric monitoring results in different parts of China for the same time period with and without the power plant operating, we could validate the result we reported in the paper; however, such an experiment is impossible.

Instead, we check whether our modeling result is reasonable by comparing our findings with the available monitoring data. The weekly average $PM_{2.5}$ mass concentration in Beijing has been reported to range from 37 to 357 $\mu g/m^3$, with ammonium, SO_4, and NO_3 accounting for 25–30% of the mass (He et al. 2001). We calculate that ammonium sulfate should account for about 15–20% of the total $PM_{2.5}$ mass, based on the average weekly $PM_{2.5}$ mass and chemical concentrations reported in the paper. Therefore, the average weekly ammonium sulfate concentration would range from 6 μg/m^3 to 71 μg/m^3.

Assuming that SO_2 emitted from power plants accounts for 35% of the ambient sulfate concentration (because power plant emissions account for 35% of SO_2 emissions from all industrial sectors), an average breathing rate of 20 m^3/day, a total population of 1.2 billion, and taking the lower bound for SO_4 concentration of 6 μg/m^3, the total amount ingested is:

$$DOSE = 6 \times 10^{-6}\ \text{g/m}^3 \times 35\% \times 20\ \text{m}^3/d \times 1.2 \times 10^9 \times 365\ d/\text{year}.$$

The total SO_2 emission rate from coal-fired power plants in the country is 7 million tons per year, and thus the lower bound estimate for the *iF* (*iF* = DOSE/Emissions) is:

$$IF(SO_4) = \frac{6 \times 10^{-6}\ \text{g/m}^3 \times 35\% \times 20\ \text{m}^3/d \times 1.2 \times 10^9 \times 365\ d/\text{year}}{7 \times 10^{12}\ \text{g/year}}$$

$$= 2.6 \times 10^{-6}.$$

Similarly, the upper-bound estimate for the *iF* can be calculated by assuming a SO_4 concentration of 71 μg/m^3 and the resulting estimate is 3.1×10^{-5}.

Our *iF* estimates for SO_4, which range from 3×10^{-6} to 1×10^{-5}, are within this range. This shows good agreement between the modeling estimates and monitoring results, which provides some evidence for the validity of our modeling approach.

7.4.2 Comparison with Intake Fraction Estimates from Other Studies

The *iF* of airborne particulate matter from power plants was estimated by Smith (1993) to be roughly on the order of 10^{-5} for developing countries, and 10^{-6} for developed countries. In addition, analyses in the United States using a similar methodology as this study estimated the *iF*s of primary fine particles to be on the order of 10^{-6}, and of SO_4 and NO_3 to be on the order of 10^{-7} (Evans et al. 2002; Levy, Spengler et al. 2002). Our estimates for these three pollutants are an order of magnitude higher than those for the United States. Both the magnitude of difference in *iF*s between China and the United States and our *iF* estimate for primary particles are consistent with Smith's estimates.

There are several reasons why the *iF* estimates calculated in Beijing are higher than those reported from similar studies in the United States. First, the population density is higher in China than the United States: the total population of China is about four times the total population of the United States, whereas the areas of the two countries are similar. Second, the domain we define covers a larger area than those in the U.S. studies. For example, Levy, Spengler et al. (2002) covered an area in the midwestern U.S. of approximately 750 km × 750 km (about 0.6 million km^2), with a population of about 33 million. This study covers 6 million km^2, with a population of about 1.1 billion. Because fine particles can travel a long distance in the atmosphere, the larger domain in this study captures the full impact of fine particles, in particular the secondary ones formed in the atmosphere.

7.4.3 Limitations

In the sensitivity analysis section, we analyze the impact of parameter uncertainties (e.g., background concentration or particle size distribution) on the *iF* predictions. It is beyond the scope of this study to evaluate the validity of the CALPUFF modeling system itself. There are very few intensive tracer field experiments available for investigating CALPUFF model simulations of mesoscale transport and dispersion. Existing studies showed that although CALPUFF performed better than other models (e.g., ISCST3, CTDMPLUS) tested in tracer studies (Strimaitis et al. 1998), the consistency of CALPUFF predictions to observed tracer concentrations varied (U.S. EPA 1998). This depended on many factors, such as the complexity of the terrain and the algorithms chosen in the simulations. Nevertheless, our consistency check and comparison with estimates from other studies demonstrate that substantial bias is unlikely.

7.4.4 Conclusion

We successfully demonstrate that *iF*s of power plant emissions in China can be calculated using a state-of-the-art air-dispersion model—CALPUFF. For the power plant in Beijing, the intake fraction of primary fine particles is roughly on the order of 10^{-5}, whereas the *iF*s of SO_2, SO_4, and NO_3 are on the order of 10^{-6}.

The sensitivity analyses shows that stack height and emission rates have a negligible influence on the *iF*. The background ozone concentration has a moderate impact on the *iF* of SO_4 and NO_3 and a slight impact on that of SO_2. The primary particle size distribution has a large impact on the *iF* of primary particles, whereas the background ammonia concentration is an important factor influencing the intake fraction of NO_3.

By combining our findings on population exposure, emissions estimates, epidemiology, and willingness-to-pay information, we can estimate the human health and economic damages from point-source emissions. From the individual power plant perspective, this information can provide the basis for choosing different control technologies by comparing the benefit and cost of each option. From a policy perspective, this information can provide a basis for those attempting to prioritize among competing control strategies for pollution control in China, such as setting regional priorities in pollution control and conducting policy simulations to compare different economic incentives, such as environmental taxes and pollution permit trading. These are the subjects of the following chapters, using both our results here and those of chapters 5 and 6.

Acknowledgments

This chapter is adapted from articles by the authors published in the journals *Atmospheric Environment* (Zhou et al. 2003) and *Environment International* (Zhou et al. 2006); permission to reproduce sections of the text here was kindly granted by the editors of these journals. The research of this chapter was funded by a generous grant from the China Sustainable Energy Program of the Energy Foundation to the China Project of the Harvard University Center for the Environment and Tsinghua University. Additional funding and support were provided by the Harvard Center for Risk Analysis, the V. Kann Rasmussen Foundation's grant to the Harvard China Project, and the Ministry of Health and Welfare of Japan (for International Health Cooperation Research, C11-8). We thank John Spengler, Joseph Scire, Jennifer Godfrey, Wei Wang, and Steven Melly for providing technical support on the analysis.

Notes

1. Twenty-three vertical layers were used with the full sigma (σ) levels from bottom to top set at 1.00 (ground), 0.99, 0.98, 0.96, 0.93, 0.89, 0.85, 0.80, 0.75, 0.70, 0.65, 0.60, 0.55, 0.50, 0.45, 0.40, 0.35, 0.30, 0.25, 0.20, 0.15, 0.10, 0.05, 0.00 (model lid), where σ is a one-dimensional vertical coordinate calculated as

$$\sigma = \frac{p_0 - p_{top}}{p_{surface} - p_{top}},$$

where p is the atmospheric pressure (Dudhia et al. 2002).

2. See http://sedac.ciesin.org/china/. Last accessed June 27, 2003.

3. Tibet, Xinjiang, and Heilongjiang are not covered in the modeling domain.

4. There is no power plant in Macao in the RAINS-Asia list.

References

Bennett, D. H., T. E. McKone, J. S. Evans, W. M. Nazaroff, J. D. Margni, O. Jolliet, and K. R. Smith. 2002. Defining intake fraction. *Environmental Science and Technology* 36 (9):207A–211A.

Bertok, I., J. Cofala, F. Gyarfas, Z. Klimont, W. Schöpp, and M. Amann. 2001. Structure of the RAINS 7.5 energy and emissions database. International Institute for Applied Systems Analysis. Laxenburg, Austria.

China Health Yearbook Committee. 1998. *China health yearbook 1998*. Beijing: People's Health Publisher. In Chinese.

China Meteorology Yearbook 1997. 1999. Beijing: China Meteorology Publishing House. In Chinese.

Dockery, D. W., A. Pope, X. Xu, J. D. Spengler, J. H. Ware, M. E. Fay, B. G. Ferris, and F. E. Speizer. 1993. An association between air pollution and mortality in six US cities. *New England Journal of Medicine* 329 (24):1753–1759.

Dudhia, J., D. Gill, K. Manning, A. Bourgeois, and W. Wang. 2002. PSU/NCAR mesoscale modeling system tutorial class notes and users' guide (MM5 modeling system version 3). Available at www.mmm.ucar.edu/mm5/documents/MM5_tut_Web_notes/TutTOC.html.

Evans J. S., S. K. Wolff, K. Phonboon, J. I. Levy, and K. R. Smith. 2002. Exposure efficiency: An idea whose time has come? *Chemosphere* 49 (9):1075–1091.

Florig, K. 1997. China's air pollution risks. *Environmental Science and Technology* 31 (6):275A–279A.

He, K., F. Yang, Y. Ma, Q. Zhang, X. Yao, C. K. Chan, S. Cadle, T. Chan, and P. Mulawa. 2001. The characteristics of $PM_{2.5}$ in Beijing, China. *Atmospheric Environment* 35:4959–4970.

He, L. G., and Y. H. Wu. 1999. Experimental results on the size distribution of particles emitted from electrostatic precipitators (ESP) for boilers. *Journal of Power Sector Environmental Protection* 15 (4):4–7. In Chinese.

Hinds, W. C. 1998. *Aerosol technology: Properties, behavior, and measurement of airborne particles*. New York: John Wiley & Sons.

Kandlikar, M., and G. Ramachandran. 2000. The causes and consequences of particulate air pollution in urban India: A synthesis of science. *Annual Review of Energy and the Environment* 25:629–684.

Lai, A. C. K., T. L. Thatcher, and W. M. Nazaroff. 2000. Inhalation transfer factors for air pollution health risk assessment. *Journal of the Air and Waste Management Association* 50 (9):1688–1699.

Levy, J. I., J. D. Spengler, D. Hlinka, D. Sullivan, and D. Moon. 2002. Using CALPUFF to evaluate the impacts of power plant emissions in Illinois: model sensitivity and implications. *Atmospheric Environment* 36 (6):1063–1075.

Levy, J. I., S. K. Wolff, and J. S. Evans. 2002. A regression-based approach for estimating primary and secondary particulate matter intake fraction. *Risk Analysis* 22 (5):895–903.

Lutgens, F. K., and E. J. Tarbuck. 2000. *The atmosphere: An introduction to meteorology.* Lebanon, Ind.: Prentice Hall.

Ministry of Power (MOP). 1998. *Development and environmental protection of China's electric power industry.* Beijing: China Power Industry Press. In Chinese.

Paine, R. J., and D. W. Heinold. 2000. Issues involving use of CALPUFF for long-range transport modeling. Presented at the Air and Waste Management Association 93rd Annual Meeting and Exhibition, June 18–22, Salt Lake City, Utah.

Phonboon, K. 1996. Risk assessment of environmental effects in developing countries. Ph.D. diss., Harvard School of Public Health.

Pope, C. A., D. W. Dockery, and J. Schwartz. 1995. Review of epidemiological evidence of health effects of particulate air pollution. *Inhalation Toxicology* 7 (1):1–18.

Pope, C. A., R. T. Burnett, M. J. Thun, E. E. Calle, D. Krewski, K. Ito, and G. D. Thurston. 2002. Lung cancer, cardiopulmonary mortality, and long-term exposure to fine particulate air pollution. *Journal of the American Medical Association* 287 (9):1132–1141.

Samet, J. M., S. L. Zeger, F. Dominici, F. Curriero, I. Coursac, D. W. Dockery, J. Schwartz, and A. Zanobetti. 2000. The national morbidity, mortality, and air pollution study. Part II. Morbidity, mortality, and air pollution in the United States. Health Effects Institute, Boston, Mass.

Schwartz, J., D. W. Dockery, and L. M. Neas. 1996. Is daily mortality associated specifically with fine particles? *Journal of the Air and Waste Management Association* 46 (10):927–939.

Schwartz, J. 1994. What are people dying of on high air pollution days? *Environmental Research* 64 (January):26–35.

Scire, J. S., D. G. Strimaitis, and R. J. Yamartino. 2000. A user's guide for the CALPUFF dispersion model. Version 5. Earth Tech Inc., Concord, Mass.

Sehmel, G. A. 1980. Particle and gas dry deposition: A review. *Atmospheric Environment* 14:983–1011.

Seinfeld, J. H., and S. N. Pandis. 1998. *Atmospheric chemistry and physics.* New York: John Wiley & Sons.

Smith, K. R. 1993. Fuel combustion, air pollution exposure, and health: The situation in developing countries. *Annual Review of Energy and the Environment* 18:529–566.

Socioeconomic Data Applications Center (SEDAC). 2003. China Dimensions World Wide Web Home Page, Center for International Earth Science Information Network, Columbia University and U.S. National Aeronautics and Space Administration. Available at http://sedac.ciesin.org/china/.

Strimaitis, D. G., J. S. Scire, and J. C. Chang. 1998. Evaluation of the CALPUFF dispersion model with two power plant data sets. Preprints from 10th Joint Conference on the Application of Air Pollution Meteorology, January 11–16, Phoenix, Ariz.

Sun, Q., and M. Wang. 1997. Ammonia emission and concentration in the atmosphere over China. *Atmospheric Science* 21 (5):590–598. In Chinese.

United States Environmental Protection Administration (U.S. EPA). 1995. User's guide for the Industrial Source Complex (ISC3) dispersion models. Volume 1: User instructions. Report no. EPA-454/B-95-003a. U.S. Environmental Protection Agency, Research Triangle Park, N.C.

United States Environmental Protection Administration (U.S. EPA). 1998. Interagency Workgroup on Air Quality Modeling (IWAQM) phase 2 summary report and recommendations for modeling long-range transport impacts. December. Available at http://www.epa.gov/ttn/scram/7thconf/calpuff/phase2.pdf.

United States Environmental Protection Administration (U.S. EPA). 2005. Sulfur dioxide national emissions totals. Available at http://www.epa.gov/air/airtrends/2005/pdfs/SO2National.pdf.

Wang, X., and K. R. Smith. 1999. Secondary benefits of greenhouse gas control: Health impacts in China. *Environmental Science and Technology* 33 (18):3056–3061.

West, J. J., A. S. Ansari, and S. N. Pandis. 1999. Marginal $PM_{2.5}$: Nonlinear aerosol mass response to sulfate reductions in the eastern United States. *Journal of the Air and Waste Management Association* 49 (12):1415–1424.

Wilson, W. E., and H. H. Suh. 1997. Fine particles and coarse particles: concentration relationships relevant to epidemiologic studies. *Journal of the Air and Waste Management Association* 47 (12):1238–1249.

Wolff, S. K. 2000. Evaluation of fine particle exposures, health risks and control options. Ph.D. diss., Harvard School of Public Health.

World Bank. 1997. *Clear water and blue skies: China's environment in the new century.* Washington D.C.: World Bank.

Xu, X., J. Gao, D. W. Dockery, and Y. Chen. 1994. Air pollution and daily mortality in residential areas of Beijing, China. *Archives of Environmental Health* 49 (4):216–222.

Zhou, Y. 2002. Evaluating power plant emissions in China: Human exposure and valuation. Ph.D. diss., Harvard School of Public Health.

Zhou, Y., J. I. Levy, J. S. Evans, and J. K. Hammitt. 2006. The influence of geographic location on population exposure to emissions from power plants throughout China. *Environment International* 32 (3):365–273.

Zhou Y., J. I. Levy, J. K. Hammitt, and J. S. Evans. 2003. Estimating population exposure to power plant emissions using CALPUFF: A case study in Beijing, China. *Atmospheric Environment* 37 (6):815–826.

8

The Economic Value of Air-Pollution-Related Health Risks in China: A Contingent Valuation Study

Ying Zhou and James K. Hammitt

8.1 Introduction

Many cities in developing countries are experiencing severe levels of air pollution, with the rapid increase in fossil-fuel consumption for industrial, residential, and motor-vehicle use. Decision makers in these countries are faced with the difficult task of mitigating air pollution while supporting continued economic growth. Benefit-cost analysis can help balance concerns about the cost of pollution control with the benefits of improved environmental quality, as applied in chapters 9 and 10. To conduct such analyses, it is necessary to quantify the value of improvements in longevity and health quality in monetary terms. Although there exist numerous studies of the value of reducing risks of mortality and other health effects in the United States and other industrialized countries, there are relatively few such estimates for developing countries. As an alternative, estimates of the value of health improvements in developing countries are often obtained by transferring estimates from industrialized countries. These benefit transfers attempt to account for differences in income, though the rate at which benefits vary with income is highly uncertain. Because of the paucity of estimates for developing countries, it is virtually impossible to identify and account for any differences in preferences that may result from disparities in economic opportunities, health care systems, and cultural characteristics.

Most estimates of the monetary value of reductions in health risk are obtained using either contingent valuation or wage-differential approaches (described below). There are a few papers that have estimated the monetary value of reductions in air pollution or related health effects using contingent valuation in China. Two studies surveyed respondents about their willingness to pay to reduce air pollution, by 50% (Li et al. 2001) and to meet local air-quality standards (Zhang et al. 2004). To determine the valuation of health risk implied by these studies, one would need to understand survey respondents' beliefs about the effects of these changes in air quality on health. A third study (Peng and Tian 2003) asked patients who had been admitted to a Shanghai hospital for treatment of respiratory disease about their willingness to pay to relieve or cure their disease. Although this population is of interest because of its familiarity with respiratory disease, it is not representative of the general population. Estimates of the value of reducing mortality risk based on the study of wage differentials have been obtained for Hong Kong, Taiwan, South Korea, and India (see Viscusi and Aldy 2003, for a review). In addition, there are a number of contingent valuation studies of the value of reductions in health risk carried out in developing countries, including one in Thailand (Chestnut et al. 1997) and several in Taiwan (e.g., Alberini et al. 1996; Fu et al. 1999; Hammitt and Liu 2004; Liu et al. 2000; Liu et al. 2005).

The monetary value of a reduction in health risk is defined as the affected individual's willingness to pay (WTP) to obtain it. WTP describes the amount of money that the individual views as equally desirable as the reduction in health risk. Taking account of his income and the other ways he could spend his money, the risk reduction and the amount of money equal to his WTP for the risk reduction have the same effect on his well-being. If the individual had to pay more than his WTP to obtain the risk reduction, he would prefer not to receive the reduction. If he could receive the risk reduction and pay nothing, or any amount less than his WTP, he would prefer to receive the reduction.

There are two methods for estimating WTP for risk reduction, "revealed preference" and "contingent valuation" (CV). Revealed-preference methods are based on the assumption that people choose the alternative, from among those available to them, that they most prefer. By observing the choices people make in settings where they are choosing between different levels of risk and money, an analyst can estimate his or her WTP for risk reduction. For example, people who purchase home smoke detectors or safer automobiles may be assumed to value the improved safety more than the price of the smoke detector or the additional price of the more expensive automobile. Most of the revealed-preference estimates of the value of mortality risk are based on comparing wages and occupational fatality risks. Among the jobs for which an individual is qualified, the more dangerous jobs offer higher wages (if not, employers could not attract workers to take these jobs). Workers who choose a lower-paying but safer job are interpreted as valuing the incremental safety more than the reduction in pay, and workers who accept the higher-paying but more dangerous jobs are interpreted as valuing the higher pay more than the incremental safety.

The alternative method, CV, is based on asking survey respondents how they would act in hypothetical settings involving choices between alternatives that differ in risk and financial consequences (CV and other survey-based methods are sometimes called "stated-preference" methods). Compared with revealed-preference methods, CV is much more flexible, because respondents can be asked about their choices in hypothetical settings. In contrast, revealed-preference estimates can only be obtained in cases where the analyst can determine (or can reasonably assume) not only what option each individual chose, but also what alternatives were rejected. CV also offers the advantage that respondents can be informed of the risks and monetary consequences associated with alternative choices. With revealed preference, the analyst does not know how well the individuals knew the consequences of alternative choices. In contrast, revealed-preference methods offer the advantage of relying on decisions with real consequences, unlike CV, where there is less incentive for the individual to carefully consider his decision.

Contingent valuation has been viewed with skepticism by many economists, largely because it relies on statements about hypothetical choices and because results sometimes conflict with economic theory (see, e.g., Diamond and Hausman 1994; Hanemann 1994; Hammitt and Graham 1999; Carson 2000). Nevertheless, estimates produced by high-quality CV studies are broadly consistent with revealed-preference estimates (Hanemann 1994; Carson 2000). Estimates of the value of reducing mortality risk obtained using CV are similar to estimates from wage-differential studies, though CV estimates are often somewhat smaller (Hammitt 2000). For example, U.S. EPA 1999 adopted the average estimated VSL from twenty-six wage-differential and CV studies to estimate the benefits of the 1990 Clean Air Act amendments in the United States. The average of the five CV estimates, $2.9 million, is just over half the average of the twenty-one wage-differential estimates, $5.2 million (1990 dollars), and the CV estimates are less variable, with a standard deviation of $1 million, compared with the standard deviation of $3.5 million for the wage-differential studies.

We conduct a CV study in three locations in China to value adverse health effects associated with air pollution, as well as study regional differences in WTP within China. We include three health endpoints in our study: the common cold, chronic bronchitis (CB), and premature fatality. These endpoints represent a range of severity of health endpoints related to air pollution. Moreover, these endpoints have been valued in other countries, so it is possible to compare the estimates we obtain in China to estimates from other societies. The survey was conducted in the cities of

Beijing and Anqing (Anhui province) and the rural areas near Anqing. These three locations were chosen to represent the preferences of residents in a big city, a small city, and the rural areas of China.

Under this approach, representative individuals were surveyed and asked how they would act in a hypothetical setting. Estimates were obtained using standard CV questions. Respondents were asked whether they would be willing to purchase medicines or other treatments that would reduce the risk that they would suffer each of three adverse health effects. Following recommended practice (e.g., Arrow et al. 1993; Hammitt and Graham 1999), we test whether estimated WTP differs significantly between subsamples of respondents offered larger and smaller risk reductions. Except when some interaction terms were included in the model, WTP does not significantly differ between subsamples, which suggests that respondents may not have adequately considered the size of the risk reduction described in the questionnaire and that our estimates of the value of preventing health effects are sensitive to the arbitrary choice of the size of the risk reduction stated in the questionnaire.

This study makes two main contributions to the literature on benefit valuation. First, we present estimates of the value of three distinct health risks in three regions of China. Second, we test different statistical models to study their impact on estimated WTP and the relationships between WTP and individual characteristics.

8.2 Survey Description

8.2.1 Data Collection

In-person interviews of about 3,600 people (1,200 people in each of three locations in China) were conducted in June and July 1999. Respondents were selected by stratified random sampling, the details of which vary by location.

Beijing, the national capital, with an estimated population of more than 16 million,[1] was chosen to represent the preferences of residents of large cities. Beijing is divided into ten administrative districts, and each district is further divided into residential communities. One urban district (Dongcheng) and two suburban districts (Shijingshan and Haidian) were chosen to represent residential, industrial, and cultural areas, respectively. About 400 people were interviewed in each district. For each of the three districts, several communities were randomly chosen. The number of communities chosen in each district depends on the size of the communities. Households within the chosen communities were randomly selected from the registration of the community. Interviewers entered the selected households and interviewed every adult between the ages of 18 and 65 years in the household on that day.

The city of Anqing in Anhui province has a population of 540,000 and was chosen to represent the preferences of people residing in small cities. It has two urban and one suburban district. A randomization and interviewing process similar to that used in Beijing was carried out in the two urban districts, with 600 people interviewed in each.

Rural areas around Anqing, with a population of about 5.5 million, were included in the survey to represent the preferences of rural residents. Among the eight rural counties near Anqing, three were randomly chosen, and villages were randomly selected within these counties. Because the registration system in rural areas is not as good as those in the cities, and houses are not as close to each other, households were chosen with the help of representatives from the villages. Interviewers were led into different households in each village by these representatives. The households were chosen by the representatives to try to cover the entire geographical area of the village and to represent households with different income levels.

In Beijing, we recruited about twenty interviewers from Beijing Medical University, with help from faculty members. For urban Anqing and the rural areas, we recruited approximately twenty interviewers from local hospitals, with the help of the Anhui Biomedicine and Environmental Health Institute.

8.2.2 The Questionnaire

The CV questionnaire included three valuation questions, concerning:

- Colds, in which people were asked how much they were willing to pay to prevent suffering an episode of minor illness similar to their most recent cold (i.e., to reduce the risk of suffering a cold in the next few days from one to zero). (We note that the analysis of colds is not used in the integrated assessment of the rest of the book but stands here as an additional analysis that is also useful for comparison with results for the other two health risks below.)
- Chronic bronchitis, in which people were asked how much they were willing to pay to reduce their lifetime chance of developing CB by 1% or by 5% from an initial risk of 18% (i.e., to 17% or to 13%). The date or age at which they might develop CB was not specified.
- Mortality, in which people were asked how much they were willing to pay to reduce their probability of death in the next year by 1 in 1,000 or 2 in 1,000, from an initial level of 7 in 1,000.

The scale of risk reduction chosen for contingent valuation is a key consideration, because it can influence estimates of the value of preventing health effects. This is a subject to which we return in the results section.

The subject's WTP was elicited using a double-bounded, dichotomous-choice format (Hanemann et al. 1991). The respondent was asked whether she would purchase the medicine or other treatment offering the stated risk reduction at a specified price. This price—the initial bid amount—was randomly selected from a set of five values. The respondent was then asked a follow-up dichotomous-choice question, where the bid amount depended on her response to the initial bid. If she responded "yes" to the initial bid, the follow-up bid was twice as large as the initial bid; if she responded "no" to the initial bid, the follow-up bid was half as large as the initial bid. This process yields interval-censored data. A respondent's exact WTP is not observed, but it is assumed to lie above any bid amount to which she replied "yes" and below any bid amount to which she replied "no." For respondents who said "no" to both initial and follow-up questions, the lower bound is assumed to be zero. For respondents who said "yes" to both initial and follow-up questions, the upper bound is infinity. Following the two dichotomous-choice questions, we asked respondents to state their maximum WTPs for the risk reduction in an open-ended follow-up.

It is well known that individuals have difficulty in comprehending the small increments of probability concerned (Hammitt and Graham 1999). To help in communicating these small probability changes, respondents were shown a card with an array of 10,000 dots. The appropriate number of dots was colored to convey the magnitude of risk change.

To capture the full range of the WTP distribution, the set of initial bids was adjusted during the survey. The adjustment was planned because we had little information on the WTP distribution before beginning the survey. Our goal in adjusting bid amounts was to have about half the respondents who received the medium initial bid respond "yes" to it and to spread the other bid amounts to capture the tails of the WTP distribution.

8.3 Methods

8.3.1 Regression Model

The WTP is estimated using regression methods to characterize its relationship to respondent characteristics (Alberini 1995). Assuming that WTP follows a specified distribution, we estimate the distribution's parameters by maximum likelihood techniques. For the equations describing WTP, we test normal, Weibull, and lognormal distributions. On the basis of both the χ^2 goodness-of-fit test and the log likelihood of the fitted model, both the lognormal and Weibull distributions fit the data much better than the normal distribution.[2] We report results using the lognormal distribution below.

In all three regression models, independent variables are factors that are likely to influence respondents' WTP. To avoid overfitting the model to random variations in the data, we use the same set of independent variables in the models describing WTP for cold, CB; and mortality risk reduction. Variables included in each model are listed later in this chapter, in tables 8.3, 8.4, and 8.5.

Three types of models were tested:

- A standard model, in which WTP is assumed to be strictly positive. This is the typical model used for analyzing dichotomous-choice CV data (Hanemann et al. 1991; Alberini 1995).
- A two-part model, in which it is assumed that different processes drive the zero and positive responses (Duan et al. 1984). The two-part model predicts whether an individual has a nonzero WTP and, if so, its magnitude. It uses an open-ended, follow-up question to distinguish true zeros from other no/no responses.
- A mixed model, in which the respondents who answered "no" to both initial and follow-up questions are assumed to be an unknown mixture of those having zero WTP and others having WTP between zero and the follow-up bid, with no information on which respondents are of each type (Werner 1999).

For the three health endpoints surveyed, between 4.4% and 34.6% of respondents in the three locations indicated in the open-ended, follow-up question that their maximum WTP is zero. It is unclear whether these respondents truly have a zero WTP or were confused, objected to the survey, or responded "no" for other reasons.

In both the standard and mixed models, we ignore the open-ended follow-up question. Respondents who reported zero maximum WTP are included with those who responded "no" to both dichotomous-choice questions but reported a positive WTP in response to the open-ended question. The assumption of the standard model is that every respondent, including those who stated that their maximum WTP is zero, has a positive albeit potentially small WTP. For each model, each respondent's mean and median WTP are predicted using his or her characteristics and the lognormal distribution. The sample mean and median WTP using the standard model are calculated by averaging each individual's mean and median WTP using the following formula, where F is the lognormal distribution function with geometric mean $\psi(Zj)$ and geometric standard deviation θ, and Zj are the individual covariates:

$$\text{MEAN} = \frac{1}{N}\sum_{j=1}^{N} E[F(\psi(Zj), \theta)]$$

and

$$\text{MEDIAN} = \frac{1}{N}\sum_{j=1}^{N} \text{MEDIAN}j,$$

where $F[\text{MEDIAN}j, \varphi(Zj), \theta] = 0.5$.

In the mixed model, no/no respondents are assumed to include two sets of respondents, those who have a zero WTP, and those who have a positive WTP that is less than the bid amount in the follow-up question. There is a probability $p(Zj)$ that respondent j has a positive WTP and complementary probability $1 - p(Zj)$ that she or he has zero WTP. A logistic distribution was used to predict probability $p(Zj)$. The estimated sample mean and median WTP are calculated as follows,

$$\text{MEAN} = \frac{1}{N}\sum_{j=1}^{N} p(Zj) \times E[F(\psi(Zj), \theta)]$$

and

$$\text{MEDIAN} = \frac{1}{N}\sum_{j=1}^{N} \text{MEDIAN}j,$$

where $1 - p(Zj) + p(Zj) \times F[\text{MEDIAN}j, \psi(Zj), \theta] = 0.5$, noting that when the predicted probability that respondent j's WTP is zero exceeds one half (i.e., $p[Zj] < 0.5$) MEDIANj is set to zero.

In the two-part model, respondents with WTP equal to zero are distinguished from those with positive WTP. A logistic regression is run on all the respondents to predict whether an individual will have a zero WTP. A second model is used to describe the WTP of respondents with positive values, assuming these values are lognormally distributed. Each respondent's estimated mean and median WTP are obtained by multiplying each respondent's predicted probability of having a nonzero WTP by his or her predicted mean or median WTP conditional on a positive value. For respondents who reported zero maximum WTP, the predicted mean and median WTP are set to zero, that is,

$$\text{MEAN} = \frac{1}{N}\sum_{j=1}^{N} p(Zj) \times E\{F[\psi(Zj), \theta]\}$$

and

$$\text{MEDIAN} = \frac{1}{N}\sum_{j=1}^{N} p(Zj) \times \text{MEDIAN}j,$$

where $F[\text{MEDIAN}j, \varphi(Zj), \theta] = 0.5$.

8.4 Results

8.4.1 Descriptive Statistics

The variables used in the analysis are defined in table 8.1. Table 8.2 provides descriptive statistics for the three samples. The average age of respondents in the rural

Table 8.1
Definition of variables

Variable	Definition
OWNCOST	Dummy variable equal to 1 if respondent paid out of his/her own pocket for the last episode of cold
HEALTH	Respondent's perception of own current health on a scale between 1 (perfect health) and 5 (worst possible health)
INCOME	Monthly household income divided by 100 (yuan)
PEOPLE	Number of people living in the household;
EXERCISE	Dummy variable equal to 1 if the respondent voluntarily exercises in spare time
SEX	Dummy variable equal to 1 if male
AGE	Age in years
EDUCATION	Categorical variable from 1 (illiterate) to 6 (college or above)
INSURANCE	Dummy variable equal to 1 if the respondent has some form of insurance coverage
CURRENT SMOKER	Dummy variable equal to 1 if respondent is a current smoker
LONGER	Dummy variable equal to 1 if respondent believes he will live longer than average
MANUAL	Dummy variable equal to 1 if manual laborer
RISK REDUCTION	Dummy variable equal to 1 if respondent was randomized into the group in which higher risk reduction scenarios were provided
LESSCB	Dummy variable equal to 1 if respondent believes chance of developing chronic bronchitis is lower than average
ALLERGY	Dummy variable equal to 1 if respondent experienced allergic reaction to either medicine or food

Table 8.2
Mean and standard deviation of respondent characteristics

Variable	Rural Areas	Anqing City	Beijing City
OWNCOST	0.67 (0.47)	0.55 (0.50)	0.52 (0.50)
HEALTH	2.41 (0.85)	2.11 (0.81)	2.16 (0.91)
HOUSEHOLD INCOME (yuan/month)	598 (1,426)	1,112 (1,000)	1,800 (2,000)
PEOPLE	4.4 (1.4)	3.6 (1.26)	3.52 (1.32)
EXERCISE	0.02 (0.16)	0.28 (0.45)	0.55 (0.50)
SEX (male)	0.54 (0.50)	0.43 (0.49)	0.45 (0.50)
AGE	36.9 (11.6)	42.8 (12.63)	42.88 (10.15)
EDUCATION	2.28 (0.96)	3.34 (1.18)	3.54 (0.98)
INSURANCE	0.03 (0.17)	0.50 (0.50)	0.55 (0.50)
CURRENT SMOKER	0.34 (0.47)	0.25 (0.44)	0.35 (0.48)
LONGER (life expectancy)	0.11 (0.31)	0.31 (0.46)	0.37 (0.48)
MANUAL	0.7 (0.46)	0.19 (0.39)	0.23 (0.42)
LESSCB	0.51 (0.50)	0.62 (0.48)	0.59 (0.49)
ALLERGY	0.05 (0.21)	0.06 (0.24)	0.15 (0.35)

areas, 37 years, was the youngest among the three locations. There were more men than women interviewed in the rural areas and more women than men in the two cities. Both the average household income and the education level were highest in Beijing and lowest in the rural areas. The average household size was largest in the rural areas and smallest in Beijing. About 50% of the respondents in Beijing and urban Anqing are covered by health insurance, offering either full or partial coverage, whereas only 4% of the respondents from rural areas are covered.

Distribution of willingness to pay As simple theory predicts, the probability that a respondent says "yes" to the initial bid declines with the bid amount. Our efforts to choose bids to capture the distribution of WTP were largely successful. For example, figure 8.1 shows the results for WTP to avoid an episode of cold. The fraction of respondents saying "yes" to the initial bid decreases with an increase in the initial bid in all three locations. Approximately 80% of respondents said "yes" to the smallest bid (4 yuan) in all three locations. The fraction saying "yes" to the largest bid (120 yuan) is less than 10% in the rural areas and Anquing, but about 35% in Beijing.

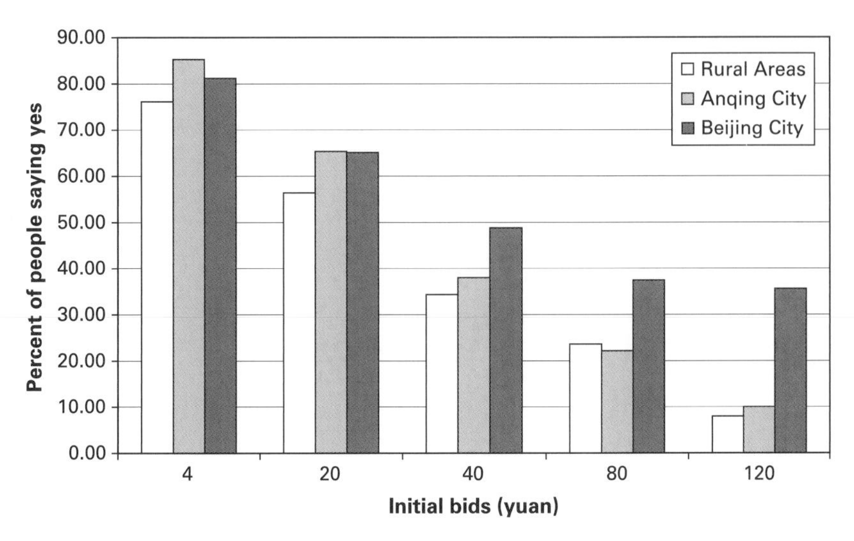

Figure 8.1
Percentage of people saying "yes" to initial bid for cold.

Comparison among regression models Figures 8.2 and 8.3 show a graphical comparison between the standard, two-part, and mixed models for cold and mortality data, respectively, for urban Anqing. To further compare the three models, χ^2 goodness-of-fit tests are used to compare the fit of the three models when no independent variables are included. Of the nine models—the cold, CB, and mortality models, each for the three locations—the mixed model fits best in six cases, the two-part model in two cases, and the standard model in one case.[3]

Regression coefficient estimates The signs and magnitudes of the parameter estimates in the standard, two-part, and mixed models are similar. Only results from the mixed model are reported here, because it is the best-fitting model overall. There are two components of the mixed model, which will be referred to in the following discussion as the part indicating if WTP is positive, and the part predicting the level of nonzero WTP.

Willingness to pay to prevent a cold Table 8.3 shows the characteristics of the most recent episode of cold that the respondents experienced. The duration ranged from 4.7 to 7.4 days, the number of work-loss days ranged from 0.2 to 1.2 days, and the number of symptoms ranged from 3.2 to 3.7.

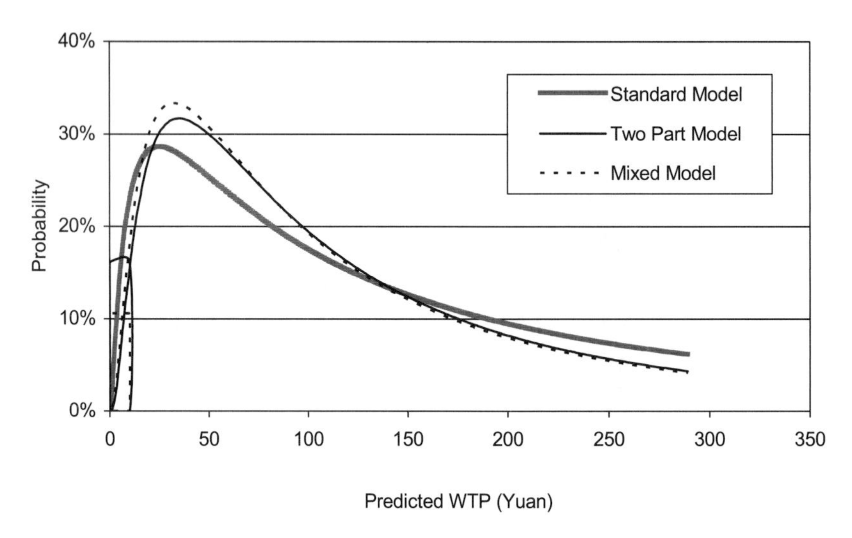

Figure 8.2
Comparison of predicted willingness to pay among three models for Anqing City cold data.

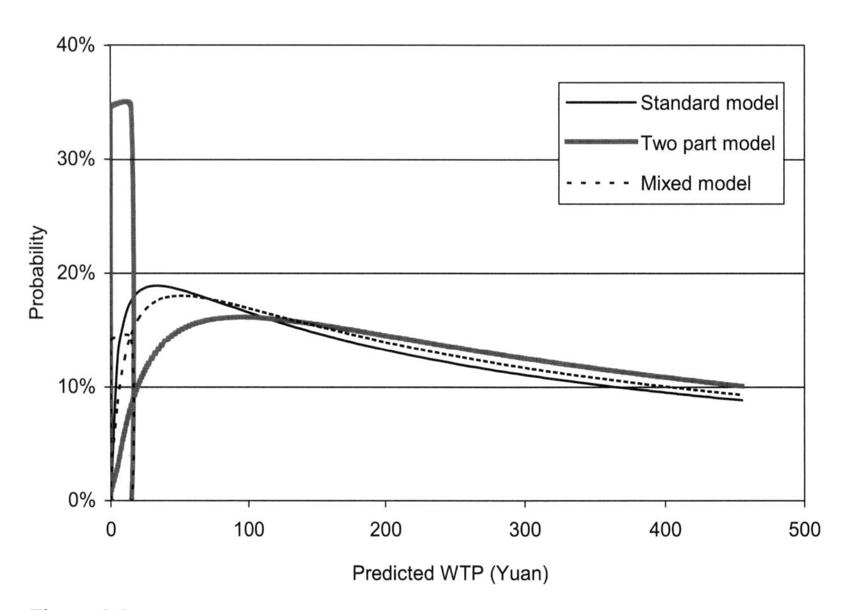

Figure 8.3
Comparison of predicted willingness to pay among the three models for Anqing City mortality risk data.

Table 8.3
Some characteristics of the most recent episode of cold

	Duration	Number of Work-Loss Days	Number of Symptoms
Rural areas	4.7	1.2	3.7
Anqing City	5.6	0.2	3.2
Beijing City	7.4	0.9	3.3

Table 8.4 shows the results of WTP to prevent an episode of cold. The results are similar in the three locations. Several variables in the part predicting the level of nonzero WTP are significantly different from zero in each of the three locations, whereas none is significant in the part indicating whether WTP is positive.

In the part predicting the level of nonzero WTP, *EXERCISE* and *OWNCOST* are significant in all three locations. Respondents who paid out of their own pockets for the last cold episode they experienced were willing to pay more to avoid a cold. In urban Anqing and rural areas, respondents who exercised voluntarily in their spare time had greater willingness to pay to avoid an episode of cold, but in Beijing such respondents had lower WTP. WTP was generally larger for younger respondents with higher income, more education, and poorer self-reported health, though these relationships are not statistically significant in all three locations. The estimated income elasticity of WTP to avoid cold is 0.27, 0.38, and 0.40 in the rural areas, urban Anqing, and Beijing, respectively.

Willingness to pay to reduce the risk of developing chronic bronchitis Table 8.5 shows the estimates of WTP to reduce the lifetime risk of developing CB from 18% to 17% or to 13%. Unlike the model for predicting WTP to avoid cold, most of the variables in the part indicating if WTP is positive are significantly different from zero in both Anqing and Beijing. Respondents with higher household income, who exercise in their spare time, who have a small household size, and males were more likely to give a positive WTP. The coefficient on "risk reduction" measures the log-odds ratio of having a positive WTP among respondents in the 5% versus 1% risk-reduction scenario. The coefficient is not significantly different from zero in any of the three locations.

In the part predicting the level of WTP, conditional on it being positive, the results are similar to those of the cold model. The income elasticity of WTP to reduce the risk of developing CB is 0.06, 0.16, and 0.17 in rural areas, urban Anqing, and Beijing, respectively.

Table 8.4
Mixed model for cold

	Rural Areas			Anqing City			Beijing City		
	Coefficient	SE	*P* Value	Coefficient	SE	*P* Value	Coefficient	SE	*P* Value
Part Predicting the Level of Nonzero WTP									
Scale	1.59	0.06	0.00	1.27	0.05	0.00	2.11	0.09	0.00
Intercept	1.72	0.39	0.00	1.56	0.34	0.00	3.25	0.67	0.00
HEALTH	0.18	0.07	0.01	0.15	0.06	0.01	0.11	0.08	0.19
INCOME	0.00	0.00	0.74	0.01	0.01	0.01	0.00	0.00	0.19
PEOPLE	0.09	0.04	0.02	−0.07	0.04	0.06	−0.01	0.06	0.84
EXERCISE	1.47	0.43	0.00	0.24	0.11	0.03	−0.34	0.15	0.03
SEX	0.21	0.14	0.14	0.08	0.12	0.50	−0.48	0.19	0.01
AGE	−0.02	0.01	0.01	0.00	0.00	0.28	−0.01	0.01	0.11
EDUCATION	0.09	0.07	0.22	0.35	0.05	0.00	0.12	0.09	0.17
INSURANCE	0.23	0.33	0.50	0.19	0.10	0.05	0.00	0.15	0.97
CURRENT SMOKER	0.25	0.14	0.07	−0.09	0.13	0.47	0.15	0.20	0.44
LONGER	0.15	0.19	0.41	0.27	0.10	0.01	0.11	0.16	0.50
MANUAL	−0.13	0.14	0.33	0.10	0.12	0.40	−0.31	0.19	0.09
OWNCOST	0.85	0.12	0.00	0.39	0.09	0.00	1.13	0.15	0.00

Part Indicating Whether WTP is Greater than Zero									
Intercept	15.69	211.92	0.94	17.55	288.64	0.95	17.56	170.96	0.92
HEALTH	2.07	30.09	0.95	−0.10	44.22	1.00	−0.09	15.39	1.00
INCOME	1.09	15.73	0.94	−0.02	13.00	1.00	0.07	10.09	0.99
PEOPLE	0.02	5.20	1.00	−0.15	17.95	0.99	−0.09	11.07	0.99
EXERCISE	−0.05	9.01	1.00	−0.07	20.88	1.00	−0.11	6.29	0.99
SEX	0.01	10.33	1.00	−0.06	57.69	1.00	−0.08	16.52	1.00
AGE	−0.10	0.13	0.42	0.09	43.80	1.00	0.12	11.06	0.99
INSURANCE	−0.27	12.95	0.98	0.11	23.83	1.00	0.08	8.63	0.99
CURRENT SMOKER	0.32	3.87	0.93	−0.07	15.73	1.00	−0.13	7.38	0.99
LONGER	0.43	21.09	0.98	−0.09	21.41	1.00	−0.09	6.10	0.99
MANUAL	−0.01	11.72	1.00	−0.08	12.53	0.99	−0.04	10.04	1.00
OWNCOST	0.01	5.16	1.00	0.99	50.65	0.98	0.98	9.13	0.91
EDUCATION	0.02	5.20	1.00	−0.08	23.08	1.00	0.08	9.85	0.99
Number of observations	1,219			1,206			1,156		

Table 8.5
Mixed model for chronic bronchitis

	Rural Areas			Anqing City			Beijing City		
	Coefficient	SE	*P* Value	Coefficient	SE	*P* Value	Coefficient	SE	*P* Value
Part Predicting the Level of Nonzero WTP									
Scale	1.59	0.08	0.00	1.47	0.07	0.00	1.69	0.08	0.00
Intercept	4.84	0.55	0.00	2.50	0.57	0.00	4.75	0.67	0.00
HEALTH	0.14	0.10	0.14	0.28	0.10	0.01	−0.04	0.09	0.64
INCOME	0.01	0.00	0.10	0.01	0.01	0.15	0.00	0.00	0.56
PEOPLE	−0.03	0.06	0.60	−0.04	0.06	0.49	−0.03	0.06	0.69
EXERCISE	1.03	0.44	0.02	0.04	0.17	0.80	0.10	0.16	0.52
SEX	0.47	0.19	0.01	0.06	0.20	0.76	0.18	0.25	0.46
AGE	−0.02	0.01	0.01	−0.01	0.01	0.06	0.00	0.01	0.99
EDUCATION	0.02	0.10	0.86	0.34	0.08	0.00	0.00	0.09	1.00
INSURANCE	0.11	0.40	0.79	−0.03	0.15	0.82	−0.08	0.17	0.64
CURRENT SMOKER	0.06	0.18	0.75	0.34	0.22	0.11	−0.34	0.24	0.16
LONGER	0.30	0.23	0.19	0.28	0.16	0.08	0.02	0.18	0.90
MANUAL	−0.18	0.17	0.31	−0.38	0.19	0.05	−0.31	0.20	0.11
LESS CB	−0.02	0.15	0.88	0.50	0.16	0.00	0.06	0.16	0.70
RISK REDUCTION	−0.07	0.15	0.62	0.14	0.14	0.33	0.20	0.16	0.21

Part Indicating Whether WTP Is Greater than Zero									
Intercept	11.30	97.86	0.91	31.17	14.86	0.04	11.58	5.48	0.03
HEALTH	2.27	35.86	0.95	13.23	7.32	0.07	−0.76	0.90	0.40
INCOME	1.76	20.07	0.93	3.29	1.65	0.05	2.54	0.46	0.00
PEOPLE	−1.64	27.94	0.95	−6.78	3.39	0.05	−2.76	0.33	0.00
EXERCISE	0.52	13.27	0.97	5.14	3.13	0.10	4.35	1.50	0.00
SEX	−1.18	23.43	0.96	13.60	6.80	0.05	4.04	2.48	0.10
AGE	0.02	0.86	0.99	0.16	0.12	0.20	0.24	0.11	0.04
EDUCATION	−0.58	9.46	0.95	9.50	4.95	0.05	−4.10	2.07	0.05
INSURANCE	0.03	13.13	1.00	−3.33	2.27	0.14	−9.17	2.27	0.00
CURRENT SMOKER	0.04	5.23	0.99	−4.19	2.10	0.05	6.18	0.96	0.00
LONGER	−0.01	8.14	1.00	−1.03	2.55	0.69	4.74	1.70	0.01
MANUAL	−0.02	12.93	1.00	−9.57	4.84	0.05	−6.04	1.18	0.00
LESS CB	−0.10	19.03	1.00	1.52	2.22	0.49	4.84	1.09	0.00
RISK REDUCTION	0.00	6.19	1.00	−3.30	2.24	0.14	−0.14	0.32	0.66
Number of observations	705			748			712		

In addition, we test for sensitivity of estimated WTP to magnitude of risk reduction by including the variable "risk reduction." The coefficient of this variable estimates the incremental WTP for a 5% reduction rather than a 1% reduction in the probability of developing CB. The estimated coefficient is small in absolute value with inconsistent signs and not significantly different from zero, which suggests respondents did not recognize the magnitude of the reduction.

Willingness to pay to reduce current mortality risk Table 8.6 reports the results of WTP to reduce mortality risk in the next year, by 1/1,000 or by 2/1,000. The estimated coefficients of most of the variables in both the part indicating whether WTP is positive and the part describing the level of nonzero WTP are significantly different from zero. Results across the three locations are similar. In the part indicating whether WTP is positive, *HEALTH*, *PEOPLE*, *CURRENT SMOKER*, and *MANUAL* are significant and with the same signs in all three locations. Respondents with poor self-perceived health or small household size, nonsmokers, and those performing manual labor were more likely to provide a positive WTP. In addition, those with higher household income, males, and those covered by health insurance appear to have a higher chance of positive WTP, though these relationships are not statistically significant in any of the three locations. The estimated coefficient of *RISK REDUCTION* is significantly positive in urban Anqing, which implies that respondents offered the larger risk reduction are more likely to provide a positive WTP.

In the part predicting the level of nonzero WTP, none of the variables is significant in all three locations. Variables that are significant with the same signs in two of the three locations are *INCOME*, *EXERCISE*, *AGE*, *EDUCATION*, *INSURANCE*, and *MANUAL*. These coefficients imply that WTP was higher for younger respondents with higher income, more education, insurance coverage, those doing manual labor and those who exercise in their spare time. The estimated coefficient of *RISK REDUCTION* in the lognormal distribution part is again small in absolute value and not significantly different from zero, which suggests respondents did not recognize the magnitude of the reduction specified in the survey.

The income elasticity of WTP to reduce mortality risk in the next year is 0.25 in urban Anqing, 0.10 in Beijing, and 0.06 in the rural areas.

Estimated willingness to pay Table 8.7, panels a–c, summarize the estimated mean and median WTP from the standard, two-part, and mixed models for the three health endpoints. Values are converted using an exchange rate of US$1 = 8 yuan to facilitate comparisons with results of other international studies.

The value of preventing a cold is estimated directly from the regression models. The values for CB and mortality estimate the value per statistical case. These are calculated by dividing WTP by the risk reduction. For example, on the basis of the standard model, median WTP to reduce mortality risk in the next year in Beijing is about US$22. The value per statistical life (VSL) is calculated using the following formula:[4]

$$VSL = \frac{1}{2} \times \left(\frac{WTP1}{1/1000} + \frac{WTP2}{2/1000} \right) = \frac{WTP}{2} \times \left(\frac{1000}{1} + \frac{1000}{2} \right) = \frac{WTP \times 3000}{4}$$
$$= WTP \times 750 = \$22 \times 750 = \$16{,}500.$$

The median WTP to prevent a cold in our study ranged from $3 to $6, which is about ten times smaller than median estimates for similar illness in the United States and also in Taiwan. For Taiwan, Alberini et al. (1996) and Liu et al. (2000) estimated values of about US$40 to prevent a cold. U.S. studies have typically valued one-day avoidance of various symptoms between US$10 and US$150 (see Liu et al. 2000). Chestnut et al. (1997) estimated median WTP for one-day avoidance of different symptoms as between $4 and $24 for adults surveyed in Bangkok.

We know of two previous U.S. studies to value prevention of chronic bronchitis. Viscusi et al. (1991) estimated the mean and median WTP for a statistical case as US$883,000 and US$457,000, respectively. Krupnick and Cropper (1992) administered a modified version of the Viscusi et al. questionnaire to persons who had a relative with chronic lung disease. They estimated the mean and median WTP for a statistical case of CB as US$2,080,000 and US$1,070,000.[5] The values we estimate for China, with means ranging from $1,600 to $3,400 and medians from $500 to $1,000, are about 1,000 times smaller. The large difference between our results and those of the U.S. studies may be due in part to the somewhat more severe set of symptoms described in the U.S. studies.

There are many estimates of VSL in the U.S. and other industrialized countries. These range between about US$100,000 and US$10 million. Reviewers of the literature based on compensating wage differentials for occupational hazards conclude that the most reasonable values are US$1.6–8.5 million (Fisher et al. 1989), US$1.5–2.5 million (Mrozek and Taylor 2002), and US$4–9 million (Viscusi and Aldy 2003). Recent CV estimates for the U.S. are US$2.4–5.7 million (median WTP; Corso et al. 2001) and US$1.5–4.8 million (mean) and US$700,000–1.1 million (median) (Alberini et al. 2004). Two wage-differential estimates for Taiwan are US$400,000 and US$600,000 (Liu et al. 1997; Liu and Hammitt 1999). Our

Table 8.6
Mixed model for mortality

	Rural Areas			Anqing City			Beijing City		
	Coefficient	SE	*P* Value	Coefficient	SE	*P* Value	Coefficient	SE	*P* Value
Part Predicting the Level of Nonzero WTP									
Scale	2.22	0.10	0.00	1.96	0.08	0.00	1.64	0.08	0.00
Intercept	4.46	0.69	0.00	2.59	0.27	0.00	6.21	0.84	0.00
HEALTH	−0.15	0.12	0.20	0.46	0.09	0.00	−0.02	0.40	0.95
INCOME	0.01	0.01	0.11	0.02	0.01	0.01	0.01	0.00	0.02
PEOPLE	0.13	0.07	0.06	−0.06	0.05	0.21	−0.12	0.07	0.08
EXERCISE	1.62	0.91	0.08	0.31	0.14	0.03	0.08	0.48	0.87
PEOPLE	0.35	0.12	0.00	0.02	0.06	0.80	−0.29	0.79	0.71
AGE	−0.05	0.01	0.00	−0.03	0.01	0.00	−0.01	0.01	0.31
EDUCATION	0.33	0.12	0.01	0.26	0.05	0.00	−0.06	0.29	0.83
INSURANCE	0.77	0.19	0.00	0.44	0.12	0.00	0.07	0.12	0.56
CURRENT SMOKER	0.41	0.21	0.05	0.16	0.12	0.19	0.05	0.38	0.89
LONGER	0.45	0.42	0.29	0.15	0.10	0.13	0.20	0.15	0.18
MANUAL	0.09	0.40	0.83	−0.31	0.14	0.03	−0.38	0.22	0.08
RISK REDUCTION	−0.06	0.17	0.72	0.11	0.16	0.49	−0.02	0.35	0.96

Part Indicating Whether WTP Is Greater than Zero									
Intercept	13.65	4.78	0.00	13.96	2.32	0.00	13.35	1.82	0.00
HEALTH	3.74	0.75	0.00	1.36	0.39	0.00	4.06	1.23	0.00
INCOME	1.74	0.12	0.00	3.33	0.32	0.00	1.90	1.49	0.20
PEOPLE	−4.73	0.63	0.00	−8.78	1.19	0.00	−5.35	1.09	0.00
EXERCISE	−0.77	0.32	0.02	6.86	2.33	0.00	0.07	0.06	0.25
PEOPLE	7.00	2.55	0.01	4.33	0.57	0.00	6.91	5.82	0.23
AGE	0.40	0.08	0.00	0.08	0.06	0.19	0.37	0.02	0.00
EDUCATION	−7.44	1.84	0.00	−5.87	0.68	0.00	−7.86	7.92	0.32
INSURANCE	−7.21	2.62	0.01	−10.41	1.28	0.00	−6.38	0.46	0.00
CURRENT SMOKER	2.52	1.05	0.02	−3.17	0.80	0.00	3.31	9.28	0.72
LONGER	2.08	0.31	0.00	0.94	0.21	0.00	1.69	0.52	0.00
MANUAL	−2.30	1.18	0.05	3.77	0.89	0.00	−3.05	2.39	0.20
RISK REDUCTION	0.32	0.65	0.62	1.65	0.49	0.00	−0.49	1.24	0.69
Number of observations	1,195			1,213			830		

Table 8.7
Estimated WTP per health effect

	Rural Areas		Anqing City		Beijing City	
	Mean/SD	Median/SD	Mean/SD	Median/SD	Mean/SD	Median/SD
a. Estimated WTP to Prevent Cold (in US$)						
Mixed model	10	3	8	4	51	6
	8	2	6	3	34	4
Standard model	10	3	8	4	51	6
	8	2	6	3	34	4
Two-part model	6	3	6	4	18	6
	6	3	10	6	13	4
b. Estimated Value per Statistical Case of Chronic Bronchitis (in US$)						
Mixed model	2,930	824	1,590	535	3,350	797
	2,040	574	1,050	364	1,030	255
Standard model	2,930	824	1,670	501	3,430	786
	2,040	574	1,190	356	1,230	284
Two-part model	2,810	873	1,570	615	2,810	1,000
	1,873	582	954	375	907	324
c. Estimated Value per Statistical Life (VSL; in US$)						
Mixed model	105,000	8,810	28,700	4,200	63,200	16,300
	231,000	19,400	30,400	4,480	32,000	8,310
Standard model	101,000	7,830	29,400	4,220	64,300	16,300
	175,000	13,600	41,200	5,910	36,100	9,170
Two-part model	178,000	40,500	14,900	4,780	45,500	16,900
	4,560,000	1,040,000	21,500	6,890	29,000	10,800

Table 8.8
Comparison of median WTP estimates from different studies

	China	United States	Taiwan
COLD	\$3–6/episode	\$12–154[a]	\$40/episode[b]
CB	\$500–\$1,000	\$0.8–\$1 million	
MORTALITY	\$4,200–\$16,900	\$1.6–\$8.5 million	\$413,000–\$624,000
INCOME per year per person	\$200–\$700	\$35,000	\$4,000–\$18,000

[a] Values are for one-day avoidance of various symptoms.
[b] An average episode has about 5.3 illness days and 2.2 symptoms.

estimates for China have an average mean between \$15,000 and \$178,000, and average median between \$4,200 and \$16,900.[6]

Table 8.8 summarizes the WTP estimates and annual average income from several studies. Taking the mid-points of the ranges in this table, average annual income per person in the U.S. studies is about eighty times larger than in our China sample, and average income in Taiwan is about twenty-five times larger than in the China sample. The differences in estimated VSL are more than proportionate to these income differences. The U.S. estimates are about 500 times larger, and the Taiwan estimates are about 50 times larger, than our estimates for China. In contrast, the differences in WTP to avoid a cold are much smaller. Both the U.S. and Taiwan estimates are about ten times larger than our estimates for China. One hypothesis for the seemingly low estimates of VSL for China is that the mortality risk reduction we offered in the CV questions (1 or 2 per 1,000) is much larger than the risk reduction typically presented in U.S. CV studies (typically parts per 10,000). Because our results suggest that estimated WTP is not very sensitive to the stated magnitude of risk reduction, it is possible that we would have estimated approximately the same WTP had we offered respondents a much smaller risk reduction of 1 or 2 per 10,000. If so, our estimated VSL would have been about ten times larger. This possibility suggests that our estimates of VSL for China should be interpreted cautiously, perhaps as lower-bound estimates.

8.5 Discussion

8.5.1 Comparison between Regression Models

Among all of the models for different health outcomes in different locations, there is a consistent trend in the predicted median WTP. The two-part model always produces the highest median, whereas the medians predicted by the mixed and standard models are similar.[7] The two-part model yields the smallest predicted mean WTP in eight of the nine cases.

As a significant proportion (between 4.4% and 34.6%) of respondents indicated that their maximum WTP was zero, the two-part and mixed models are more compatible than the standard model with the data. The standard model assumes that WTP is strictly positive. The difference between the two-part model and the mixed model is that the former makes use of additional information from the follow-up maximum WTP question, whereas the latter does not. If answers to the open-ended question are reliable, the two-part model has an advantage over the mixed model of making use of additional information; however, if respondents cannot reliably distinguish between zero and very small but positive values of WTP, the mixed model has an advantage over the two-part model by making use of only the dichotomous-choice questions. Dichotomous-choice questions are believed to be easier for respondents to answer and to yield unbiased responses. In contrast, estimates of WTP from open-ended questions are believed to be biased downward (e.g., Hammitt et al. 2001).

8.5.2 Scope Sensitivity

As discussed above, we varied the magnitude of risk reduction between respondents in order to test for respondent understanding of the CV questions. Respondents offered a larger risk reduction should have a greater WTP. If not, this suggests respondents did not adequately appreciate the magnitude of the risk reduction presented to them (Hammitt and Graham 1999). The coefficient for the variable *RISK REDUCTION* is never significantly different from zero in the part of the mixed model predicting the level of nonzero WTP. In the part indicating whether WTP is positive, the coefficient is significant only once, in the urban Anqing mortality mixed model, with a positive sign.

To further explore scope sensitivity, interaction terms between *RISK REDUCTION* and other variables were added. For the urban Anqing mortality model, when the interaction between the number of people in the household and *RISK REDUCTION* was added to the standard and the mixed models, *RISK REDUCTION* is significant with an estimated value of 0.91 and a P value of 0.05. The estimated coefficient of the interaction term is −0.22 with a P value of 0.08. Thus, WTP of respondents who were offered the higher risk-reduction scenario is larger than those offered the lower risk reduction by exp(0.91 − 0.22 × number of people in the household). This provides some evidence of scope sensitivity for small households.

For the standard model of WTP to reduce mortality risk in Beijing and the rural areas, when the interaction between income and *RISK REDUCTION* is added to the model, the parameter estimates for *RISK REDUCTION* are 0.41 and 0.32, respectively, with P values of 0.06 and 0.10. Parameter estimates for the interaction terms are negative (−0.026 and −0.06) with P values less than 0.01. That is, estimated WTP for respondents who were offered the larger risk reduction is larger than for those offered the smaller risk reduction in Beijing and the rural areas by $\exp(0.41 - 0.026 \times \text{income})$ and $\exp(0.32 - 0.06 \times \text{income})$, respectively. Although the income effects in both locations are positive and significant, the negative sign for the interaction term suggests that scope sensitivity declines as income increases.

8.6 Conclusion

We have presented results from a large CV study of the value of reducing the risk of adverse health effects in China. Although estimates of WTP are not significantly related to many of the explanatory variables, they do exhibit a degree of internal consistency. The estimated values per effect vary widely, with prevention of premature fatality valued at more than ten times the value of prevention of chronic bronchitis. This, in turn, is valued at 40–200 times the value of preventing a cold; however, the insensitivity of estimated WTP to the magnitude of the risk reduction in the main models for CB and fatality risks (without interaction terms) suggests that respondents did not clearly understand the risk reductions they were asked to value, and casts some doubt on the results. Moreover, because we observe little difference in WTP in response to differences in stated risk reduction, it is possible that the difference between the estimated value per statistical fatality and value per statistical case of CB may be artifacts, at least in part, of the different risk reductions offered for these health endpoints.

Comparing among the three locations, our estimates appear to be more similar between Beijing and the rural areas than between these areas and Anqing. Except for colds, the estimates of WTP in urban Anqing are substantially smaller than the estimates for the other locations. Because the income level in the Anqing sample is intermediate to those of the rural and Beijing samples, an intermediate valuation of health would be more plausible. This pattern might reflect social changes that are taking place in China. For example, the reform of state-owned enterprises may have influenced small cities such as Anqing more substantially than big cities such as Beijing, where there are many job opportunities outside the state-owned enterprises, and more than the rural areas, where most people are self-employed. If respondents in Anqing believe their future income is at greater risk than respondents in the other communities, their WTP for health could be smaller (Eeckhoudt and Hammitt 2001).

Our estimates of willingness to pay to reduce health risk are between about 10 and 1,000 times smaller than estimates for the United States and for Taiwan. The large income difference between China and the United States and Taiwan is surely an important explanatory factor. There may be other factors, however, including differences in culture, health care systems, and experience with a market economy and with surveys, that contribute to estimated differences in WTP. In future work, it would be useful to identify and investigate some of these differences.

This study demonstrates that it is possible to conduct a large-scale contingent valuation in China, not only in major cities like Beijing, but also in smaller cities and rural areas as well. We found that respondents were able to answer double-bounded dichotomous-choice questions, and the pattern of responses satisfies the minimal requirement that the probability of accepting a bid is larger for small bids than for large ones. The failure to find a difference in WTP between respondents offered larger and smaller risk reductions is disappointing, though not unanticipated. Many CV studies of health risk in other countries have found that WTP is either insensitive, or only modestly sensitive, to the stated risk reduction (Hammitt and Graham 1999). This insensitivity suggests that the respondents did not understand or adequately consider the magnitude of the risk reduction stated in the survey. In future work, it would be useful to devote additional attention to understanding how Chinese understand health risks, probability, and changes in risk, and to determine whether other methods of communicating changes in probability would be more effective. In addition, it might be useful to provide respondents with additional training in thinking about changes in risk, potentially including practice questions as part of the contingent-valuation instrument.

As noted above, the apparent lack of sensitivity of estimated willingness to pay to the stated magnitude of the risk reduction suggests that we would have estimated similar values of WTP, and thus larger values per statistical case, had we asked about smaller risk reductions. This concern may be particularly important for our estimates of VSL, because we asked about risk reductions of a few parts per thousand per year, whereas CV studies in other countries have often asked about risk reductions ten times smaller. In future work, it would be valuable to test for differences in WTP between substantially different stated magnitudes of risk reduction to investigate how sensitive the estimated value per statistical case is to the stated magnitude of risk reduction.

Acknowledgments

The research of this chapter was generously funded by a grant to the China Project of the Harvard University Center for the Environment from the V. Kann Rasmussen Foundation, with additional funding from the Bedminster Foundation and the Dunwalke Trust. We thank Xiping Xu, Binyan Wang, Beijing Medical University and Anhui Biomedical Research Institute for help administering the survey and Scott Venners for advice and help with the data. Another analysis of these data is reported in *Environmental and Resource Economics* (Hammitt and Zhou 2006).

Notes

1. This total includes an estimate of unregistered, floating population and is for 2000 (Beijing Municipality Statistical Bureau 2000). See appendix A for more information.
2. The log likelihoods of the three distributions (lognormal, Weibull, and normal) without independent variable are as follows:

	Lognormal	Weibull	Normal
Cold (Anqing City)	−1300	−1278	−1695
CB (Anqing City)	−719	−729	−1444
Mortality (Anqing City)	−1120	−1124	−2460
Cold (Beijing City)	−1312	−1298	−1993
CB (Beijing City)	−806	−803	−1262
Mortality (Beijing City)	−954	−963	−1523
Cold (Rural areas)	−1360	−1354	−1990
CB (Rural areas)	−783	−791	−1336
Mortality (Rural areas)	−1227	−1227	−2478

3. The χ^2 goodness-of-fit test statistics for each sample and critical value of the 95% significance level are as follows:

	Standard Model	Two-Part Model	Mixed Model	Critical Value	Degrees of Freedom
Cold (Anqing City)	111	24	28	33	21
CB (Anqing City)	46	60	46	33	21
Mortality (Anqing City)	91	97	80	33	21
Cold (Beijing City)	100	63	53	38	25
CB (Beijing City)	32	40	28	28	17
Mortality (Beijing City)	165	208	162	38	25
Cold (Rural areas)	56	54	39	33	21
CB (Rural areas)	77	58	77	33	21
Mortality (Rural areas)	55	642	339	33	21

4. WTP1 corresponds to WTP of respondents in the 1/1000 risk reduction scenario. WTP2 corresponds to WTP of respondents in the 2/1000 risk reduction scenario. WTP1 and WTP2 are assumed to be the same as represented by WTP in the VSL calculation, since the coefficient for the "risk reduction" variable is not statistically significant.

5. The sample size is seventy. In addition, it is noted in the paper that "those familiar with the disease reveal a significantly higher mean WTP for chronic-bronchitis risk reductions." This suggests that these estimates may not be representative of the general population.

6. The median VSL estimate for the rural areas from the two-part model is not included in the range, because this model provides a very poor fit. The estimated VSL from the two-part model is much larger than those from the other models in all locations.

7. In the cold models, the median estimates from the mixed, standard, and two-part models are the same.

References

Alberini, Anna. 1995. Efficiency vs. bias of willingness-to-pay estimates: Bivariate and interval-data models. *Journal of Environmental Economics and Management* 29:169–180.

Alberini, Anna, Maureen Cropper, Tsu-Tan Fu, Alan Krupnick, Jin-Tan Liu, Daigee Shaw, and Winston Harrington. 1996. Value of reduced morbidity in Taiwan. In *The economics of pollution control in the Asian Pacific*, ed. Robert Mendelsohn and Daigee Shaw. London: Edward Elgar Publishing.

Alberini, Anna, Maureen Cropper, Alan Krupnick, and Nathalie B. Simon. 2004. Does the value of a statistical life vary with age and health status? Evidence from the US and Canada. *Journal of Environmental Economics and Management* 48:769–792.

Arrow, K., R. Solow, P. Portney, E. Leamer, R. Radner, and H. Schuman. 1993. Report of NOAA panel on contingent valuation. *Federal Register* 58 (January 11):4601–4614.

Beijing Municipality Statistical Bureau. 2000. *Beijing social and economics statistical yearbook: 2000*. Beijing: China Statistics Press. In Chinese.

Carson, Richard A. 2000. Contingent valuation: A users guide. *Environmental Science and Technology* 34 (8):1413–1418.

Chestnut, Lauraine G., Bart D. Ostro, and Nuntavarn Vichit-Vadakan. 1997. Transferability of air pollution control health benefits estimates from the United States to developing countries: Evidence from the Bangkok study. *American Journal of Agricultural Economics* 79:1630–1635.

Corso, Phaedra S., James K. Hammitt, and John D. Graham. 2001. Valuing mortality-risk reduction: Using visual aids to improve the validity of contingent valuation. *Journal of Risk and Uncertainty* 23 (2):165–184.

Diamond, Peter A., and Jerry A. Hausman. 1994. Contingent valuation: Is some number better than no number? *Journal of Economic Perspectives* 8 (4):45–64.

Duan, Naihua, Willard G. Manning Jr., Carl N. Morris, and Joseph P. Newhouse. 1984. Choosing between the sample selection model and the multi-part model. *Journal of Business and Economic Statistics* 2 (3):283–289.

Eeckhoudt, Louis R., and James K. Hammitt. 2001. Background risks and the value of a statistical life. *Journal of Risk and Uncertainty* 23 (3):261–279.

Fisher, Anne, Lauraine G. Chestnut, and Daniel M. Violette. 1989. The value of reducing risks of death: A note on new evidence. *Journal of Policy Analysis and Management* 8 (1):88–100.

Fu, Tsu-Tan, Jin-Tan Liu, and James K. Hammitt. 1999. Consumer willingness to pay for low-pesticide fresh produce in Taiwan. *Journal of Agricultural Economics* 50 (2):220–233.

Hammitt, James K. 2000. Valuing mortality risk: Theory and practice. *Environmental Science and Technology* 34 (8):1396–1400.

Hammitt, James K., and John D. Graham. 1999. Willingness to pay for health protection: Inadequate sensitivity to probability? *Journal of Risk and Uncertainty* 18 (1):33–62.

Hammitt, James K., and Jin-Tan Liu. 2004. Effects of disease type and latency on the value of mortality risk. *Journal of Risk and Uncertainty* 28 (1):73–95.

Hammitt, James K., Jin-Tan Liu, and Jin-long Liu. 2001. Contingent valuation of a Taiwanese wetland. *Environment and Development Economics* 6 (2):259–268.

Hammitt, James K., and Ying Zhou. 2006. The economic value of air-pollution-related health risks in China: A contingent valuation study. *Environmental Resource Economics* 33 (3):399–423.

Hanemann, W. Michael. 1994. Valuing the environment through contingent valuation. *Journal of Economic Perspectives* 8 (4):19–43.

Hanemann, Michael, John Loomis, and Barbara Kanninen. 1991. Statistical efficiency of double-bounded dichotomous choice contingent valuation. *American Journal of Agricultural Economics* 73 (4):1255–1261.

Krupnick, Alan, and Maureen L. Cropper. 1992. The effect of information on health risk valuations. *Journal of Risk and Uncertainty* 5 (1):29–48.

Li, Ying, Mo Bai, Kai-zhong Yang, and Xue-jun Wang. 2001. Study on residents' willingness to pay for improving air quality in Beijing. *Urban Environment and Urban Ecology* 14 (5):6–8. In Chinese.

Liu, Jin-Tan, and James K. Hammitt. 1999. Perceived risk and value of workplace safety in a developing country. *Journal of Risk Research* 2 (3):263–275.

Liu, Jin-Tan, James K. Hammitt, and Jin-Long Liu. 1997. Estimated hedonic wage function and value of life in a developing country. *Economics Letters* 57 (3):353–358.

Liu, Jin-Tan, James K. Hammitt, Jung-Der Wang, and Jin-Long Liu. 2000. Mother's willingness to pay for her own and her child's health: A contingent valuation study in Taiwan. *Health Economics* 9 (4):319–326.

Liu, Jin-Tan, James K. Hammitt, Jung-Der Wang, and Meng-Wen Tsou. 2005. Valuation of the risk of SARS in Taiwan. *Health Economics* 14 (1):83–91.

Mrozek, Janusz R., and Laura O. Taylor. 2002. What determines the value of life? A meta-analysis. *Journal of Policy Analysis and Management* 21 (2):253–270.

Peng, Xi-zhe, and Wen-hua Tian. 2003. Study on willingness to pay due to air pollution in Shanghai. *World Economic Forum* 2:32–44. In Chinese.

U.S. Environmental Protection Agency (U.S. EPA). 1999. The benefits and costs of the Clean Air Act, 1990 to 2010. Report no. EPA 410-R-99-001. Washington, D.C.

Viscusi, W. Kip, and Joseph E. Aldy. 2003. The value of a statistical life: a critical review of market estimates throughout the world. *Journal of Risk and Uncertainty* 27 (1):5–76.

Viscusi, W. Kip, Wesley A. Magat, and Joel Huber. 1991. Pricing environmental health risks: Survey assessment of risk-risk and risk-dollar trade-offs for chronic bronchitis. *Journal of Environmental Economics and Management* 21 (1):32–51.

Werner, Megan. 1999. Allowing for zeros in dichotomous-choice contingent-valuation models. *Journal of Business and Economic Statistics* 17 (4):479–486.

Zhang, Ming-jun, Jian-feng Fan, Chen-xia Hu, and Bo Zhang. 2004. Assessment of total economic value of improving atmospheric quality of Lanzhou. *Journal of Arid Land Resources and Environment* 18 (3):28–32. In Chinese.

9

Sector Allocation of Emissions and Damage

Mun S. Ho and Dale W. Jorgenson

Many of the previous estimates of damages from air pollution in China have focused on total damages. For example, World Bank (1997) calculates the damage to health due to the observed levels of ambient pollution concentrations in Chinese cities, that is, the pollution coming from all sources: local human activity, desert dust, and imported pollution. Such an aggregate number does not estimate the contributions of different source types, for instance how much can be attributed to the electric power industry, the cement industry, motor vehicles, and so forth. Without such sector estimates, policy makers are unable to prioritize pollution control efforts or efficiently allocate the environmental protection budget. For example, if they chose to put a national cap on sulfur dioxide (SO_2) emissions, how much of a quota should be allocated to the cement industry? Or if they decide to auction off SO_2 emission rights, how would they estimate the health benefits of such a policy or the effects on the economy? Answering these questions requires knowledge of effects at the industry level.

Other studies have focused in detail on particular sectors or cities, for instance estimating the damage due to a particular power plant or a particular industry (e.g., Feng 1999; Hirschberg et al. 2003; Wang and Mauzerall 2006). That important research gives an indication of the damage incurred, so many millions of yuan annually for a particular plant, say, and this figure may suggest the appropriate level of pollution control effort for that plant. Such studies, however, do not give a systematic view of all the sources and do not allow us to answer the above questions.

In this chapter, we estimate the health damage due to air pollution of each sector of the economy, including households. We allocate the share of total damages to each of thirty-three industries and also estimate the damages per unit of industry output—for example, the damage done per yuan of cement produced. The results of these estimates are needed to examine the relative benefits of various pollution control policies, as we shall show in the next chapter.

To allocate total pollution damage to the various sources, one must first identify the emissions from each sector and second relate how these emissions contribute to the ambient concentration of pollutants. It is probably obvious that different sectors produce different levels of emissions, both in absolute terms as well as emissions per yuan of output. What may be less obvious is that each ton of, say, particulates emitted from different sectors may cause different amounts of health damage.

The discussions of the intake fraction (*iF*) in chapters 4–7 show why. First, the location of the source of emissions is important, because the proximity to population centers and the meteorological conditions determine the amount of damage to a large degree. Second, source characteristics such as stack height, emission temperature, and velocity matter (e.g., higher chimneys allow emissions to be blown away from population centers). Third, in the case of primary particulate matter, the particle size distribution is important, because the transportation rates and the health impacts are different for different particle diameters. Other factors include those that affect the rate of chemical reactions that generate secondary particles.

It is therefore important to identify the major sources of *health damage* rather than just the sources of emissions to develop efficient pollution control strategies. In the sections below, we describe our estimates of sector emissions, how they relate to industry characteristics such as output and fuel use, and how these are translated to estimates of health damages. We concentrate on two pollutants, total suspended particulates (TSP) and SO_2, for now, recognizing that other emissions such as mercury and nitrogen oxides also affect human health.

9.1 Output and Emissions by Sector

The policy analysis in chapter 10 is based on a model of the Chinese economy that is built around the 1997 input-output table, the most recent official accounts of interindustry flows. That model divides the economy into thirty-three production sectors, plus one nonproduction sector (households). The emission and fuel use data here are constructed to be consistent with that framework.

Table 9.1 gives the emissions of TSP and SO_2 for the thirty-four sectors in 1997. These estimates are modified from data in the *China Environmental Yearbook* (1998–2000) and the *China Energy Yearbook* (Fridley et al. 2001), which provide estimates for fourteen sectors at our level of classification. (Estimates for finer particle sizes such as PM_{10} [particles less than 10 microns in diameter] are not available at the national level.) We divide these aggregate estimates into our more detailed sectors using fuel emission factors provided by Lvovsky and Hughes (1997) and the fuel use data in the input-output table. The details of the disaggregation procedures are described below.

The official emission data were massively revised in 1995 and subsequent years to incorporate information for town and village enterprises that were previously ignored.[1] There is also a dramatic decline in the TSP estimates for the nonindustrial sectors, from 6.4 million tons (mt) in 1995 to 3.1 in 1997 and 2.1 in 1999. These are official estimates based on emission factors, not actual measurements of total emissions. We believe that the newer figures reflect more accurate estimation methods rather than changes in actual emissions, and so we use the 1998 and 1999 data to adjust our 1997 estimates when possible (as detailed below).

The Chinese emission data are divided into two types: combustion and process. The former describe emissions from the burning of fossil fuels and the latter from mechanical or other chemical processes. The data for the industrial sectors (mining, manufacturing, utilities) are estimated for twenty sectors, whereas only one total is provided for the remaining nonindustrial sectors (agriculture, construction, transportation, services, and households). We allocate these official estimates to our thirty-four sectors using equations to be described (equations [9.1]–[9.3]), and the estimates for the nonindustrial sectors are obviously quite rough. We should also note that the household sector is widely believed to be getting much cleaner with the substitution of gas for coal in urban residential heating and cooking; there is, however, no systematic national data on this.

Table 9.1 shows that the scales of combustion and process emissions of TSP are similar. The major sectors contributing to the national total of 11.4 million tons (Mt) of combustion TSP are electricity (4.0 Mt), nonmetal mineral products (mostly cement, 2.5 Mt), chemicals (0.7 Mt), and metals smelting (chiefly iron and steel, 0.6 Mt).[2] For process TSP, 92% is from cement and metals smelting. One can say that the widespread use of coal for manufacturing purposes, as well as for heating, means that most sectors contribute significant amounts of pollution, and there were only a few relatively "clean" sectors in 1997.

Emissions of SO_2 are predominantly from combustion, which accounted for 17.4 Mt nationally. Again the major sources are electricity (7.8 Mt), nonmetal mineral products (cement, 1.5 Mt) and chemicals (1.2 Mt). The sector pattern is similar to that of combustion TSP, except that the cement sector emits proportionately less SO_2.

Table 9.1
Sector emissions, 1997

Sector	TSP (kiloton)			SO_2 (kiloton)		
	Combustion	Process	Total	Combustion	Process	Total
1 Agriculture	160	0	160	367	0	367
2 Coal mining and processing	181	65	245	163	29	192
3 Crude petroleum mining	57	51	108	78	23	101
4 Natural gas mining	0	3	3	1	1	2
5 Metal ore mining	22	31	53	21	14	35
6 Nonferrous mineral mining	44	60	104	45	27	72
7 Food products, tobacco	310	16	326	467	11	478
8 Textile goods	166	3	169	332	2	334
9 Apparel, leather	16	1	17	24	0	24
10 Sawmills and furniture	90	46	136	114	4	118
11 Paper products, printing	201	92	292	254	8	261
12 Petroleum refining and coking	72	76	148	112	51	163
13 Chemical	672	200	872	1,216	214	1,429
14 Nonmetal mineral products	2,549	10,756	13,305	1,525	654	2,178
15 Metals smelting and pressing	632	1,244	1,876	739	925	1,664
16 Metal products	39	84	123	26	8	34
17 Machinery and equipment	125	24	150	185	4	189
18 Transport equipment	43	16	59	65	3	67
19 Electrical machinery	28	15	43	44	3	47
20 Electronic and telecommunications equipment	8	13	22	13	2	16
21 Instruments	6	3	9	9	0	10
22 Other manufacturing	83	99	182	181	55	235
23 Electricity, steam, hot water	3,953	72	4,025	7,846	49	7,895
24 Gas production and supply	16	3	19	24	2	26
25 Construction	121	0	121	414	0	414
26 Transport and warehousing	361	0	361	609	0	609
27 Post and telecommunication	0	0	0	6	0	6
28 Commerce and restaurants	118	0	118	273	0	273
29 Finance and insurance	13	0	13	28	0	28
30 Real estate	72	0	72	124	0	124

Table 9.1
(continued)

Sector	TSP (kiloton)			SO_2 (kiloton)		
	Combustion	Process	Total	Combustion	Process	Total
31 Social services	222	0	222	441	0	441
32 Health, education, other services	416	0	416	714	0	714
33 Public administration	112	0	112	216	0	216
Households	462	0	462	778	0	778
Total	11,369	12,974	24,343	17,452	2,089	19,541

Source: *China Environment Yearbook* (various issues) and authors' calculations. Estimates are for 1997, based on information in 1997–1999.

How do these emissions relate to the fuel mix and output levels of the different sectors? In table 9.2 we provide estimates of output (in 1997 yuan), the coal, oil, and gas inputs of each sector, along with the combustion TSP from table 9.1. The fuel inputs for each sector are derived from the yuan values in the input-output table and converted to quantities by assuming that all sectors pay a common price. The price is chosen such that the quantity of national coal, oil, and gas outputs are equal to the official total output.[3] Output is the sum of domestic purchases, net exports, and changes in inventories.

Even a casual examination will show that there is a wide variation of emissions of TSP per ton of coal or oil combusted. We are more specific below, but for now we note that the largest emitters are the largest users of coal: electricity, nonmetal mineral products, and metals smelting. The transportation sector is a small user of coal, but the second-largest user of oil and is thus also a large emitter of TSP. At the other end, the cleanest major sectors are communications, finance, electronic equipment, and apparel. Some of the nonmanufacturing sectors, including households, have surprisingly high estimated emissions because of their high use of coal for heating.

In table 9.3 we present a different angle of the output-emission relationship. In addition to gross output, which is the value of the industry's goods to its customers, we also give the value added, which is the output minus intermediate inputs, essentially the value of labor and capital. The column marked "Energy Value" is the amount paid for purchases from the coal, oil (crude and refined), gas, and electricity sectors (i.e., both primary and secondary energy inputs, including feedstocks). Finally, the column marked "Energy Use" lists the official estimates of the energy inputs in standard coal equivalents derived from quantity data on coal, oil, gas, and

Table 9.2
Emissions, fuel use, and output, 1997

Sector	Combustion TSP (kiloton)	Gross Output (billion yuan)	Coal Use (million tons)	Oil Use (million tons)	Gas Use (million m^3)
1 Agriculture	159.7	2,467.7	12.7	9.61	0.0
2 Coal mining and processing	180.7	238.0	40.8	2.72	7.2
3 Crude petroleum mining	56.9	188.1	12.7	15.09	5.0
4 Natural gas mining	0.3	10.9	0.1	0.24	627.6
5 Metal ore mining	22.0	115.2	5.0	1.06	2.4
6 Nonferrous mineral mining	44.0	222.4	9.9	3.32	3.1
7 Food products, tobacco	310.4	1,371.3	33.2	1.74	44.3
8 Textile goods	166.1	913.0	21.8	1.06	31.5
9 Apparel, leather	16.4	627.0	6.7	0.76	0.0
10 Sawmills and furniture	90.2	222.5	8.1	0.57	0.0
11 Paper products, printing	200.5	445.5	18.1	1.34	14.5
12 Petroleum refining and coking	71.6	308.5	20.0	8.75	421.7
13 Chemical	671.8	1,488.8	148.5	32.08	4,541.7
14 Nonmetal mineral products	2,548.8	874.9	206.7	10.34	1,442.4
15 Metals smelting and pressing	632.0	750.2	183.2	8.81	968.9
16 Metal products	38.9	477.8	18.3	1.80	69.1
17 Machinery and equipment	125.4	882.1	36.2	3.81	366.8
18 Transport equipment	42.9	570.7	12.4	1.88	247.6
19 Electrical machinery	28.0	563.3	8.1	1.93	156.1
20 Electronic and telecommunications equipment	8.3	488.1	2.4	0.68	260.3
21 Instruments	6.0	96.0	1.7	0.26	1.6
22 Other manufacturing	82.8	286.9	13.7	1.08	6.4
23 Electricity, steam, hot water	3,953.0	380.8	384.9	14.80	538.5
24 Gas production and supply	15.5	12.4	4.4	1.40	25.4
25 Construction	120.8	1,738.6	9.4	20.63	0.0
26 Transport and warehousing	361.0	506.6	17.5	23.07	2.9
27 Post and telecommunication	0.1	195.9	0.0	0.54	0.0
28 Commerce and restaurants	118.2	1,412.6	9.4	7.19	111.7
29 Finance and insurance	13.4	359.5	1.1	0.51	0.0
30 Real estate	72.3	185.5	5.8	0.19	0.0

Table 9.2
(continued)

Sector	Combustion TSP (kiloton)	Gross Output (billion yuan)	Coal Use (million tons)	Oil Use (million tons)	Gas Use (million m^3)
31 Social services	221.8	564.5	17.8	6.55	36.3
32 Health, education, other services	416.0	658.5	33.5	1.15	38.5
33 Public administration	111.7	443.4	9.0	2.76	1.5
Households	461.7		37.1	2.11	1,990.8
Total	11,369.1	20,067	1,350.3	189.8	11,963.6

Note: Fuel use is combustion, excluding the transformation to secondary fuels and products.
Source: Input-output table, authors' calculations.

electricity used. These exclude transformation to secondary energy products (NBS 2001, table 5.5; NBS 2000, table 7.3).

From these data we calculate the energy and emission ratios given in table 9.4. The first column is the value share of energy in output (i.e., the share of total production costs devoted to energy inputs, including those used for feedstocks). The sectors with the largest energy shares are the energy-transformation sectors: refining and coking, gas production, and electricity. Next are the mining industries, transportation, metals smelting (iron and steel), and nonmetal mineral products (cement) industries with shares greater than 10%. In the third column is the ratio of total energy use expressed in coal equivalents to output, that is, the use of coal, oil, gas, and electricity converted to a common heat unit and covering only combustion (i.e., excluding their use as feed stocks). The sectors with the largest energy-to-output ratios are gas production, metals smelting, electricity and coal mining, with ratios greater than 0.24 kg of coal equivalent per yuan of output. The petroleum refining sector transforms a large portion of its purchased energy inputs into secondary products and so has an energy use-to-output ratio much smaller than the value share of total energy input.

In the last column we calculate the emissions of TSP per yuan of energy input purchased. The refining and chemicals sectors purchase a lot of fuels as feedstock and thus have low emission-to-energy purchased ratios. For the other sectors that do little transformation, their ratios of emissions-to-energy-purchased give convenient summaries of the separate information for coal, oil, and gas inputs given in table 9.2. The highest emissions per yuan of energy purchased is for nonmetal mineral

Table 9.3
Output and energy use

Sector	TSP Emissions (kton)	Gross Output (billion yuan)	Value Added (billion yuan)	Energy Value (billion yuan)	Energy Use (million tons of coal equivalent)
1 Agriculture	160	2,467.7	1,474.2	42.1	59.05
2 Coal mining and processing	245	238.0	118.2	23.2	57.91
3 Crude petroleum mining	108	188.1	123.0	27.2	33.25
4 Natural gas mining	3	10.9	5.9	1.4	2.40
5 Metal ore mining	53	115.2	39.9	12.2	8.18
6 Nonferrous mineral mining	104	222.4	96.6	24.4	13.49
7 Food products, tobacco	326	1,371.3	381.4	21.4	38.43
8 Textile goods	169	913.0	258.1	14.4	30.80
9 Apparel, leather	17	627.0	193.1	5.6	4.37
10 Sawmills and furniture	136	222.5	63.1	6.2	5.01
11 Paper products, printing	292	445.5	139.1	16.7	21.99
12 Petroleum refining and coking	148	308.5	71.9	156.4	73.89
13 Chemical	872	1,488.8	402.4	136.0	192.80
14 Nonmetal mineral products	13,305	874.9	276.0	97.0	123.17
15 Metals smelting and pressing	1,876	750.2	159.2	87.3	214.48
16 Metal products	123	477.8	116.3	23.0	10.45
17 Machinery and equipment	150	882.1	282.4	30.5	24.48
18 Transport equipment	59	570.7	158.2	13.4	15.19
19 Electrical machinery	43	563.3	130.2	12.1	6.47
20 Electronic and telecommunications equipment	22	488.1	124.4	5.4	4.93
21 Instruments	9	96.0	29.3	2.0	0.83
22 Other manufacturing	182	286.9	125.5	7.8	13.25
23 Electricity, steam, hot water	4,025	380.8	163.8	107.7	100.76
24 Gas production and supply	19	12.4	3.3	5.8	4.27
25 Construction	121	1,738.6	499.7	63.0	11.79
26 Transport and warehousing	361	506.6	279.8	64.2	68.43
27 Post and telecommunication	0	195.9	112.6	6.2	7.00
28 Commerce and restaurants	118	1,412.6	726.3	31.0	23.94

Table 9.3
(continued)

Sector	TSP Emissions (kton)	Gross Output (billion yuan)	Value Added (billion yuan)	Energy Value (billion yuan)	Energy Use (million tons of coal equivalent)
29 Finance and insurance	13	359.5	219.5	3.4	1.82
30 Real estate	72	185.5	140.8	2.4	2.63
31 Social services	222	564.5	224.3	26.0	14.55
32 Health, education, other services	416	658.5	314.8	21.5	19.41
33 Public administration	112	443.4	200.0	14.7	8.61
Households	462			62.2	163.68
Total	24,343	20,067	7,653.1		1,381.73

Notes: Energy yuan value is total purchases, including feedstocks for secondary products, from input output table and authors' calculations. Energy use excludes transformation (*China Statistical Yearbook*, 1999).

products with 137 g of TSP per yuan of energy purchased, followed by electricity with 37 g/yuan.

We also express energy purchased as a share of value added, because this is a more readily available quantity that is used by many other researchers. It is the value of energy divided by the sum of the value of labor and capital inputs, reported in the second column in table 9.4. The rankings of energy share of value added and energy share of gross output are similar, but not identical. The sectors with the three highest energy shares of output (column 1) also have the three highest energy shares of value added. The chemicals sector, however, is only twelfth in terms of energy-to-output ratio despite having the sixth-highest energy-to-value-added ratio. We believe the output measure is a more useful one when describing energy conservation in production functions of the sort used in the next chapter, and this highlights one of the sources of different results from studies that use the more easily available value-added measure.

9.2 Pollutant Emissions and Emission Factors

We now turn to the factors that contribute to the sector emissions given in table 9.1. We gave output and the estimated use of energy—coal, oil, and gas—in tables

Table 9.4
Energy and emission ratios

Sector	Value Share of Energy in Output	Value Share of Energy in Value Added	Energy: Output (kg coal equivalent per yuan)	TSP Emissions: Energy Value (gram per yuan)
1 Agriculture	0.0171	0.0286	0.0239	3.80
2 Coal mining and processing	0.0975	0.1963	0.2433	10.56
3 Crude petroleum mining	0.1446	0.2211	0.1768	3.97
4 Natural gas mining	0.1284	0.2373	0.2202	2.14
5 Metal ore mining	0.1059	0.3058	0.0710	4.34
6 Nonferrous mineral mining	0.1097	0.2526	0.0607	4.26
7 Food products, tobacco	0.0156	0.0561	0.0280	15.23
8 Textile goods	0.0158	0.0558	0.0337	11.74
9 Apparel, leather	0.0089	0.0290	0.0070	3.04
10 Sawmills and furniture	0.0279	0.0983	0.0225	21.94
11 Paper products, printing	0.0375	0.1201	0.0494	17.49
12 Petroleum refining and coking	0.5070	2.1752	0.2395	0.95
13 Chemical	0.0913	0.3380	0.1295	6.41
14 Nonmetal mineral products	0.1109	0.3514	0.1408	137.16
15 Metals smelting and pressing	0.1164	0.5484	0.2859	21.49
16 Metal products	0.0481	0.1978	0.0219	5.35
17 Machinery and equipment	0.0346	0.1080	0.0278	4.92
18 Transport equipment	0.0235	0.0847	0.0266	4.40
19 Electrical machinery	0.0215	0.0929	0.0115	3.55
20 Electronic and telecommunications equipment	0.0111	0.0434	0.0101	4.07
21 Instruments	0.0208	0.0683	0.0086	4.50
22 Other manufacturing	0.0272	0.0622	0.0462	23.33
23 Electricity, steam, hot water	0.2828	0.6575	0.2646	37.37
24 Gas production and supply	0.4677	1.7576	0.3444	3.28
25 Construction	0.0362	0.1261	0.0068	1.92
26 Transport and warehousing	0.1267	0.2294	0.1351	5.62
27 Post and telecommunication	0.0316	0.0551	0.0357	0.00
28 Commerce and restaurants	0.0219	0.0427	0.0169	3.81

Table 9.4
(continued)

Sector	Value Share of Energy in Output	Value Share of Energy in Value Added	Energy: Output (kg coal equivalent per yuan)	TSP Emissions: Energy Value (gram per yuan)
29 Finance and insurance	0.0095	0.0155	0.0051	3.82
30 Real estate	0.0129	0.0170	0.0142	30.00
31 Social services	0.0461	0.1159	0.0258	8.54
32 Health, education, other services	0.0326	0.0683	0.0295	19.35
33 Public administration	0.0332	0.0735	0.0194	7.62

Source: Calculated from table 9.3.

9.2 and 9.3 and now describe the relationship between emissions, output, and fuel use in greater detail. Recall that total emissions are made up of noncombustion (process) and combustion parts. The combustion component is ascribed to the burning of coal, oil, and gas, ignoring minor contributions from sources such as wood. Let us then specify total emissions of pollutant x from sector j at time t, as:

$$EM_{jxt} = EM_{jxt}^{NC} + EM_{jxt}^{C}, \tag{9.1}$$

$$EM_{jxt}^{NC} = \sigma_{jx} QI_{jt}, \tag{9.2}$$

and

$$EM_{jxt}^{C} = \sum_{f} (\psi_{jxf} AF_{jft}), \tag{9.3}$$

where $x = \text{TSP}, \text{SO}_2$, $f = \text{coal, oil, gas}$, $j = \text{sectors (from } 1, \ldots 33, H)$, EM_{jxt} is the annual emissions of pollutant x (in kilotons, or ktons), and the NC and C superscripts denote noncombustion and combustion, respectively. QI_{jt} is the output of industry j (in billion constant 1997 yuan), and AF_{jft} is the quantity of fuel f combusted (billion tons of oil equivalent, toe). σ_{jx} represents the process emission rates (kton/billion yuan) and ψ_{jxf} the fuel emission factors (ktons/billion toe). These coefficients are the average rates of emissions, and by the linearity assumption of equation (9.1), they are also the marginal rates. That is, although it is very likely that the last ton of coal used in the newest plant will have an emission rate lower than the average of all existing plants, we ignore this distinction for now. The j index runs over the thirty-three production sectors and the household sector ($j = H$).

Equations (9.1)–(9.3) and the following sections on emission factors, concentrations, and dose-response follow closely Lvovsky and Hughes 1997, the basis of chapters 2 and 3 of the World Bank report *Clear Water, Blue Skies* (1997), which estimated environmental damages and the benefits of pollution control policies. We also use their data, as described in more detail in appendix G.[4]

The fuel inputs in oil equivalents (billion toe) are derived from the quantities, which in turn are derived from the purchased values:

$$AF_{jft} = \theta_f FT_{fjt}, \tag{9.4}$$

and

$$FT_{jft} = \lambda_{fj} \xi_f A_{fjt}. \tag{9.5}$$

$FT_{j,coal,t}$ and $FT_{j,oil,t}$ are the physical quantities of coal and oil (billion tons), while $FT_{j,gas,t}$ is natural gas (billion cubic meters, m^3). The θ_f coefficients translate the physical units to oil equivalents. A_{fjt} is the constant yuan measure from the input-output matrix, and this is converted to physical units by the ξ_f coefficient. $1/\xi_f$ can be interpreted as the average price of the fuel. A_{fjt} represents the total purchases of fuels, and for the refining/coking and chemical sectors, part of the fuel inputs are converted to secondary products. We therefore multiply A_{fjt} by a loss ratio, λ_{fj}, to obtain only the combusted portion.

The official estimates of TSP and SO_2 emissions, EM_{jxt}, are available for thirteen industrial sectors, some of which are aggregates of our sectors in table 9.1. The combustion component is computed using an equation similar to table 9.1, with more detail on fuel types. Unfortunately these detailed data are not readily available. To make estimates for all of our sectors, we turn to the emission factors provided by Lvovsky and Hughes. Their coefficients for coal, oil, and gas were scaled evenly so that the second term on the right side of table 9.1 matches the 1997 official combustion emissions. The details are given in appendix G, with the scaled ψ_{jxf}'s reported in appendix tables G.1 and G.2. The sectors that produce the most TSP per unit of fuel combusted are the nonindustrial sectors, with their small boilers and poorly controlled emissions.

For the process emissions, the σ_{jx} coefficient is derived by dividing the official emissions by the value of gross output given in the input-output table. The procedure and results are reported in appendix table G.3. This source of pollution is essentially dominated by the nonmetal mineral products (cement) and the metals

Table 9.5
Concentration of pollutants in major cities 1997

City	Concentration of TSP (μg/m^3)	Assumed Concentration of PM_{10} (μg/m^3)	Concentration of SO_2 (μg/m^3)
Beijing	377	204	124
Tainjing	318	172	75
Shenyang	369	199	82
Harbin	310	167	26
Shanghai	229	124	68
Jinan	420	227	141
Wuhan	241	130	42
Guangzhou	217	117	70
Chongqing	200	108	208
Chendu	248	134	60
Xian	385	208	91
National urban average	298	160	93

Source: Fridley 2001 and authors' calculations.

smelting (iron and steel) sectors. The other manufacturing sectors have low estimated process emissions.

The result of these emissions is the high levels of pollution seen in many Chinese cities. The concentrations of TSP and SO_2 observed in eleven cities in 2000 are given in table 9.5. The national urban average was 251 μg/m^3 of TSP. Although this is much improved from the early 1990s, it is very high compared to conditions in richer countries, for example TSP concentrations in Los Angeles averaging less than 100 μg/m^3. The trends since 1992 are shown in table 9.6. Total national combustion TSP emissions have fallen considerably despite an enormous increase in output. In 1992, they were probably more than 20 million tons (Mt), whereas in 2002 they were around 10 Mt at a time when GDP was 240% higher. We say "probably" because the changes in data coverage make the older data difficult to compare with more recent figures. As shown in the last two rows of the TSP data of table 9.6, based on the old narrow definition, emissions in 1995 were 14.5 Mt, where based on the new definition they were 24.7 Mt (Sinton et al. 2004, table 8B.2). Total combustion TSP has fallen faster than total SO_2 but note that the reverse is true for some industries.

Table 9.6
Trends in combustion emissions (kilotons) for major sectors

Pollutant and Sector	1992	1995[a]	1997[b]	2000	2002
TSP					
13 Chemicals	749	784	672	677	562
14 Nonmetal mineral products	812	757	2,549	2,753	1,307
15 Metals smelting and pressing	606	551	632	679	602
23 Electricity	3,883	4,443	3,953	3,827	3,284
26 Transport and Warehousing	1,060	969	361	357	
34 Households	2,165	3,175	462	479	
Total (pre-1997 definitions)	14,002	14,509			
Total (post-1997 definitions)		24,710	15,730	11,472	10,120
SO_2					
13 Chemicals	1,172	1,191	1,216	1,040	762
14 Nonmetal mineral products	714	521	1,525	1,736	1,550
15 Metals smelting and pressing	518	562	739	651	536
23 Electricity	5,598	6,946	7,846	8,107	7,501
Total (pre-1997 definitions)	16,382	17,206			
Total (post-1997 definitions)		21,410	20,870	17,169	16,990

Notes: National totals and 2002 data are from Sinton (2004).
Official SO_2 data for 2002 is for total emissions, allocated to combustion using 2000 shares.
[a] The data prior to 1997 excludes township and village enterprises.
[b] The official data in 1997 is adjusted to incorporate later information by the authors.

9.3 Methodologies for Estimating Emissions, Concentrations, and Health Damages

In this section we describe our methods to estimate damages to human health from these emissions using the intake fraction (*iF*) results from chapters 5–7. Although the use of national average *iF*s is highly simplified, it is nevertheless a methodological advance, including over our previous effort in Ho, Jorgenson, and Di 2002. There we employed the approach of Lvovksy and Hughes 1997 that underlies *Clear Water, Blue Skies* (World Bank 1997), which used a simple linear expression linking emissions to concentrations and differentiated emissions into three categories on the basis of release height: low, medium or high. These concentrations multiplied by dose-response coefficients gave the health effects. That approach only considered primary particulates and ignored, for example, the formation of secondary sulfate particles from SO_2.

9.3.1 Intake Fractions

As described in chapter 4 and adapting the notation to the needs of this chapter, the *iF* for pollutant *x*, from source *r*, iF_x, gives the fraction of all emissions from *r* that is ingested by someone within a given domain and is calculated as follows:

$$iF_{xr} = \frac{BR \sum_d C_{xd} POP_d}{EM_{xr}}, \tag{9.6}$$

where C_{xd} is the change in concentration at location *d* due to emissions EM_{xr}, POP_d is the population at *d*, and *BR* is the breathing rate. The summation is over all locations in the domain where the concentration is estimated to be positive. To recall, *iF*(primary TSP) denotes the fraction of particulate matter that is emitted that is estimated to be inhaled given changes in TSP concentration, population distribution, and breathing rate; *iF*(primary SO_2) is analogous; and *iF*(SO_4^-/SO_2) denotes the inhaled fraction of secondary sulfate *particles* (chiefly ammonium sulfate, $[NH_4]_2SO_4$) converted in the air from each gram of SO_2 *gas* emitted. Below we denote the first two as *iF*(SO_2) and *iF*(TSP) for brevity, but readers should keep in mind that these refer to inhalation of primary pollutants, while *iF*(SO_4^-/SO_2) concerns inhalation of a secondary one, sulfate particles.

Chapters 5–7 report how *iF*s are calculated from a sample of emission sources for each of the highly polluting industries: electricity, chemicals, nonmetal mineral products (cement), metals smelting (iron and steel), and transportation. We need to make several adjustments to these sample *iF* estimates. First, the values reported in table 5.10 are for the 50-km domain, and we make a simple adjustment for the whole country by multiplying these manufacturing industry *iF*s by 3, following the recommendation in section 5.4.1.[5] Second, we have to make estimates for the other industries. We take the average of the three manufacturing industries (cement, iron and steel, chemicals) and apply it to all the other manufacturing industries. For the agriculture and service sectors, which have mostly low-height chimneys, we use the *iF*(TSP) for transportation (because of its ground-level emissions) and adjust it by considering only the agriculture or service output in urban areas. That is, we assume that emissions in rural areas are less damaging and can be ignored. These simple rules parallel the treatment in Ho, Jorgenson and Di 2002, where all manufacturing emissions were assigned to medium-height stacks and all services to low-height ones.

Because no estimates were made for $iF(SO_2)$ for the transportation sector, we apply the value from the manufacturing sectors. This is a conservative estimate given the lower emission heights. For the transportation sector only, chapter 5 makes an estimate of the *iF* for secondary nitrates.[6]

For the electric power sector, the Liu and Hao analysis in chapter 6 of 278 plants gives a mean $iF(TSP) = 2.16 \times 10^{-6}$ and $iF(SO_2) = 2.38 \times 10^{-6}$. In this case we do not have much guidance about adjusting for exposures beyond the modeled domain of 50 km, and we follow the conservative recommendation of chapter 5 and multiply by 2, after which the *iF*s become 4.32×10^{-6} and 4.76×10^{-6}, respectively. The Zhou et al. analysis in chapter 7 uses a more sophisticated air-dispersion model covering most of China, albeit with emission characteristics of only one power plant, which indicates that this adjustment is reasonable. Table 7.4 reports mean $iF(SO_2) = 4.8 \times 10^{-6}$ and $iF(PM) = 6 \times 10^{-6}$ (weighting over various particle sizes). These remarkably similar figures from such different models give us confidence about our adjustment, erring on the conservative side. The results of these adjustments to the primary intake fractions from chapters 5–7 for all industries are given in table 9.7.

The third adjustment is for secondary particles. Chapter 7 reports sulfate-SO_2 and nitrate-NO_X *iF*s for the electric power industry. Because the sensitivity analysis in section 7.3.5 shows that stack height is not very important for secondary particles, we apply the average value given in table 7.4 to all sources of SO_2. That is, we assume $iF(SO_4/SO_2, j) = 4.4 \times 10^{-6}$ for all j as shown in table 9.7.[7] For industry emissions of nitrogen oxides (NO_X), there unfortunately are sparse data and we ignore them, with the exception of one key NO_X-emitting sector, transportation, by using data described in appendix D. For this sector we use the $iF(NO_3^-/NO_X) = 3.1 \times 10^{-6}$ value given in table 5.5.

9.3.2 Health Effects

The next step is to estimate the damage to human health due to these pollutants. World Bank 1997 and Lvovsky and Hughes 1997 identify eight separate health effects for PM_{10} and two for SO_2. These effects are listed in table 9.8. The most important in terms of the total value of damages are chronic bronchitis, mortality, and restricted activity days (also known as "work days lost" in some studies). In those studies the authors decided that the evidence of mortality effects from SO_2 is too uncertain and chose to combine the mortality effects of PM and SO_2, as we shall explain in more detail below. We, however, make separate premature mortality attributions to the two pollutants, hence the eleven health effects given in table 9.8.

Although health damage is believed to be primarily from very fine particles and studies of health effects often use PM_{10}, the emission data in the official environmental statistics that we presented in section 9.1 are in terms of TSP. We therefore first need to convert the TSP measures into PM_{10}. The average ratio of PM_{10}:TSP for six cities in 2001 is given in table 4.4 as 0.54. We therefore assume

$$EM_{j,PM10,t} = 0.54 EM_{j,TSP,t}. \tag{9.7}$$

This means that we are assuming the same factor for all industries because there is little information on particle size distribution at the industry level. The lack of data also means that there is no point in specifying damages in terms of finer particles, such as $PM_{2.5}$. The *iF*s calculated in chapters 5 and 6 were for TSP, whereas the analysis in chapter 7 included small particles. We make the conservative assumption that $iF(PM_{10}) = iF(TSP)$, though the smaller particles travel farther and have higher *iF*s.

With the *iF*s for primary TSP, primary SO_2, and secondary sulfates for each industry, we can now estimate the health effects by sector. Let us denote the national *iF*s in the last three columns of table 9.7 by iF_{xj}^N to distinguish them from the source-specific ones in equation (9.6). We interpret the product of iF_{xj}^N and the total national emissions from sector j as the total dosage (i.e., the total amount of x inhaled by all people):

$$DOSE_{xj} = iF_{xj}^N EM_{xj}, \tag{9.8}$$

where x = primary PM_{10}, primary SO_2, secondary sulfates. We use "PM_{10}" to denote total primary and secondary particles, for example,

$$DOSE_{PM10,j} = DOSE_{primaryPM10,j} + DOSE_{sulfates,j}.$$

To emphasize how to interpret equation (9.8), substitute in the definition of intake fraction from equation (9.6):

$$DOSE_{xj} = iF_{xj}^N EM_{xj} = BR \sum_d C_{xd,j} POP_d. \tag{9.9}$$

The dosage can thus be interpreted as the quantity of x inhaled by the population in location d exposed to concentration C because of all emissions from industry j, summed over all locations in the country. Note that for sulfates, equation (9.8) is written as $DOSE_{SO4,j} = iF(SO4/SO2,j)EM_{SO2,j}$.

The number of cases of health effect h from pollutant x, HE_h, (e.g., number of cases of chronic bronchitis [CB]) is usually derived by multiplying the concentration (C_x) by the dose-response coefficient (DR_{hx}) and population (POP):

Table 9.7
Intake fractions used to estimate primary and secondary health effects

	iF (50-km domain)		*iF* (entire domain)		Secondary PM
Sector	TSP	SO_2	TSP	SO_2	$iF(SO_4^-/SO_2)$
1 Agriculture			1.54E-06	3.32E-07	4.40E-06
2 Coal mining and processing			1.05E-05	1.66E-05	4.40E-06
3 Crude petroleum mining			1.05E-05	1.66E-05	4.40E-06
4 Natural gas mining			1.05E-05	1.66E-05	4.40E-06
5 Metal ore mining			1.05E-05	1.66E-05	4.40E-06
6 Nonferrous mineral mining			1.05E-05	1.66E-05	4.40E-06
7 Food products and tobacco			1.05E-05	1.66E-05	4.40E-06
8 Textile goods			1.05E-05	1.66E-05	4.40E-06
9 Apparel, leather			1.05E-05	1.66E-05	4.40E-06
10 Sawmills and furniture			1.05E-05	1.66E-05	4.40E-06
11 Paper products, printing			1.05E-05	1.66E-05	4.40E-06
12 Petroleum processing and coking			1.05E-05	1.66E-05	4.40E-06
13 Chemical	3.28E-06	6.61E-06	9.84E-06	1.98E-05	4.40E-06
14 Nonmetal mineral products	3.46E-06	3.99E-06	1.04E-05	1.20E-05	4.40E-06
15 Metals smelting and pressing	3.75E-06	6.03E-06	1.13E-05	1.81E-05	4.40E-06
16 Metal products			1.05E-05	1.66E-05	4.40E-06
17 Machinery and equipment			1.05E-05	1.66E-05	4.40E-06
18 Transport equipment			1.05E-05	1.66E-05	4.40E-06
19 Electrical machinery			1.05E-05	1.66E-05	4.40E-06
20 Electronic and telecommunications equipment			1.05E-05	1.66E-05	4.40E-06
21 Instruments			1.05E-05	1.66E-05	4.40E-06
22 Other manufacturing			1.05E-05	1.66E-05	4.40E-06
23 Electricity, steam, and hot water	2.16E-06	2.38E-06	4.32E-06	4.76E-06	4.40E-06
24 Gas production and supply			1.05E-05	1.66E-05	4.40E-06
25 Construction			5.79E-05	1.66E-05	4.40E-06
26 Transport and warehousing	7.72E-05		7.72E-05	1.66E-05	4.40E-06

Table 9.7
(continued)

Sector	iF (50-km domain) TSP	iF (50-km domain) SO_2	iF (entire domain) TSP	iF (entire domain) SO_2	Secondary PM $iF(SO_4^-/SO_2)$
27 Post and telecommunication			3.76E-05	1.66E-05	4.40E-06
28 Commerce and restaurants			5.49E-05	1.66E-05	4.40E-06
29 Finance and insurance			3.86E-05	1.66E-05	4.40E-06
30 Real estate			3.86E-05	1.66E-05	4.40E-06
31 Social services			3.86E-05	1.66E-05	4.40E-06
32 Health, Education, other services			3.86E-05	1.66E-05	4.40E-06
33 Public administration			3.86E-05	1.66E-05	4.40E-06
Households			2.31E-05	1.66E-05	4.40E-06
Three-manufacturing average	3.50E-06	5.54E-06	1.05E-05	1.66E-05	
Electricity (chap 5, 1,680 km)			6.0E-06	4.8E-06	4.4E-06
Transportation ($iF[NO_3^-/NO_X]$)					3.1E-06

$$HE_h = DR_{hx} \times C_x \times POP$$
$$(\text{number}) = \left(\frac{\text{number per million}}{\text{people per } \mu g/m^3}\right) \times (\mu g/m^3) \times (\text{millions}). \tag{9.10}$$

In light of the definition of dosage in equation (9.8) in terms of concentrations, we can rewrite this health effect due to emissions from sector j as:

$$HE_{hj}^S = \sum_x \left(DR_{hx} \frac{DOSE_{xj}}{BR} \right) = \sum_x \left(DR_{hx} \frac{iF_{xj}^N EM_{xj}}{BR} \right), \tag{9.11}$$

where h = mortality, respiratory hospital admissions, and so forth. That is, the sector health effect (denoted by the superscript S) is given by multiplying the dose-response coefficient by the dosage due to j's emissions, dividing by the breathing rate (to give dosage per unit time) for each pollutant, and then summing over $x = PM_{10}, SO_2$.

We should note that $DOSE/BR$ on the right side of equation (9.11) has the same units as $C \times POP$ that appears on the right side of equation (9.6). This formula

Table 9.8
Dose-response and valuation estimates for PM_{10} and SO_2, base case

Health Effect	Dose-Response (cases per million people per μg/m³ increase) WB (1997)	Present Study	Valuation WB (1997) in US$	WB (1997) updated to yuan 1997	Our base valuation yuan 1997
Due to PM					
1 Mortality	6	1.95	60,000	585,000	370,000
2 Respiratory hospital admissions	12	12	284	2,770	1,751
3 Emergency room visits	235	235	23	220	142
4 Restricted activity days	57,500	57,500	2	23	14
5 Lower respiratory infection/child (asthma)	23	23	13	130	80
6 Asthma attacks	2,608	2,608	4	40	25
7 Chronic bronchitis	61	61	8,000	78,000	48,000
8 Respiratory symptoms	183,000	183,000	0.6	6	3.7
Due to SO_2					
9 Mortality	0	1.95	60,000	585,000	370,000
10 Chest discomfort	10,000	10,000	0.6	10	6.2
11 Respiratory symptoms/child	5	5	13	10	6.2

Notes: WB (1997) refers to World Bank (1997) and Lvovsky and Hughes (1997).

shows that the assumed breathing rate, *BR*, is immaterial. In the calculation of *iF* in chapters 5–7, a particular *BR* was used; in equation (9.11) the same *BR* appears in the denominator, canceling out the one embedded in the numerator of *iF*.

How should these industry effects be aggregated to obtain national health damages? In light of the way the simple air-dispersion model calculates concentrations, we can sum the changes in concentration, that is, assume that the emissions from industry k do not affect the emissions from j. The formation of secondary particles is more complicated, but lacking any clear guidance we shall assume that they can also be summed. That is, the total health effect for the whole economy is assumed to be the sum over all industry effects:

$$HE_{ht} = \sum_{j} HE^{S}_{hj}. \tag{9.12}$$

The final step is to put a yuan value on these health damages to compare with the costs of other environmental damages and compare the benefits of different policies. For those wishing to perform benefit-cost analysis, valuation is an essential ingredient. Let V_{ht} denote the value of one case of health effect h (e.g., yuan per case of CB). The national value of damage due to all cases of h in year t is given by

$$D_{ht} = V_{ht} HE_{ht}. \tag{9.13}$$

The value of total national health damages is then simply the sum over all of the health effects:

$$TD_t = \sum_h D_{ht}. \tag{9.14}$$

The value of damages from industry j only is simply the sum of all h health effects from emissions in that industry:

$$D_{jt}^S = \sum_h V_{ht} HE_{hjt}^S. \tag{9.15}$$

9.3.3 Comparison with Previous Method to Estimate Damages

Since the above method of *iF*s is quite different from the procedures in Garbaccio, Ho, and Jorgenson 2000 and Ho, Jorgenson, and Di 2002, we should briefly note the major changes. As noted earlier, there we followed the method of Lvovsky and Hughes 1997 for World Bank 1997, which expresses concentration as a linear function of industry emissions classified by three emission heights, $c =$ low, medium, high:

$$C_{xt}^N = \gamma_{low,x} E_{low,xt}^C + \gamma_{medium,x} E_{medium,xt}^C + \gamma_{high,x} E_{high,xt}^C, \tag{9.16}$$

where $x = PM_{10}$, SO_2; E_{cx}^C is the emissions from all industries in urban areas at height class c; and C_x^N is the average national urban concentration. The health effects were then derived from an equation such as (9.10).

This simple method was easier to implement because Lvovsky and Hughes provided the concentration coefficients, γ_{cx}. This, however, attributes all observed PM concentrations to primary emissions, in contrast to the *iF* method above that distinguishes primary PM from those derived from SO_2 and NO_X. That is, our previous method attributed too much damage to primary TSP emissions and too little to SO_2 emissions. It also attributed naturally occurring pollutants like wind-blown dust to industrial activity. Our current method can be thought of as replacing the three γ_{cx} coefficients with our set of *iF*s for each industry and adding a secondary particle dimension. Finally, we should emphasize that the empirical basis of the *iF*s is much deeper and stronger, with a large sample of detailed stack characteristics and population distributions from original Chinese sources.

9.3.4 Dose-Response Coefficients and Health Effect Valuations

Dose-response coefficients The issues involved in defining and estimating appropriate dose-response coefficients are discussed by Levy and Greco in chapter 4.[8] These include differentiating impacts by age, the ambiguity of results when both TSP and SO_2 are included, and the difference between the coefficients estimated in China compared to those estimated in cleaner, more developed countries.

The most important effect is mortality, and here we face a difficult choice of whether to include an estimate of chronic mortality. No cohort studies of chronic mortality have been conducted in China and importing coefficients from studies in wealthier countries introduces large uncertainties. These uncertainties, compared to those for acute mortality, are discussed in chapter 4 and also in ECON 2000 (chapter 3) and Lvovsky et al. 2000. Recall that the dose-response coefficient (β) for acute mortality is of the order of 0.06% per μg $(PM_{10})/m^3$, whereas the more uncertain one for chronic mortality is 0.4%. In light of the Levy and Greco discussions, we conclude that we shall take a more conservative approach and consider only acute mortality in the base case. Later, when we report on the total health damages in table 9.11, we also give a calculation that includes the chronic effect based on non-China research, which allows analysts who prefer that approach to make the appropriate interpretation. ECON 2000 and Wang and Mauzerall 2006 both include chronic mortality, whereas World Bank 1997 and Wang and Smith 1999a do not.

As summarized in table 4.5, the mean coefficients for acute mortality from the HEI 2004 and Aunan and Pan 2004 "meta-analyses" of existing literature are about 0.04–0.05% per μg/m^3 for both PM_{10} and SO_2, when attributed to one pollution component. Levy et al. 1999 summarizes twenty-four studies and gives separate pooled estimates for single- and multipollutant models. Controlling for copollutants, the pooled estimate for PM_{10} in these studies of cleaner developed countries is 0.06% ([0.02, 0.09] 95% confidence interval), and for SO_2 it is 0.004% (0, 0.008). These other estimates are summarized in table 9.9. In two of the China studies referenced in chapter 4 (Xu et al. 1994, 2000), when both pollutants are included in the model one of them becomes statistically insignificant, a problem often encountered when dealing with these two highly correlated series.

This problem has left the question of how to aggregate the two effects with no obviously satisfactory solution in most cases. We do not have statistically significant coefficients to include both, and some studies may have double counted using estimates from separate, single-pollutant models. Most studies use one of the concentration-response coefficients and set the other to zero. For example, World Bank 1997 applies a $\beta(PM_{10})$ of 0.1% for the combined effect and sets $\beta(SO_2) = 0$.[9] This combination is also used by Wang and Smith 1999b, (table 5; see also 1999a) and Abeygunawardena et al. 1999. The practical effect of this type of choice is to estimate the benefits of reducing SO_2 emissions at very low values, especially if the model does not include secondary sulfate formation.

The recommendation of Aunan and Pan 2004 and ECON 2000 (chapter 3) is to calculate both effects and to use the higher of the two. This is applicable for city-specific estimates when the policy does not radically change the TSP-SO_2 mix; however, this is not easily implemented for our goal of making a simple estimate of national health effects without city-level data.

Here we include the mortality effects from both pollutants by making some simple adjustments to the coefficients from the single pollutant models. If we apply the 0.04–0.05% range from HEI 2004 for PM_{10} to the national average of measured urban concentrations, 160 $\mu g/m^3$ (table 9.5), and the urban population of 1997, we get 160,000–210,000 premature deaths (ignoring threshold effects and background concentrations). Applying the 0.04–0.05% range for SO_2 we get 95,000–120,000 cases.[10] Without doing a city-by-city calculation and summing up, if we apply the "take the higher of the two" recommendation of Aunan and Pan 2004 and ECON 2000 to the national averages, we should be aiming for a pair of coefficients that would deliver 160,000–210,000 mortality cases. If we apply a concentration-response of 0.03% for both PM_{10} and SO_2 to the national average concentrations in 1997 and add them up, we get 195,000. In our use of these estimates in chapter 10 we are interested in the marginal damages, the effect of reducing emissions by 1 ton, and thus the absolute totals are not our main concern. The use of 0.03% for both pollutants will give the "correct" marginal effect if the pollution-control policy delivers reductions in PM_{10} and SO_2 concentration in proportion to the levels given in table 9.5. The quotation marks around "correct" are to emphasize that this method of applying the coefficient for one pollutant to estimate total damages is rough but gives a figure familiar to experts in the field.[11]

This extended discussion is partly to be transparent about the complicated and uncertain nature of such exercises in estimating the health effects of air pollution. To summarize, we wish to take into account the benefits of reducing particulates

Table 9.9
Alternative dose-response and valuation estimates for PM_{10} and SO_2

Estimates of Dose-Response		
	Change in Mortality Risk per $\mu g/m^3$ (%)	Range Given
World Bank (1997), citing literature		
Acute mortality, PM_{10}	0.1	
ECON (2000), Guangzhou study and citing literature		
Chronic mortality, PM_{10}	0.4	(0.0, 0.6)
Acute mortality, SO_2	0.12	(0.09, 0.16)
Levy et al. (1999), pooled estimates of 24 international studies (controlled for copollutants); 95% confidence interval		
Acute mortality, PM_{10}	0.06	(0.02, 0.9)
Acute mortality, SO_2	0.004	(0, 0.008)
Aunan and Pan (2004), pooled estimate of six China studies (single pollutant); (1 SE)		
Acute mortality, PM_{10}	0.03	($\pm$0.01)
Acute mortality, SO_2	0.04	($\pm$0.01)
HEI (2004), summary of developing country studies (single pollutant); 95% confidence interval		
Acute mortality, PM_{10} (random-effects model)	0.049	(0.023, 0.076)
Acute mortality, SO_2 (random-effects model)	0.052	(0.030, 0.074)

Table 9.9
(continued)

Estimates of Value of Statistical Life			
	Valuation in Study Country (US$)	Valuation in China (1997 yuan)	Ratio GDP per Capita, China to Study Country
World Bank (1997)	3,000,000	585,000	0.0235
U.S. EPA (2000), meta-analysis for U.S.[a]	6,100,000	1,090,000	0.0235
Zhou and Hammitt (chapter 8), China CV study	29–63,000	230–506,000	1
Liu, Hammitt and Liu (1997), Taiwan wage study	413–624,000	186–282,000	0.054
ECON (2002), recommendations for China	77,000	610,000	
Range up to 200 times GDP/capita		1,210,000	

Notes: The ratio of GDP per capita is derived using nominal exchange rates for 1997. For valuations given in US$, these are converted to yuan values using the ratio of GDP per capita in 1997 and the nominal rate of 8.29 yuan/US$.
[a] U.S. EPA (2000) gives US$6.1 million in 1999 dollars, which is based on the US$4.8 million in 1990 dollars of U.S. EPA (1997).

and SO_2 separately because they are often targeted with distinct policies. In our base case we are going to use 0.03% as the concentration-response coefficient for both PM_{10} and SO_2 and add the two effects. We do not apply it to national concentrations but to the *iF*s and emissions. The resulting total number of mortality cases, as discussed later in table 9.11, is smaller than World Bank 1997, which was calculated using the smaller urban population of the early 1990s. It is also a bit lower than applying the Levy et al. 1999 summary coefficient derived from developed countries for PM_{10} only. It is thus a conservative base case.

Applying this 0.03% coefficient, the number of excess deaths per million people per year due to an increase in the concentration of PM_{10} of 1 $\mu g/m^3$, is simply:

$$\begin{aligned} DR_{PM10} &= \beta \times \text{death rate} \times 10^6 \\ &= 0.0003 \times 0.00651 \times 10^6 \\ &= 1.95, \end{aligned}$$

where *DR* denotes "dose-response" and we use the death rate in 1997 (NBS 1999a, table 4.2). Similarly, $DR_{SO2} = 1.95$ deaths per million per $\mu g/m^3$. These base case coefficients are given in table 9.8.

We should note that we have used PM_{10} as the measure for fine particulate effects. In equation (9.8) above we have simply added primary PM_{10} to sulfates to obtain the dosage of PM_{10}, and this $DOSE_{PM10}$ will be multiplied by DR_{PM10}. Sulfates, however, are very fine particles and should ideally be multiplied by $DR_{PM2.5}$. We ignore this for practicality and merely note that our *iF* procedure gives a conservative estimate of the health effects.

For the other health effects, because they are few China estimates available, we shall use the DR_{hx} coefficients taken from World Bank 1997 (annex 2.1) and reproduced in table 9.8. The most important effect other than mortality is chronic bronchitis and the coefficient chosen here is 61 per million, which is similar to the 8% figure in table 4.5 of chapter 4.

In light of the uncertainty surrounding these dose-response estimates, we also use some low and a high alternative values in our calculations to help readers make their own judgments on imbedded assumptions. As shown in table 9.9, the Aunan and Pan 2004 meta-analysis gives a one-standard-error range around their 0.04% estimate of ± 0.01. HEI 2004 gives a 95% confidence interval of $(0.023, 0.076)$ for the PM_{10} coefficient and a similar range for SO_2. In light of this, we use 0.02% as the lower end for both PM_{10} and SO_2, and 0.04% as the upper end for both PM_{10} and SO_2. For respiratory symptoms, ECON (2000, table 8.1) recommends a figure of 49,820 cases per million people per $\mu g/m^3$, compared to 183,000 in World Bank 1997. We take the ECON figure as the lower end value for the dose-response. Similarly, we use the ECON 2000 figure for asthma attacks of 1,770 cases per million as the low end, and 18,400 for restricted activity days.

These base, low, and high alternative values of dose response are expressed in cases per million people and summarized in table 9.10. For all health effects other than those specifically mentioned where we have some clear advice, we shall simply use the base case values as the low and high alternatives when we perform the sensitivity analysis. As is shown later in table 9.11, the dominant effects (i.e., those that comprise most of the total values) are chronic bronchitis, mortality, and restricted activity days.

Valuations of health effects Valuations derived from willingness-to-pay (WTP) studies were discussed in chapter 8. Let us emphasize again that these are the valuations of people who might suffer the health effect. This is not the same as calculat-

Table 9.10
Alternative dose-response coefficients and valuations for sensitivity analysis

	Dose-Response (cases per million per μg/m^3)			Valuation per Case (yuan 1997)		
Health Effect	Base	Low	High	Base	Low	High
Due to PM_{10}						
1 Mortality (deaths)	1.95	1.30	2.60	370,000	130,000	950,000
2 Respiratory hospital admissions	12	12	12	1,750	1,751	2,769
3 Emergency room visits (cases)	235	235	235	142	142	224
4 Restricted activity days (days)	57,500	18,400	57,500	14	14	23
5 Lower respiratory infection/child asthma (cases)	23	23	23	80	80	127
6 Asthma attacks (cases)	2,608	1,770	2,608	25	25	39
7 Chronic bronchitis (cases)	61	61	61	48,000	19,760	78,000
8 Respiratory symptoms (cases)	183,000	49,820	183,000	3.7	3.7	5.9
Due to SO_2						
9 Chest discomfort (cases)	10,000	10,000	10,000	6.2	6.2	10
10 Respiratory symptoms/child (cases)	5	5	5	6.2	6.2	10
11 Mortality (deaths)	1.95	1.30	2.60	370,000	130,000	950,000

Sources: Base coefficients and values are from table 9.8. The low and high cases are described in the text.

ing the medical costs, the cost of lost output of sick workers, the cost of parents' time to take care of sick children, and so forth. The personal WTP may or may not include these costs; it is presumably lower in a system of publicly provided medical care where individuals do not pay the entire health care cost.

Valuation of health damages is a controversial and difficult exercise, with debates over how to do time-discounting, how to aggregate the WTP (Pratt and Zeckhauser 1996), and how well the contingent valuation method works (Hammitt and Graham 1999). The U.S. EPA (2000, chapter 7) gives consideration to these and other issues, and offers guidelines for using valuation in benefit analyses. In our previous studies such as Ho, Jorgenson, and Di 2002, we took a commonly used estimated WTP for the United States and scaled it to China by the ratio of per capita incomes in the two countries. This has been the approach of most other studies in China because little valuation research has been conducted in China until recently.

Here we shall use the Zhou and Hammitt estimates from chapter 8 for our base case and low alternative, and scaled U.S. valuations for our high alternative. The Zhou and Hammitt study covers Beijing and Anqing, and the estimated values of a statistical life (VSL) are very different for different locations. The mean and median valuations are given in table 8.7. We shall follow their recommendation to ignore the surprisingly high rural estimates and use the estimates from the "mixed model" for Beijing and urban Anqing. The mean VSL for Beijing using the mixed model is US$63,200 (506,000 yuan), and for Anqing city is US$28,700 (230,000 yuan). The estimates for the "standard model" are very close to these values. It is difficult to derive a good national mean from this small sample. Beijing residents are much richer than the average Chinese but have a lower VSL than the rural observations. For simplicity, we decided to take a simple average of the means for Beijing (506,000 yuan) and Anqing (230,000 yuan), that is, 370,000 yuan, as our base case VSL.

All of the other studies of pollution damage in China have to similarly choose valuations based on only a few studies. Another, almost contemporaneous, study of WTP in Chongqing in 1998 by Wang and Mullahy (2003) produced a median VSL of 286,000 yuan. In light of the skewed nature of the assumed distribution of values, where the mean is higher than the median, our choice of 370,000 yuan is quite comparable to the Wang and Mullahy estimates. In our earlier study (Ho, Jorgenson, and Di 2002) we used a U.S. VSL of US$3.6 million and scaled that linearly by the differences in per capita incomes for 1997 to obtain a China VSL of 702,000 yuan. That was an update of the World Bank (1997) estimate of US$60,000 (330,000 yuan) for 1992 using the same procedure and a US$3.0 million figure. Our proposed valuation here is smaller than this U.S. value transferred using unit income elasticities. In the bottom portion of table 9.9, we summarize the valuation estimates for premature mortality from the various studies for comparison. The first column gives the original study estimates, whereas the second gives the conversions to Chinese incomes and currency for 1997.

The difficulty in deciding how to scale VSL estimates from the United States to such poorer countries as China is discussed in many places, including ECON 2000 and Lvovsky et al. 2000. Direct estimates of income elasticities using cross-section data within a particular country usually give elasticities less than 1. However, applying an elasticity of, say, 0.6, to translate U.S. values to China results in counterintuitive values, as Lvovsky et al. 2000 points out.[12] Comparing international studies suggests a wide range of elasticity estimates. In a study of wage differentials and risk in Taiwan, Liu, Hammitt, and Liu (1997) estimate a range for VSL of US$413,000–624,000. Compared to the U.S. figure of US$3.6 million, this would imply an international income elasticity between 1.2 and 2 (see chapter 8, table 8.8). Scaling the US$3.6 million with an elasticity of 1.2 and GDP per capita in 1997, adjusted for purchasing power parity (PPP), generates an estimate of Chinese VSL of 440,000 yuan instead of 702,000.[13] This happens to be close to our choice of 370,000 yuan as the base case.

Lvovsky et al. 2000 starts with a U.S. estimate of US$1.62 million in 1990 dollars, which is about half that used in World Bank 1997. They recommend an income elasticity of 1.0 or slightly higher. Translated to 1997 units for China this would come also to about 350,000 yuan.[14] After surveying the literature, ECON 2000 recommended a simple rule of 100 times the gross domestic product (GDP) per capita as a conservative estimate of the VSL.[15] Again, this uses a unit income elasticity assumption, and for China this would be 610,000 yuan for 1997. That simple formula is defensible where no other estimates are available; however, given the estimates of Zhou and Hammitt in chapter 8 and Wang and Mullahy 2003, we believe our 370,000-yuan figure is a more sensible, conservative base number to focus on. It is also a figure consistent with international estimates translated using plausible income elasticities.

From this discussion it should be clear that there is a substantial range of uncertainty around risk-valuation estimates, both within particular surveys and across them.[16] We thus again report low and high alternatives to our base values to assist readers in making judgments about the cost:benefit ratios of pollution-reduction efforts. The U.S. EPA (2000, p. 90) currently recommends a VSL of US$6.1 million (1999 dollars) in their analyses, which is about 180 times the U.S. per capita GDP. Using this ratio (i.e. implicitly assuming unit income elasticity), the figure for China in 1997 would be 1.09 million yuan. This is similar to the high-end recommendation of ECON 2000 (p. 48), which suggests using "100 times income" as the central estimate, with a "subjective confidence" interval of (50 times, 200 times). If, instead of using unit income elasticity, we apply the elasticity of 1.1 recommended by Lvovsky et al. (2000) to this US$6.1 million valuation for the United States and scale it with PPP-adjusted GDP per capita estimates from World Bank 2001, we would arrive at 950,000 yuan.[17] This is about two and a half times our central estimate of 370,000 yuan and serves as our high alternative VSL.

For the low alternative, examine again the Zhou and Hammitt estimates in chapter 8, table 8.7. In addition to the means, they report the medians, and for the two cities these are around US$16,300 and US$4,200 (or 130,000 and 33,600 yuan) for all three models that they estimated. In describing the international comparisons in table 8.8, Zhou and Hammitt discuss how these estimates of median VSLs are low compared to those from U.S. and Taiwanese studies. They suggest that these can serve as lower bounds. Because our base estimate is already lower than the values used in other studies, we decide to use the median value for Beijing as our low-end alternative—that is, 130,000 yuan. The Liu, Hammitt, and Liu 1997 study of Taiwan gives a range for VSL of US$413,000–624,000. If we scale the lower end of that range to the differences in PPP-adjusted per capita incomes between Taiwan and China using an income elasticity of 1.1, it would be 162,000 yuan,[18] close to the Beijing median. (The valuations given in the second column of table 9.9 are converted using an elasticity of 1.) The base, low, and high alternative valuations are also given in table 9.10, together with the choices of DR coefficients.

Our base value of 370,000 yuan with a range of (130,000:950,000) may strike some readers as problematically vague. It has a lower end somewhat smaller than that implied by the (50 times, 200 times) income interval suggested by ECON (2000) but similar upper end. This range of uncertainty, however, is quite common in these kinds of analyses. The U.S. EPA 1997 report on the Clean Air Act includes a detailed discussion of valuation uncertainties and for its central estimate of the benefits of US$22.2 trillion, the (5%, 95%) confidence interval was (US$5.6 trillion, US$49.4 trillion).[19] That is, it produced a high-end value more than twice the base estimate, and a low-end value around a quarter of the base. The predominant contributor to this uncertainty was the VSL uncertainty.

In addition to premature mortality, World Bank 1997 included seven other health effects for particulate matter and we reproduce their valuations in table 9.8 under the column marked "Valuation in US$." As noted above, the VSL chosen by World Bank 1997 is US$3.0 million in the United States, which scaled to Chinese income gives the US$60,000 in table 9.8.[20] For a case of chronic bronchitis (CB), the valuation is US$8000. We update these estimates to 1997 income levels and convert to yuan values in the column marked "Updated to yuan 1997." This procedure adjusts the World Bank estimates to 585,000 yuan for premature mortality and 78,000 yuan for chronic bronchitis in 1997. The parallel Lvovsky and Hughes 1997 study also includes two minor effects of SO_2, and these are included in table 9.8. As explained above in discussing mortality dose-response of PM versus SO_2, World Bank 1997 combines the two into a single DR coefficient for PM and sets the SO_2 effect to zero. We calculate distinct mortality effects for the two pollutants and thus report two mortality estimates in both the SO_2 and TSP sections of table 9.8.

For CB, the Zhou and Hammitt estimates in table 8.7 give an average value for the two cities of 19,800 yuan. As they point out, however, their survey specified CB in a less severe manner compared to other studies, and thus these estimates should be regarded as a lower estimate. For example, the World Bank 1997 estimate, as recalculated in table 9.8, gives a valuation of 78,000 yuan, equivalent to 13% of the study's VSL. The more recent U.S. EPA recommendation for CB is US$260,000 in 1990 dollars, equivalent to 5.4% of its recommended VSL (U.S. EPA 1997, chapter 6, table 13). If this figure is transferred to China assuming unit income elasticity, it would be 65,000 yuan. ECON 2000 recommends using two-thirds of the U.S. EPA 1997 value, or US$5800 per Chinese case, which is 48,000 yuan. Given this range of estimates, we choose to use the ECON 2000 value as the base case, the Zhou and Hammitt value as the low alternative, and the World Bank 1997 value as the high alternative. This base value of 48,000 yuan also happens to be 13% of our base VSL of 370,000.

Because there are fewer valuation estimates for the other health effects listed in table 9.8, we shall make use of the World Bank 1997 and ECON 2000 figures. The most important of these are restricted activity days (or work days lost) and respiratory symptoms. For the base case, we simply adjust the World Bank 1997 valuations given in the column labeled "Updated to yuan 1997" in table 9.8 by the ratio of our VSL to the World Bank's VSL (i.e., by $370/585 = 0.63$). This results in our base case values in the last column of table 9.8 for effects other than mortality and CB. This is a somewhat arbitrary procedure, but we believe it provides a reasonable, conservative estimate.

There is little guidance for choosing reasonable low- and high-end alternatives for health effects other than mortality and CB. Because our procedure gives a conservative base valuation for these effects we keep them as the low alternatives. For the high alternative, we simply use the World Bank 1997 estimates, rescaled to 1997 levels as reported in table 9.8.

These base, low, and high alternatives are given in table 9.10. To summarize, we use the mean estimates derived from Zhou and Hammitt for our base-case mortality valuation, ECON 2000 recommendation for CB and scaled estimates from World Bank 1997 for the other health effects. The median valuation in Zhou and Hammitt is used as our lower-case VSL. For the upper-case VSL, we use a value transferred from U.S. EPA 2000, assuming an income elasticity of 1.1.

In estimating total health damages, we add up the eleven health effects. This is a procedure generally used in the literature, but we should note that there may be an element of double counting. For example, the response to respiratory symptoms may have already been included in restricted-activity days. There is little analysis of this problem, and we merely note that it reinforces the reasons to choose a conservative base case.

9.4 Air Pollution Damage Estimates

We now report the results of applying the two methodologies for calculating industry damages to the data for 1997. We begin with the total economic damages.

9.4.1 Total National Damages from Air Pollution

The total national health damages using the *iF* method are given by equation (9.12) for the number of cases of the various health effects, and by equation (9.14) for yuan values. These total estimates are not the objective of our study, and we briefly note them for the record. The results for each of the eleven health effects, calculated by using the central parameter values and the *iF* estimates, are given in the first two columns of table 9.11. The number of cases of excess mortality due to PM and SO_2 is around 94,000, and for CB it is 1.4 million. The value of national health damages due to air pollution is 137 billion yuan, or 1.8% of GDP in 1997. Of this total value, CB accounts for 65 billion yuan, premature mortality 34 billion, and restricted activity days 18 billion yuan.

If the lower-end values for dose-response and valuations given in table 9.10 are used, the total value of damages would be considerably lower, around 49 billion yuan, or just 0.65% of GDP. This is mostly due to using the lower mortality and CB valuations. If we use the upper-end alternatives instead, the total value is 283 billion yuan, or about twice the base case. Finally, we have discussed in section 9.3.4 above how the estimate for chronic mortality is very uncertain and that we

Table 9.11
Total health damages in 1997, using alternative values of DR and valuation.

	Base-Case Parameters		Low-End Value (millions yuan)	High-End Value (millions yuan)	Including Chronic Mortality (millions yuan)
	Number of Cases	Value (millions yuan)			
Due to PM					
1 Mortality	43,165	15,971	3,742	54,691	207,623
2 Respiratory hospital admissions	265,222	464	465	735	464
3 Emergency room visits	5,202,763	738	738	1,167	738
4 Restricted activity days	1,270,853,429	18,182	5,820	28,758	18,182
5 Lower resp. infection/child asthma	507,236	41	41	64	41
6 Asthma attacks	57,641,491	1,422	965	2,249	1,422
7 Chronic bronchitis	1,352,630	64,926	26,735	105,535	64,926
8 Respiratory symptoms	4,044,629,173	14,965	4,075	23,670	14,965
Due to SO_2					
9 Mortality	50,557	18,706	4,382	64,039	18,706
10 Chest discomfort	258,866,940	1,596	1,596	2,524	1,596
11 Respiratory symptoms/child	129,433	0.8	0.8	1.3	0.8
Total		137,012	48,560	283,432	344,635
% of GDP		1.84	0.65	3.81	4.28
Total due to PM		116,709			
Due to primary PM		56,093			
Due to secondary PM		60,616			
Total due to SO_2 and NO_X		80,919			

do not include it in our policy analysis. For those who wish to consider it, we include this effect in the last column of table 9.11 using the base-case valuations. Such a high mortality effect raises the total estimated damages to 345 billion yuan, or 4.3% of GDP, greatly exceeding the damages from CB and the other health effects, and even higher than the upper-end alternative estimate for acute effects.

Of the 137 billion yuan damage for the base case, 117 billion is due to particulate matter (PM). When we separately calculate the primary PM damage versus secondary PM, the primary PM contributes 56 billion, or 48% of this total PM, as shown in the last two rows of table 9.11. This also means that SO_2 and nitrogen oxide[21] gas emissions account for 81 billion yuan of damages when we include the secondary effects, equivalent to 59% of total damages. Hirschberg et al. (2003) also conclude that the majority of damages to health are caused by secondary PM.

We have noted above that the official emission data include noncombustion (process) emissions of TSP and SO_2. We discuss in appendix G.2 why we use only the combustion information in light of the epidemiological evidence discussed in chapter 4. For those who believe that process emissions should be included, however, we provide an additional calculation. We should note that we multiply all TSP emissions by 0.54 to convert to PM_{10}, including process emissions; this procedure likely understates the effect of combustion emissions and overstates that for noncombustion. In table 9.12 we reproduce the base case results in the first two columns, and in the columns marked "Combustion + Process" we include the noncombustion emissions, EM_{jx}^{NC}, in the calculation of equation (9.8). This raises the health effects by nearly half; the total value of damages is now 200 billion yuan instead of 137 billion, equivalent to 2.7% of GDP.

We noted in section 9.3.3 above that the method to estimate damages used earlier in Garbaccio, Ho, and Jorgenson 2000 relies on observed concentrations, which are due to both human activity and nature (such as wind-blown dust). If we use that method, the total damages are 70% higher than our current base case, 231 billion yuan versus 137 billion. Care should thus be taken when comparing the current estimates to our earlier ones.

9.4.2 Damage per Unit Output and Industry Allocation of Total Damages

We now turn to the health damages by industry and begin with the calculation of marginal damages per unit output of each sector. The marginal damage per unit of pollutant x (using equation 9.11) is:

$$MDX_{jx} = \sum_{h} V_{ht} DR_{hx} iF_{xj}^{N} / BR. \tag{9.17}$$

Table 9.12
Total health damages (1997) including noncombustion emissions

	Combustion Only		Combustion + Process	
	Number of Cases	Value (million yuan)	Number of Cases	Value (million yuan)
Due to PM				
1 Mortality	43,165	15,971	65,146	24,104
2 Respiratory hospital admissions	265,222	464	400,284	701
3 Emergency room visits	5,202,763	738	7,852,239	1,114
4 Restricted activity days	1,270,853,429	18,182	1,918,027,901	27,441
5 Lower respiratory infection/child asthma	507,236	41	765,543	61
6 Asthma attacks	57,641,491	1,422	86,995,074	2,146
7 Chronic bronchitis	1,352,630	64,926	2,041,449	97,990
8 Respiratory symptoms	4,044,629,173	14,965	6,104,332,276	22,586
Due to SO_2				
9 Mortality	50,557	18,706	59,426	21,988
10 Chest discomfort	258,866,940	1,596	304,282,645	1,876
11 Respiratory symptoms/child	129,433	0.8	152,141	0.9
Total		137,012		200,007
% of GDP		1.84		2.69
Total due to PM		116,709		176,142
Due to primary PM		56,093		108,970
Due to secondary PM		60,616		67,172
Total due to SO_2 (including secondary sulfate PM) and NO_X		80,919		91,038

The total emissions from producing output in sector j are given by equation (9.1). The simple form of this equation means that the marginal emission rate (emissions per unit output) is equal to the average rate. This emission rate for pollutant x, in the base year tb, is given simply by $EM_{jx,tb}/QI_{j,tb}$. The value of the marginal damage due to x from the last unit of output is

$$MDX^{O}_{jx,tb} = MDX_{jx}\frac{EM_{jx,tb}}{QI_{j,tb}}, \tag{9.18}$$

where the O superscript denotes that it is the damage per unit output. The total damage due to one unit of output is obtained by summing over all pollutants:

$$MD^{O}_{j,tb} = \sum_{x} MDX^{O}_{jx,tb}. \tag{9.19}$$

This also means that the industry damages given in equation (9.15) can also be expressed as

$$D^{S}_{jt} = MD^{O}_{jt}QI_{jt}. \tag{9.20}$$

The results of this calculation are given in table 9.13, where only combustion emissions are considered. For the chemical industry, for example, $MDX(SO_2) = 0.00434$ means that primary SO_2 emissions and secondary sulfates cause damages equal to 0.434 fen per yuan of chemical output (a fen is a Chinese cent, a hundredth of a yuan). To keep the notation simple, here the "SO_2" label refers to SO_2 plus the NO_X from the transportation sector. The total damage due to one yuan of chemical output is $MD^{O} = MDX(PM) + MDX(SO_2) = 0.00607$ yuan. The output of this industry in 1997 was 1,489 billion yuan (table 9.2) and thus the total damages due to this sector were $0.00607 \times 1489 = 9.04$ billion yuan.

The industry with the highest marginal damage is electricity, 9.36 fen per yuan, followed by transportation (2.67 fen), nonmetal mineral products (1.96 fen), health-education-services (1.49 fen), and gas production (1.47 fen). Electricity and cement have high damages because of their emission rate (tons of pollutant per unit output), whereas transportation and services have high damages because of their low emission heights. At the other end, sectors with the lowest marginal damages are post and telecommunications (0.01 fen), electronic and telecommunications equipment (0.02 fen), and apparel (0.03 fen), because they have trivial estimated emissions, followed by gas mining (0.04 fen) and agriculture (0.05 fen). The latter two industries have low damages, given our assumption that rural emissions have zero *iF*s.

Recall from table 9.11 that primary PM contributes about 41% of total economy damages and the other 59% is from SO_2 and NO_X, both as primary pollutants and precursors of secondary ones. This division varies over the industries as shown by the marginal damage coefficients, $MD(primary\ PM)$ and $MD(SO_2)$. In the chemicals case about 29% of the industry damages are due to primary PM. The nonmanufacturing sectors have the highest shares because of primary PM (about 60%) because they have low SO_2 emissions; these include construction, commerce, social services, and so forth. Nonmetal mineral products (cement) also have a high primary PM share of 60%. At the other end, the industry with the highest share of SO_2- and NO_X-related damages is agriculture, where they contribute 92%, followed by electricity, 81%. The post and telecommunications industry has a higher share but is a very clean sector with trivial damages.

Outdoor air pollution health damages are the product of the marginal damage per yuan and the level of industry output. Recall from equations (9.14) and (9.15) that the sector damages are D_{jt}^S and total national damages are TD_t. These sector damages are given in the fourth column of table 9.13, and the sector shares of national damages, D_{jt}^S/TD_t, are in the last column. Of the total national damages of 137 billion yuan, the electricity sector contributes the most by far, 36 billion, or 26%. It is the greatest user of coal and has very high SO_2 emissions. Its ratio of SO_2 to primary TSP is higher than average.

The next most damaging industry is nonmetal mineral products (mostly cement, 13% of the total), which also is a big user of coal, followed by transportation (10%), the greatest burner of oil, with damaging low-height emissions. Next is health-education-services (7%), which is not a very big user of coal but also has very damaging low-height emissions. Following these are chemicals (6.6%), households (5.9%), and metals smelting (4.8%). Recall that cement, transportation, chemicals, and metals smelting are industries included in the detailed *iF* study in chapter 5, and electric power was the subject of chapters 6 and 7.

To make clear that damages and fuel use are not correlated 1:1, because of varying pollutant, dispersion, and exposure characteristics (introduced into our estimates by *iF*), in table 9.14 we rank the top ten industries by both health damage caused and coal use. Health-education-services is the seventh-ranking user of coal but has the fourth-highest damages, because of a low-height high-*iF* designation, as well as high-emission coefficients (TSP emitted per unit fuel). These same comments apply to households, social services, construction, and commerce. Metals smelting (iron and steel) is the third-largest user of coal but only ranks seventh in damages

Table 9.13
Marginal sector health damage (yuan of damage per yuan of output), 1997

Sector	MDX^{O}_{xj}(yuan/yuan)		MD^O	Value of Damages (million yuan)	Share of Total Damages (%)
	x = primary	$x = SO_2$			
	PM	NO_X	yuan/yuan		
1 Agriculture	0.00004	0.00048	0.00052	1,278	0.93
2 Coal mining and processing	0.00311	0.00341	0.00652	1,551	1.13
3 Crude petroleum mining	0.00124	0.00207	0.00331	623	0.45
4 Natural gas mining	0.00012	0.00033	0.00045	5	0.00
5 Metal ore mining	0.00078	0.00091	0.00170	195	0.14
6 Nonferrous mineral mining	0.00081	0.00100	0.00181	403	0.29
7 Food products and tobacco	0.00093	0.00169	0.00262	3,594	2.62
8 Textile goods	0.00075	0.00181	0.00255	2,332	1.70
9 Apparel, leather	0.00011	0.00019	0.00030	186	0.14
10 Sawmills and furniture	0.00166	0.00254	0.00420	935	0.68
11 Paper products, printing	0.00184	0.00283	0.00467	2,082	1.52
12 Petroleum processing and coking	0.00095	0.00181	0.00276	851	0.62
13 Chemical	0.00173	0.00434	0.00607	9,041	6.60
14 Nonmetal mineral products	0.01181	0.00779	0.01960	17,148	12.52
15 Metal smelting and pressing	0.00370	0.00505	0.00875	6,563	4.79
16 Metal products	0.00033	0.00027	0.00060	288	0.21
17 Machinery and equipment	0.00058	0.00104	0.00162	1,432	1.04
18 Transport equipment	0.00031	0.00056	0.00087	497	0.36
19 Electrical machinery	0.00020	0.00039	0.00059	333	0.24
20 Electronic and telecommunications equipment	0.00007	0.00013	0.00020	100	0.07
21 Instruments	0.00026	0.00047	0.00073	70	0.05

Table 9.13
(continued)

Sector	MDX^{O}_{xj}(yuan/yuan) x = primary PM	$x = SO_2$ NO_X	MD^{O} yuan/yuan	Value of Damages (million yuan)	Share of Total Damages (%)
22 Other manufacturing	0.00118	0.00313	0.00432	1,238	0.90
23 Electricity, steam and hot water	0.01751	0.07610	0.09362	35,654	26.02
24 Gas production and supply	0.00515	0.00952	0.01467	181	0.13
25 Construction	0.00157	0.00118	0.00275	4786	3.49
26 Transport and warehousing	0.01070	0.01598	0.02668	13,519	9.87
27 Post and telecommunication	0.00001	0.00014	0.00015	30	0.02
28 Commerce and restaurants	0.00180	0.00096	0.00276	3,892	2.84
29 Finance and insurance	0.00056	0.00039	0.00095	341	0.25
30 Real estate	0.00588	0.00332	0.00920	1,707	1.25
31 Social services	0.00593	0.00388	0.00981	5,539	4.04
32 Health, education, other services	0.00953	0.00539	0.01492	9,824	7.17
33 Public administration	0.00380	0.00243	0.00623	2,761	2.02
Households				8,035	5.86
Total				137,012	100.00

Note: MDX^{O}_{xj} is the yuan value of damage to human health due to the emission of pollutant x from producing one yuan of output from sector j. "Value of damages" is the total value due to sector j emissions.

Table 9.14
Ranking of sector damages versus fuel consumption

Sector	Value of Damages (million yuan)	Damage Rank	Coal Use (million tons)	Coal Use Rank
23 Electricity, steam, hot water	35,654	1	384.9	1
14 Nonmetal mineral products	17,148	2	206.7	2
26 Transport and warehousing	13,519	3	17.5	
32 Health, education, other services	9,824	4	33.5	7
13 Chemical	9,041	5	148.5	4
Households	8,035	6	37.1	5
15 Metals smelting and pressing	6,563	7	183.2	3
31 Social services	5,539	8	17.8	
25 Construction	4,786	9	9.4	
28 Commerce and restaurants	3,892	10	9.4	
7 Food products, tobacco	3,594		33.2	8
8 Textile goods	2,332		21.8	9
12 Petroleum refining and coking	851		20.0	10
17 Machinery and equipment	1,432		36.2	6

because of lower emission coefficients and higher smoke stacks. Note also that, whereas electricity ranks first in both categories, its 26% damage share is smaller than its share of coal use (29%) and much smaller than its shares of national combustion emissions of TSP (35%) and SO_2 (45%). Pollution-control priorities should thus be focused on damages and not on fuel use or emissions per se.

Transportation uses little coal but lots of oil and is a big source of damages because of its low-height emissions. We should emphasize again that the *iF* for transportation is based on a much weaker set of estimates because of lack of data and sample size compared to the other sectors, as described in chapter 5. An *iF* was estimated for primary PM and for secondary nitrates for road transportation only in the sample cities, and we make a rough adjustment for rural transportation and other forms of transportation. We also make an assumption that the secondary sulfate *iF* is the same as that for the electricity sector. From these estimates we calculate the total dosage of pollutants and arrive at the estimate that transportation contributes 10% of the national damages, almost as large as the worst manufacturing sector, nonmetal mineral products. Note that our assessment has not tackled the more chemically complex mobile-source pollutant ozone, which also has major health effects.[22] These large damages certainly encourage an emphasis on finding effective transportation policies, both as a means to ease congestion and to reduce air pollution damage.

Finally, if one wishes to also consider noncombustion emissions as in World Bank 1997, the industry results are affected substantially, too. This is done in appendix table G.4. The greatest difference is that the marginal damage rate of the cement sector, with its large noncombustion emissions, then becomes a strikingly large 7.3 fen per yuan of output, instead of 1.96 fen.

Comparing to Ho, Jorgenson, and Di 2002 In our earlier work where we used the Lvovsky-Hughes method of equation (9.16), the distribution of damages was very different. There the electricity sector was assumed to have only high-height emissions and thus low damages per unit emissions. The share of total damages due to electricity was thus much smaller. Our current method of explicitly accounting for secondary particles gives a greater weight to all industries with relatively high SO_2 emissions per unit output, not just electric power but also, for instance, chemicals relative to nonmetallic mineral products. Both methods give high weight to transportation damages.

9.4.3 Damage per Unit Fossil Fuel

We next calculate the damage due to primary fuels, to provide guidance for fuel price policies. In equation (9.1) we are assuming that the fuel emission factors, ψ_{jxf}, are constant. In reality, they are functions of the choice of control strategies. These choices are different for different enterprises in any given year, and they also change over time. This means that the last ton of coal burned by the cleanest enterprise will have marginal emissions smaller than the sector average. We have to ignore this for research practicality, but the reader should keep the distinction in mind.

By use of the same reasoning that was used to derive equation (9.18), the marginal damage from a unit of fuel in sector j is the sum over pollutants of the marginal damage per unit of emissions from fuel f:

$$MDF_{fj} = \sum_{x}(MDX_{jx}\psi_{jxf}\theta_f), \tag{9.21}$$

where $f =$ coal, oil, gas.

Table 9.15
Health damage from fuels, 1997

Fuel	Average Marginal Damage (yuan per unit fuel)	t_f^{xv} (yuan of damage per yuan of fuel)
Coal	93.84 yuan/ton	0.5751
Oil	54.28 yuan/ton	0.0249
Gas	0.08 yuan/1,000 m^3	0.0002

Note: Average marginal damage (AMD_f) is the yuan value of damage per physical unit of fuel (averaged over all sectors). t_f^{xv} is the yuan value of damage per yuan of fuel combusted nationally.

As described in the next chapter (on pollution control policy), we shall examine a simple national tax on fuels, a tax that is proportional to the damage caused by each fuel. In appendix table G.1 we show that different sectors produce different emissions per ton of fuel burned. We do not try to estimate the damage done by a ton of fuel for each sector but estimate only the average national damage per ton of coal, ton of oil, or cubic meter of gas. This is obtained by averaging over all sectors in the base year, *tb*:

$$AMD_f^{IF} = \frac{\sum_j MDF_{fj}^{IF} FT_{jf,tb}}{\sum_j FT_{jf,tb}}. \tag{9.22}$$

The values of the average damage are given in table 9.15. The estimated damage from coal is very high, 94 yuan per ton in 1997, or 0.58 yuan of damage per yuan of coal.[23] The damage from oil is a lot less, only 0.025 yuan per yuan of oil, while the damage from gas is negligible.

Again, these damages per unit fuel estimates are quite different from Ho, Jorgenson, and Di 2002, where we used the Lvovsky-Hughes method that used observed concentrations and included natural particulate matter. Using that method would have produced a damage rate per unit fuel that is 70% higher than those given in table 9.16.

9.5 Sensitivity Analysis

Throughout this volume we have noted the uncertainty surrounding all the components of damage analysis, including the one done here. In tables 9.9 and 9.10 we tabulated the range of estimates for the DR coefficients and valuations. We also noted our simple proxies for estimating *iF*s for the industries not studied in chapters 5–7. For the *iF*s in the five industries that are estimated from detailed samples, they too are subject to quite a bit of uncertainty, as noted in chapter 5, tables 5.8 and 5.9.

As a simple summary indicator of these uncertainties, we calculate the industry damages (D_{jt}^{S}) again but using the low and high alternatives for the DR and valuation parameters given in table 9.10. The results are given in table 9.16, where the column marked "Base Parameters" is reproduced from table 9.13. As we can see from the industry share of total damages, there is very little change going from the low-end to high-end parameters. The high-end alternative parameters allocate a slightly larger responsibility to the manufacturing industries and a slightly lower one to electricity. This is because of the differences in the SO_2:TSP emission ratio among the industries.

One important uncertainty that we discussed in section 9.3.4 regards the attribution of mortality to PM versus SO_2. In our base case, we assigned a DR of 0.03% for 1 $\mu g/m^3$ for both PM_{10} and SO_2. We noted that the World Bank 1997 study chose to set DR(PM) = 0.1% and DR(SO_2) = 0%, and that Lvovsky et al. 2000 uses DR(PM) = 0.084 and DR(SO_2) = 0. To examine how this affects our results we recalculated the total damages based on the Lvovsky et al. 2000 choices. The result is to change total excess mortality from PM to 120,000 compared to 43,000 in table 9.12, and mortality from SO_2 to zero compared to 51,000. This raises the value of total damages to 147 billion yuan compared to 137 billion. The industry shares of the total are little changed, with electricity contribution rising to 27.0% from 26.0% and nonmetal mineral products rising to 13.0% from 12.5%. These small changes in totals are not important; what is very important, however, is the different implication for PM control policy versus SO_2 control policy.

We should also reemphasize that including noncombustion emissions into the analysis changes the industry rankings considerably, as shown in appendix table G.4. The effect of this on total damages has already been discussed in table 9.12.

One comment is in order here. The large range of uncertainty reflects the use of many parameters in this analysis, each of which is not well measured. This highlights a need for more data and research; however, some of these uncertainties affect every sector in the same way. The ranking of sector damages, or their relative damages, are not affected by the uncertainty in DR or valuation, and these, rather than aggregated national damages, provide the main interpretive power of this assessment.

Table 9.16
Sensitivity analysis of sector damages

Sector	Damage Value (million yuan)			Share of Total Damages (%)	
	Base Param-eters	Low Alter-native	High Alter-native	Low Alter-native	High Alter-native
1 Agriculture	1,278	465	2,392	0.96	0.84
2 Coal mining and processing	1,551	545	3,295	1.12	1.16
3 Crude petroleum mining	623	217	1,356	0.45	0.48
4 Natural gas mining	5	2	11	0.00	0.00
5 Metal ore mining	195	69	417	0.14	0.15
6 Nonferrous mineral mining	403	141	862	0.29	0.30
7 Food products, tobacco	3,594	1,252	7,862	2.58	2.77
8 Textile goods	2,332	809	5,176	1.67	1.83
9 Apparel, leather	186	65	405	0.13	0.14
10 Sawmills and furniture	935	327	2,026	0.67	0.71
11 Paper products, printing	2,082	727	4,511	1.50	1.59
12 Petroleum refining and coking	851	296	1,867	0.61	0.66
13 Chemical	9,041	3,116	20,474	6.42	7.22
14 Nonmetal mineral products	17,148	6,117	34,642	12.60	12.22
15 Metal smelting and pressing	6,563	2,293	14,232	4.72	5.02
16 Metal products	288	102	602	0.21	0.21
17 Machinery and equipment	1,432	499	3,129	1.03	1.10
18 Transport equipment	497	173	1,087	0.36	0.38
19 Electrical machinery	333	116	729	0.24	0.26
20 Electronic and telecommunications equipment	100	35	219	0.07	0.08

Table 9.16
(continued)

Sector	Damage Value (million yuan)			Share of Total Damages (%)	
	Base Parameters	Low Alternative	High Alternative	Low Alternative	High Alternative
21 Instruments	70	24	152	0.05	0.05
22 Other manufacturing	1,238	429	2,759	0.88	0.97
23 Electricity, steam, hot water	35,654	12,723	71,936	26.20	25.38
24 Gas production and supply	181	63	397	0.13	0.14
25 Construction	4,786	1,694	9,941	3.49	3.51
26 Transport and warehousing	13,519	4,866	26,721	10.02	9.43
27 Post and telecommunication	30	10	69	0.02	0.02
28 Commerce and restaurants	3,892	1,385	7,922	2.85	2.79
29 Finance and insurance	341	121	705	0.25	0.25
30 Real estate	1,707	607	3,486	1.25	1.23
31 Social services	5,539	1,965	11,410	4.05	4.03
32 Health, education, other services	9,824	3,494	20,062	7.19	7.08
33 Public administration	2,761	980	5,679	2.02	2.00
Households	8,035	2,833	16,902	5.83	5.96
Total	137,012	48,560	283,432		

9.6 Conclusion

As the introductory chapter 1 noted, many studies have shown that the total health damages from air pollution in China are very high. In this chapter we focused on the sector contributions to this total damage. We estimated the damage incurred per unit of output of the various sectors, and the damage per unit of fossil fuel burned. We find that the electricity sector, as the greatest burner of coal with high emissions of SO_2, is the greatest source of air pollution damage. This is in contrast to our previous result in Ho, Jorgenson, and Di 2002, which gives the power sector a lower damage ranking and is due chiefly to the explicit inclusion of secondary particles. The next-most-damaging sources, as ranked in this chapter, are cement, transportation, services, and chemicals. For the economy as a whole, outdoor air pollution damage is estimated to be some 1–4% of GDP, depending on various assumptions.

The first point to emphasize, in reviewing our results, is that one should focus on damages and not on energy consumption or quantity of emissions per se. There is not a strict 1:1 link between damages and fuel combustion. The service industries and households use little coal but are estimated to have high damages because of their low smoke stacks and high emission coefficients.

Second, there is a great deal of uncertainty surrounding any exercise to estimate damages such as those we have done here and those that we have cited. Let us summarize here the key uncertainties and missing information:

- Foremost is the uncertainty regarding concentration-response of mortality and morbidity to PM and SO_2. The epidemiological literature debates the absolute size of these effects, and the relative importance of PM and SO_2. There are only a few studies in China to help resolve these matters for the high pollutant concentrations that we see in China. This is relevant for deciding on the priorities for pollution control policy, and thus more epidemiological research is essential.
- Secondary PM is estimated here to be a big contributor to air pollution, almost half of the total damages; however, this is based on a small study of power plants. We have pointed out how the methods of World Bank 1997, by not explicitly taking secondary pollutants into account, wrongly attribute the source of damage to primary PM instead of SO_2. This spotlights the need for more air-dispersion research and more sophisticated modeling of chemical transformation.
- Although *iF*s improve damage estimates by differentiating dispersion and exposure dimensions of different source types, the *iF*s of many sectors have not been explicitly estimated, especially those with low stack heights. The range of values explicitly estimated in the surveyed sample is large and thus our use of the estimates for the industries that are not covered in the survey may be quite noisy.

Third, although the estimates for the transportation sector are weak because of the poor database, they are very large and may be the second- or third-largest source of damage today given the rapid growth of this sector. As the next chapter indicates, it is likely that it will be the biggest sector source of air pollution damages in the future. This suggests that pollution control efforts give due attention to this sector and that researchers give priority to improving the data and estimates of mobile-source damages (as our own program is doing in new research).

These uncertainties and limitations show the critical importance of developing further data and *iF* estimates. We highlight several additional areas for future research that are less important than those listed above, but still notable:

- The estimated damages from the nonindustrial sectors may be implausibly high and may be due to misallocating the crude aggregate emissions in the official data. More precise *iF* estimates would require industry-specific information on emissions and stack heights.
- We have mentioned the importance of sharper estimates for secondary pollutants; to make these one also would need corresponding information on NO_X, along with other precursors to ozone.
- The big difference in the damage results when noncombustion particles are included show that more research is needed on the particle-size distribution and toxicity of these emissions. This would inform the type and strength of pollution control for the cement industry and its priority compared to other sectors.

The uncertainties that we list above may be troubling, but one should keep several points in mind. Approximations can yield useful conclusions, even if they are rough. This assessment indicates that damages are far from trivial in any case. Also, there is no option for research in this field but to advance incrementally and reduce uncertainties accordingly. We provide a first systematically quantified comparison of health damage risks of Chinese air pollution by sector, anticipating refinements as research proceeds. Even a rough sense of relative sector damages by itself is informative and useful to policy, compared to no systematic information at all. We will see in the next chapter it will also permit simulation of control policies and, importantly, their direct and indirect economic effects on a national scale.

Acknowledgments

The research of this chapter was funded by a number of sources. Development of the model was originally supported by the Integrated Assessment Program, Biological and Environmental Research, U.S. Department of Energy (grant DE-FG02-95ER62133). Later research was supported by the U.S. EPA, the Task Force on

Environmental and Natural Resource Pricing and Taxation of the China Council for International Cooperation on Environment and Development, the Harvard Asia Center, and the V. Kann Rasmussen Foundation. Mun Ho was hosted as a visiting scholar by Resources for the Future, Washington, D.C., when much of this work was done.

Notes

1. The official time series have jumps in these years and make our current estimates based on 1997 quite different from our earlier work based on 1992 and 1995 data (Garbaccio, Ho, and Jorgenson 2000; Ho, Jorgenson, and Di 2002).

2. Estimates for all sectors are taken from data for 1997 except for the following. For nonmetal mineral products the emissions from 1998 were more substantially revised and we scaled up the 1997 official emissions according to changes in cement output between 1997 and 1998. For nonindustrial sectors the 1999 total of 2.06 Mt was used instead of the 1997 figure of 3.08 Mt.

3. This makes our estimates somewhat different from those in the *China Statistical Yearbook* (NBS 1999a, tables 7.4, 7.5, 7.9), which provides data on coal and oil use, and total energy consumption in tons of coal equivalents. The assumption of a common price corresponds to the abstraction in the economic model in chapter 10. The use of implied prices from the quantity data would mean different prices of the same input for different buyers. This complicates matters, and we chose another way to reconcile the fuel use to the emissions. This is described later.

4. Gordon Hughes kindly shared his estimates with us, which are used in this section.

5. Chapter 5, section 5.4.2 also discusses the effect of using county-level population data that is less refined than the data used for the five-city sample. These county data are used in combination with the chapter 5 regression results to estimate the national effect. The national *iF* is much lower than the five-city sample estimate partly as a result, for example, for TSP in the cement industry, the $iF(\text{5-city}) = 5.85 \times 10^{-6}$, while $iF(\text{national}) = 3.46 \times 10^{-6}$. In light of this our "multiply by 3" adjustment for domain effects is conservative.

6. Because there are no official estimates of emissions for transportation, the authors of chapter 5 made their own estimates as described in appendix D: 140 kt of primary PM_{10} and 3230 kt of NO_X. We make additional assumptions about the share of these total emissions that is in the urban areas and about the contribution of transportation other than motor vehicles. For these two factors, on net, we multiply the estimated emissions by 0.7.

7. The average value for $iF(\text{sulfate}/SO_2)$ of 4.4×10^{-6} given by Zhou et al. in chapter 7 is about twenty-five times greater than the U.S. estimate given in chapter 4 (tables 4.1 and 4.2) for a slightly different domain. Are the China estimates reasonable given the bigger domain? This is discussed in section 7.4.2. We believe they are plausible, in light of China's population is about 4.5 times greater and largely concentrated in the eastern third of the country, with quite different wind patterns. Alternative values are used for sensitivity analysis here.

8. HEI 2004 is a key review of health effect studies in developing countries, and Aunan and Pan 2004 gives a summary of the latest studies in China. Aunan et al. 2004 discusses the dif-

ferences between low coefficients from China-based studies and the higher ones from studies in richer countries.

9. World Bank 1997 (p. 24) discusses the effects of both PM and SO_2 on mortality, and states that "the interpretation here is that sulfur . . . is a good proxy for fine particulates, and that a portion of the fine particulates are formed from sulfur in the atmosphere. Thus we have used PM_{10} as the key measure of health impacts."

10. If we assume a threshold effect of 50 $\mu g/m^3$ for SO_2, the number of cases is 44,000–55,000; however, we are mainly concerned with the marginal effects (i.e., the effects of reducing emissions by 1 ton), and the issue of thresholds is irrelevant.

11. The 178,000 deaths estimated in World Bank 1997 (table 2.1) is derived from using a concentration response of 0.1% and assuming a threshold of 50 $\mu g/m^3$, and where the 1992 urban population is quite a bit smaller than that in 1997.

12. Lvovsky et al. 2000 (p. 52). These authors conclude by using an elasticity of 1 in their study of six developing country cities.

13. *World Development Indicators 2001* (World Bank 2001, table 1.1), gives the PPP adjusted GDP per capita in 1999 for China of US$3550 and for the United States of US$31,910. The GDP per capita in China in 1997 was 6050 yuan.

14. The Lvovsky et al. 2000 study reports the estimates for all six cities in common US$ units and converts incomes using market exchange rates, but suggests that PPP rates "may be a better approach." Because we are working with unit elasticities and only with yuans here, the issue of exchange rate conversion does not arise.

15. The approach in Aunan et al. 2004 is more detailed, calculating age-specific effects and estimating the value of life years lost. Such an approach requires age-specific exposure-response coefficients and valuations. We have to defer such intensive work for our national estimates to future research.

16. U.S. EPA 1997 (appendix I) gives a convenient survey of the literature, whereas Viscusi and Aldy 2003 provides a recent review.

17. See note 13.

18. GDP per capita in 1997 in Taiwan was US$13,400 and in China was 6,050 yuan. The PPP-adjusted value for China is US$3200 from the *World Development Indicators* (World Bank 2001), and the PPP exchange rate was 1.89 yuan/US$. With an income elasticity of 1.1, we have the Chinese VSL as 162,000 yuan $= 413{,}000 \times 1.89 \times (3200/13{,}400)^{1.1}$.

19. See appendix I U.S. EPA 1997.

20. This is from chapter 2 of World Bank 1997, which also discusses the use of willingness to pay valuation versus "human capital" valuation, the method used by many in China. The US$3.0 million figure was recommended in the early 1990s.

21. Recall that we include only the NO_X from the transportation sector, and other sources of NO_X are ignored.

22. The Harvard University Center for the Environment China Project research team has initiated a new study of urban transport pollution, which includes modeling of ozone exposures.

23. The absolute magnitudes of these ratios have to be interpreted carefully. The value of all coal used is taken from the input-output table. This consists of a mixture of raw and higher value, washed coal, thus overstating the value if the objective is just raw coal. On the other

hand, the value is taken at the mine mouth and not the delivered price, which includes the costs of trade and transportation. It is understated if one wishes to compare to the purchased value of coal.

References

Abeygunawardena, P., B. Lohani, D. Bromley, and R. Barba. 1999. *Environment and economics in project preparation: Ten Asian cases*. Manila: Asian Development Bank.

Aunan, Kristin, Jinghua Fang, Haakon Vennemo, Kenneth A. Oye and Hans Martin Seip, 2004. Co-benefits of climate policy: Lessons learned from a study in Shanxi. *Energy Policy* 32:567–581.

Aunan, Kristin, and Xiao-Chuan Pan. 2004. Exposure-response functions for health effects of ambient air pollution applicable for China: A meta-analysis. *Science of the Total Environment* 329:3–16.

China Environment Yearbook. 1998–2000. Beijing: China Environmental Science Press. In Chinese.

Dockery, D., and C. A. Pope, III. 1996. Epidemiology of acute health effects: Summary of time-series studies. In *Particles in our air: Exposures and health effects,* ed. J. D. Spengler and R. Wilson. Cambridge, MA: Harvard University Press.

ECON. 2000. An environmental cost model. Report no. 16/2000. ECON Centre for Economic Analysis. May. Oslo, Norway.

Feng, Therese. 1999. *Controlling air pollution in China: Risk valuation and the definition of environmental policy*. Cheltenham: Edward Elgar.

Fridley, D. G., J. Sinton, F. Q. Zhou, B. Lehman, J. Li, J. Lewis, and J. M. Lin, eds. 2001. *China energy databook 5.0. Lawrence Berkeley National Laboratory, China Energy Group, LBNL-47832*. Berkeley, Calif.: LBNL.

Garbaccio, Richard F., Mun S. Ho, and Dale W. Jorgenson. 2000. The health benefits of controlling carbon emissions in China. In *Ancillary Benefits and Costs of Greenhouse Gas Mitigation, Proceedings of an IPCC Co-Sponsored Workshop*. March 27–29. Washington, D.C.: Organization for Economic Co-operation and Development.

Hammitt, J. K., and J. D. Graham. 1999. Willingness to pay for health protection: Inadequate sensitivity to probability? *Journal of Risk and Uncertainty* 18 (1):33–62.

Health Effects Institute (HEI). 2004. *Health effects of outdoor air pollution in developing countries of Asia: A literature review*. Special report no. 15, Boston: International Scientific Oversight Committee, HEI.

Hirschberg, S., T. Heck, U. Gantner, Y. Q. Lu, J. V. Spodaro, W. Krewitt, A. Trukenmuller, and Y. H. Zhao. 2003. Environmental impact and external cost assessment. In *Integrated assessment of sustainable energy systems in China*, ed. B. Eliasson and Y. Y. Lee. Dordecht: Kluwer Academic Publishers.

Ho, Mun S., Dale W. Jorgenson, and Wenhua Di. 2002. Pollution taxes and public health. In *Economics of the environment in China*, ed. J. J. Warford and Y. N. Li for the China Council for International Cooperation on Environment and Development. Boyds, Maryland: Aileen International Press.

Levy, Jonathan I., James K. Hammitt, Yukio Yanagisawa, and John D. Spengler. 1999. Development of a new damage function model for power plants: Methodology and applications. *Environmental Science and Technology* 33 (24):4364–4372.

Liu, Jin-Tan, James K. Hammitt, and Jin-Long Liu. 1997. Estimated hedonic wage function and value of life in a developing country. *Economics Letters* 57 (3):353–358.

Lvovsky, Kseniya, and Gordon Hughes. 1997. An approach to projecting ambient concentrations of SO_2 and PM-10. Unpublished Annex 3.2 to World Bank, *Clear water, blue skies: China's environment in the new century*. Washington, D.C.: World Bank.

Lvovsky, Kseniya, Gordon Hughes, David Maddison, Bart Ostro, and David Pearce. 2000. *Environmental costs of fossil fuels*. World Bank Environment Department Papers, no. 78. October. Washington, D.C.

National Bureau of Statistics (NBS). 1999a, 2000. *China statistical yearbook (1999, 2000)*. Beijing: China Statistics Press. In Chinese.

National Bureau of Statistics (NBS). 1999b. *China input output table, 1997*. Beijing: China Statistics Press. In Chinese.

National Bureau of Statistics (NBS). 2001. *China energy statistical yearbook (1997–1999)*. Beijing: China Statistics Press. In Chinese.

Pratt, John W., and Richard J. Zeckhauser. 1996. Willingness to pay and the distribution of risk and wealth. *Journal of Political Economy* 104 (4):747–763.

Sinton, J., D. Fridley, J. M. Lin, J. Lewis, N. Zhou, and Y. X. Chen, eds. 2004. *China energy databook 6.0. Lawrence Berkeley National Laboratory, China Energy Group, LBNL-55349*. Berkeley, Calif.: LBNL.

U.S. EPA. 1997. The benefits and costs of the Clean Air Act, 1970–1990. Report prepared for the U.S. Congress. October. Washington, D.C.

U.S. EPA. 2000. Guidelines for preparing economic analysis. Report no. EPA 240-R-00-003. Office of the Administrator. September. Washington, D.C.

Venners, Scott, Binyan Wang, Zhonggui Peng, Yu Xu, Lihua Wang, Xiping Xu. 2003. Particulate matter, sulfur dioxide and daily mortality in Chongqing, China. *Environmental Health Perspectives* 111 (4):562–567.

Viscusi, W. Kip, and Joseph E. Aldy. 2003. The value of a statistical life: A critical review of market estimates throughout the world. *Journal of Risk and Uncertainty* 27 (1):5–76.

Wang, Hong, and John Mullahy. 2003. Willingness to pay for reducing the risk of death by improving air quality: A contingent valuation study in Chongqing, China. November 1. Yale School of Public Health. New Haven, Conn. Unpublished manuscript.

Wang, Xiaodong, and Kirk R. Smith. 1999a. Secondary benefits of greenhouse gas control: Health impacts in China. *Environmental Science & Technology* 33 (18):3056–3061.

Wang, Xiaodong, and Kirk R. Smith. 1999b. Near-term health benefits of greenhouse gas reductions: A proposed assessment method and application in two energy sectors of China. World Health Organization report no. WHO/SDE/PHE/99.01. March. Geneva.

Wang, Xiaoping, and Denise L. Mauzerall. 2006. Evaluating impacts of air pollution in China on public health: Implications for future air pollution and energy policies. *Atmospheric Environment* 40 (9):1706–1721.

World Bank. 1997. *Clear water, blue skies: China's environment in the new century.* Washington D.C.: World Bank.

World Bank. 2001. *World development indicators 2001.* Washington, D.C.: World Bank.

Xu, X., J. Gao, D. W. Dockery, and Y. Chen. 1994. Air pollution and daily mortality in residential areas of Beijing, China. *Archives of Environmental Health* 49 (4):216–222.

Xu, Z., D. Yu, L. Jing, and X. Xu. 2000. Air pollution and daily mortality in Shenyang, China. *Archives of Environmental Health* 55 (2):115–120.

10

Policies to Control Air Pollution Damages

Mun S. Ho and Dale W. Jorgenson

10.1 Introduction

In light of the severe damages caused by air pollution in China, as recounted in the previous chapters, there have been many studies examining various options to control or reduce emissions. These range from electricity-generation policies to energy efficiency policies to economic deregulation to eliminate subsidies on dirty fuels.[1] Only a few of these studies, however, make an integrated estimate of the economic costs and health benefits of these pollution-control policies.[2] Another strand of related research considers the co-benefits of policies to reduce greenhouse gas emissions, ancillary benefits in the form of reduced local pollution.[3]

The aim of this chapter is to examine several pollution-control policies and how they might affect economic performance. We shall focus on general, economywide "green tax" (or Pigovian tax) policies rather than traditional regulatory approaches such as mandatory scrubbers, other technology standards, or sector-specific policies. Although many environmental problems must be addressed at the local or industry levels, there is an important role for national policies to rationalize allocation of pollution-control effort across regions and sectors. The case for market-based environmental policies has been made in many contexts.[4] Here we focus on such policies for reasons of economic efficiency and because they can be aligned with government fiscal-reform efforts.

The estimates of sector pollution damages given in the previous chapter will be used to construct pollution fees, or green taxes, to force producers to take into account the negative pollution externalities. We examine how these taxes affect fuel use and, hence, emissions and health damages. At the same time we estimate how they affect output, allocation of resources, existing taxes, and, over time, economic growth. There is widespread recognition now of the usefulness of considering local and global pollution jointly,[5] and we also consider the effects of these pollution taxes on greenhouse gas emissions.

The chapters in this volume, and studies by many others, have discussed the substantial uncertainty underlying health-damage estimates. The sources of this uncertainty include the emissions-concentration relationship, the dose-response (DR) coefficients, and the valuation of premature mortality and other health outcomes. In performing a cost-and-benefit assessment here we are adding another layer of uncertainty: our model of the economy uses estimated functions for describing how enterprises respond to changes in prices. Nevertheless, we believe our results are instructive. They give us a sense of the relative magnitudes of the costs and benefits involved that would help prioritize mitigation efforts.

We find that a policy that imposes even moderate taxes on fuels could reduce health damages by 11%, while lowering gross domestic product (GDP) only by less than 0.1% and aggregate consumption by 0.1% in the short run. Depending on how the pollution tax revenues are used, the long-run effects could be positive for GDP. If these revenues are recycled toward investment, consumption and GDP over the longer term are both higher. The foregoing estimate of GDP loss does not include the benefit of the avoided health damages from this moderate fuel tax policy, which would be about 0.20% of GDP in the short run. Depending on how one wishes to weigh present versus future consumption, the sacrifice of consumption over time is about an order of magnitude smaller than the benefit in damage reduction. This cost-benefit ratio is in line with the conclusions of the World Bank report *Clear Water, Blue Skies* (1997).

A broader-based, but less-efficient, policy that taxes output on the basis of the amount of pollution damage caused could reduce health damages by 3.5% a year and, in the short run, lower GDP and consumption by less than 0.1%. This output tax is less effective than the fuel tax because there is little fuel-switching, but the adjustment costs are spread out instead of being concentrated on the coal sector. This policy also raises substantial new revenue for the government, which could be used to help with reform of the overall tax system.

We perform sensitivity analysis of the many uncertain parameters in our analysis, including the crucial DR and health endpoint valuations. We find that our conclusion of a high benefit-cost ratio for policies to reduce particulate and sulfur dioxide (SO_2) emissions is robust to these parameters. If one takes into account the uncertainties and adjustment costs, one might want to phase in these policies gradually. The long-run benefits, however, are clear.

10.2 Methodology for Analyzing Pollution Taxes

The use of taxes to correct for externalities has been examined by many studies. Some of them ask whether the traditional Pigovian tax (i.e., a tax equal to the marginal damage caused by the externality) is appropriate if we consider an economy that has many tax distortions already in place. This is related to the question of whether it is possible to have a double dividend (i.e., lower negative externalities and higher economic efficiency; see, for instance, Bovenberg and de Mooij 1994; Bovenberg and Goulder 1996; Goulder, Parry, and Burtraw 1997; and Metcalf 2000).

In this chapter we do not directly ask what would be the optimal system of taxes to correct air pollution externalities. We ask a simpler question: what are the effects of employing taxes that are related to the level of pollution emitted (i.e., the effects on sector prices, output, consumption, and economic growth)? The reasons for this limited inquiry are straightforward. First, the estimated externalities are large for many sectors, as discussed below. A full Pigovian tax would lead to large changes in prices, more than 100% in some scenarios. Changes of this magnitude are not reliably estimated by using marginal analysis. Second, the damage to human health is estimated using linear functions as described in chapter 9. These linear approximations are not very reliable for large changes in pollution emissions. Third, a proper optimal tax analysis should take intertemporal effects into account. This has to be done in a model that specifies a capital market for savings and investment. It is difficult to give an accurate characterization of the Chinese capital markets today, with their mix of controlled credit markets and open stock markets.[6] We therefore use a simpler approach that ignores optimization over time.

We employ a multisector economic-environment model of the Chinese economy, an earlier version of which has been used to study the local health benefits of carbon control policies (Garbaccio, Ho, and Jorgenson 2000). Our approach is to begin with the estimates, from the previous chapter, of damages to human health from air pollution due to the current patterns of output and fuel use. This forms our base case. These damages were attributed to the emissions from specific industries, and we calculated the average damage per unit output of each sector. Our input-output framework also allows us to calculate the damage per unit of coal or oil used. The health effects (e.g., the number of cases of premature deaths) are translated to yuan values. The economic model is dynamic and the value of damages is estimated each year as the projected economy expands and changes in structure.

In light of these negative externalities from production of goods and use of fuels, there is a strong economic case for imposing green taxes to ensure economic agents internalize them in their decisions. In the second step, we impose taxes in proportion to the damages and examine the new trajectory of the economy. We shall examine two sets of policies. The first is a tax on sector output, where the tax rate is proportional to the health damage caused by the production of the commodity. This tax will cause the buyers of goods to face a price that reflects the pollution externalities. For example, users of cement will pay higher prices relative to users of apparel. This tax is not the most efficient,[7] but it is relatively easy to implement. Compared to the next policy, it produces smaller changes in prices and incomes and may thus find broader political support.

The second policy is a tax on primary fuels, in which the tax rate is proportional to the average damage per unit of fuel. The estimates in chapter 9 show that a ton of coal produces different levels of emissions and damages depending on which industry burns it. An efficient externality tax would tax the sectors differently; however, an industry-specific fuel tax seems to us to be a politically and institutionally infeasible option in China, so we consider a tax that is applied equally to all users. This will cause producers to internalize the damages caused by their choice of fuels in their production decisions.

In view of the size of the estimated health damages, these pollution taxes are large. To maintain revenue neutrality, we cut other, preexisting taxes. The choice of which taxes to cut, or which sectors to compensate, affects both the mix of "winners" and "losers" in a given period, as well as the mix over time.

Of course, the most efficient green tax policy would be a direct tax on emissions; however, emissions of total suspended particulates (TSP) are not systematically measured but are merely derived from fuel inputs, boiler characteristics, and small sample measurements. The requisite economywide measurement would be extremely costly and is not implemented in any country. SO_2 from large sources may be amenable to a control policy like the U.S. trading program, but this is now only at a stage of policy experimentation.[8] Our interest here is in market-oriented policies that promote efficiency and may prove feasible, and thus we concentrate on taxes on output and fuels. Nonmarket policies such as end-of-pipe regulations are important elements in the policy toolkit, though they have to be analyzed separately using control cost information that is not well developed at this stage.[9]

10.2.1 The Environment-Health Model

Our economic model, described in the next section, generates output for each of thirty-three production sectors and their demands for inputs, including energy. Households are a nonproducing, thirty-fourth sector in the model. In chapter 9 we described how emissions of TSP and SO_2 are generated from the combustion of fossil fuels and from noncombustion processes, called process emissions. We then related these emissions to the ambient concentrations in populated areas and calculated the exposures and health effects. The *iF* method was used to estimate these health effects, and our procedures are briefly summarized here to set up the following discussion and notation and for readers who skipped chapter 9 to focus on the final policy and economic simulations of this chapter. (Careful readers of chapter 9 might proceed directly to section 10.2.2.) Note that this methodology is quite different from that employed in our earlier work in Ho, Jorgenson, and Di (2002) and Garbaccio, Ho, and Jorgenson (2000).

Total emissions of pollutant x from industry $j(EM_{jxt})$ at time t (in kilotons, kt) is the sum of noncombustion and combustion emissions, which are given by the level of output (QI_{jt}) and fuel inputs, respectively:

$$EM_{jxt} = EM_{jxt}^{NC} + EM_{jxt}^{C}$$

$$EM_{jxt}^{NC} = \sigma_{jx} QI_{jt} \tag{10.1}$$

$$EM_{jxt}^{C} = \sum_{f} (\psi_{jxf} AF_{jft})$$

$$AF_{jft} = \lambda_j \theta_f FT_{fjt} \tag{10.2}$$

$$FT_{jft} = \xi_f A_{fjt}, \tag{10.3}$$

where $x = PM_{10}, SO_2$; $f = \text{coal, oil, or gas}$; $j = \text{sector}(1, \ldots 33, H)$; σ_{jx} is the process emissions coefficient; and ψ_{jxf} is the emission factor for fuel f in sector j. AF_{jft} is the quantity of fuel f (in tons of oil equivalent, or toe) consumed. The constant yuan measure A_{ijt} derived from the input-output matrix is converted to physical units by the ξ_f coefficient (e.g., $FT_{j,coal,t}$ tons of coal). The physical units are converted to toe's by the θ_f coefficient. For the petroleum refining and coal products sectors, only part of the fuel inputs is combusted; this loss ratio is given by λ_j. The j index runs over the $n = 33$ production sectors and the household sector, H. Particulate emissions are estimated in terms of TSP and converted to PM_{10} by using equation (9.7).

The amount of emissions per yuan of output, or emissions per toe of fuel used, depends on the technology employed and will change as new investments are made.

A complete study should take into account the costs of these new technologies and how much they reduce emissions and energy use.[10] Estimates of these factors have not yet been assembled for many industries in China and we use a simple mechanism to represent such changes. Lvovsky and Hughes (1997) make an estimate of the emission levels of "new" technology and write the actual emission coefficients as a weighted sum of the coefficients from the existing and new technologies. By use of superscripts O and N to denote the old and new coefficients, we have,

$$\psi_{jxft} = k_t \psi_{jxf}^{O} + (1 - k_t)\psi_{jxf}^{N}, \tag{10.4}$$

where the weight, k_t, is the share of old capital in the total stock of capital.[11] For our purposes, the use of these old and new coefficients generates a more realistic base case but they do not affect the costs of pollution control as they should in a more general model.

The next step is to estimate the health effects due to these emissions. Eight separate health effects are identified for PM_{10} and three for SO_2, ranging from acute mortality to respiratory symptoms. These effects are given in table 10.1. The most important of them in terms of the monetary valuations are mortality and chronic bronchitis (CB). The number of cases of health effect h in period t is assumed to be a linear function of the concentration, not necessarily for all conceivable pollution levels but just across the more limited range that we will be considering.[12] This is expressed in equation (9.10) as $HE_{ht} = DR_{hx} C_x POP_t$, where DR is the dose-response relationship (number of cases of h per million people per $\mu g/m^3$), C_x is the change in the concentration of x due to the emissions under consideration, and POP_t is the population (in millions).

The use of iFs to estimate marginal damages is described in chapter 9 in equations (9.8)–(9.15). To summarize, the total national dosage, or intake, of pollutant x due to sector j is given by multiplying the national $iF(iF_{xj}^N)$ with total emissions of x from j and dividing by the breathing rate, $DOSE = iF_{xj}^N EM_{xj}/BR$.

The health effect h due to emissions from sector j is the dosage multiplied by the DR coefficient and summed over all pollutants:

$$HE_{hjt}^{S} = \sum_{x} \left(DR_{hx} \frac{iF_{xjt}^{N} EM_{xjt}}{BR} \right), \tag{10.5}$$

where $h =$ mortality, respiratory hospital admissions, and so forth. The sources of the DR estimates are described in chapter 4 and section 9.3.4 of chapter 9. The estimates are reproduced from table 9.10 in table 10.1 and include the central values and low and high alternatives. The notable features about these DR coefficients are

Table 10.1
Dose-response and valuation estimates for PM_{10} and SO_2

Health Effect	Dose-Response (cases per million per µg/m^3)			Valuation per Case (yuan 1997)		
	Base	Low	High	Base	Low	High
Due to PM_{10}						
1 Mortality (deaths)	1.95	1.30	2.60	370,000	130,000	950,000
2 Respiratory hospital admissions	12	12	12	1,750	1,751	2,769
3 Emergency room visits (cases)	235	235	235	142	142	224
4 Restricted activity days (days)	57,500	18,400	57,500	14	14	23
5 Lower respiratory infection/child asthma (cases)	23	23	23	80	80	127
6 Asthma attacks (cases)	2,608	1,770	2,608	25	25	39
7 Chronic bronchitis (cases)	61	61	61	48,000	19,760	78,000
8 Respiratory symptoms (cases)	183,000	49,820	183,000	3.7	3.7	5.9
Due to SO_2						
9 Chest discomfort	10,000	10,000	10,000	6.2	6.2	10
10 Respiratory systems/ child	5	5	5	6.2	6.2	10
11 Mortality	1.95	1.30	2.60	370,000	130,000	950,000

Source: Chapter 9, table 9.10.

that they include a positive value for mortality due to SO_2 and that these acute mortality coefficients are at the low end of the values used in many other previous China studies.

The total national health effect is obtained by summing over all industries:

$$HE_{ht} = \sum_{j} HE_{hjt}^{IF,S}. \tag{10.6}$$

As we can see from equation (9.6) the *iF* depends on the population in the vicinity of the emission source, and we project that urban populations will rise much more rapidly than total population (section F.7 of appendix F). We thus make a simple projection of future *iFs* by scaling the base year values by the projected urban population:

$$iF^N_{xjt} = iF^N_{xj,t-1}\frac{POP^u_t}{POP^u_{t-1}} \tag{10.7}$$

One important determinant of *iFs* is the pollution release height and it will be convenient to group the thirty-four sectors into high, medium, and low stack heights in discussing the results. We keep the conventions used in Ho, Jorgenson, and Di 2002, which follows Lvovsky and Hughes 1997 and *Clear Water, Blue Skies* (World Bank 1997), in classifying sectors into three categories: electric power is high height; manufacturing, mining, and utilities are medium height; and agriculture, services, and households are low height.

After estimating the health effects, the next step is to value these health damages to compare with the costs of reducing them. The value of damages due to h is given by:

$$D_{ht} = V_{ht}HE_{ht} \tag{10.8}$$

The sources of the valuations, V_{ht} (yuan per case), are described in section 9.3.4 and the estimates are reproduced in table 10.1.

Some studies of health damage valuation use a fixed V_{ht} for all years of their analysis, that is, the valuation of health risk reduction is assumed to be independent of income. In light of the large statistical uncertainty of the estimates for V_{ht}, this is often a reasonable approach. China, however, is experiencing rapid increases in real incomes. For example, if income rises at an annual rate of 5%, it would have risen 3.4 times in twenty-five years. This large change in incomes should to be accounted for, and in our model we make a simple adjustment by scaling it to lagged changes in per capita incomes, assuming unit income elasticity.[13] In the base case, our model projects an average growth rate of 4–5% in per capita incomes over the next forty years, and this produces substantial revaluations in V_{ht}.

The value of total national damages is simply the sum over all the health effects:

$$TD^N_t = \sum_h D_{ht}. \tag{10.9}$$

In analyzing the effects of green taxes, we are interested not in the total damages, but in the marginal damage. The marginal damage for each unit of emissions in sector j is derived in equation (9.17), which we repeat here:

$$MDX_{jx} = \sum_{h} V_{ht} DR_{hx} iF_{xj}^{N} / BR. \tag{10.10}$$

The marginal damage from producing one unit of sector *j*'s output is the weighted sum over all pollutants of these MDX_{jx}'s:

$$MD_{jt}^{O} = \sum_{x} \left(MDX_{jx} \frac{EM_{jxt}}{QI_{jt}} \right). \tag{10.11}$$

The weights are the industry emissions and the superscript O is to denote that it is an output measure. In most of our analysis this is calculated using only combustion emissions. As a sensitivity analysis we make an additional calculation using total combustion and noncombustion emissions.

10.2.2 The Economic Model

The environment-health model described above requires as inputs industry output and fuel use. These are generated by our model of the Chinese economy. We summarize the key features of the model here, with further details in appendix F. It is a standard, multisector Solow growth model (dynamic recursive[14]) for one country that is modified to recognize the two-tier plan-market nature of the Chinese economy. The key feature is that the "general equilibrium" effects of policies are taken into account in this model. This is unlike "partial equilibrium" analysis, which does not consider feedback effects. For example, a tax on electricity not only results in more expensive inputs for steelmaking, but more expensive steel dampens its demand and lowers its output, in turn reducing demand for steelmaking inputs such as coal and capital. This affects the coal and capital prices, which in turn affect the costs of generating electricity.

This version of our model is based on the 1997 input-output table where thirty-three sectors and commodities are identified, including six energy industries. Sector output, value added, and energy use are given in tables 9.2 and 9.3. The largest sector in terms of employment and output is agriculture; the largest user of coal is electricity; and the largest emitter of TSP, from combined combustion and noncombustion, is the nonmetal mineral products (cement) sector. The largest emitter of combustion TSP and SO_2 is electricity.

The household sector maximizes a utility function that has all thirty-three commodities as arguments. Income is derived from labor, capital, and land, supplemented by transfers. As in the original Solow model, the private savings rate is set exogenously.[15] Total national savings is made up of household savings and retained earnings of enterprises. These savings, plus allocations from the central plan, finance national investment, the government deficit, and the current account surplus. The investment in period t increases the stock of capital that is used for production in future periods.

Labor is supplied inelastically[16] by households and is mobile across sectors. The capital stock is partly owned by households and partly by the government. The plan part of the stock is immobile in any given period, whereas the market part responds to relative returns. Over time, plan capital is depreciated and the total stock becomes mobile across sectors.

The government imposes taxes on value added, sales and imports and also derives revenue from a number of miscellaneous fees. On the expenditure side, it buys commodities, makes transfers to households, pays for plan investment, makes interest payments on the public debt, and provides various subsidies. The government deficit is set exogenously and projected for the duration of the simulation period. This exogenous target is met by making government spending on goods endogenous.

Finally, the rest of the world supplies imports and demands exports. World-relative prices are set to the data in the last year of the sample period. The current account balance is set exogenously in this one-country model, and an endogenous variable for terms of trade clears this equation.

The production technology for industry output is allowed to change over time, and productivity growth (the $g[t]$ term in appendix F, equation [F.3]) is projected exogenously. We make a guess of how input requirements per unit output fall over time, including energy requirements. The latter is sometimes called the AEEI (autonomous energy-efficiency improvement). In the model, there are separate sectors for coal mining, crude petroleum, natural gas, petroleum refining, electric power, and gas supply (including coal gas). Nonfossil fuels, including hydropower and nuclear power, are included as part of the electric-power sector.

10.2.3 Marginal Damages and Green Taxes

With the above economic model and the environment-health submodel, we now have the pieces to analyze the costs and benefits of two green taxes, or in the jargon of economists, Pigovian taxes, to reduce pollution damages. The first is a tax on output proportional to the damages, and the second is a tax on fuels.

Policy of output taxes To implement corrective output taxes, one needs the marginal damage per unit output of each sector. These are derived above in equation (10.11). The values for the marginal damages from combustion emissions are reported in chapter 9, table 9.13. (A separate analysis is performed for noncombustion emissions in section 10.4.2 below.)

The sector producing the highest health damage is electricity with a rate of 0.094 yuan of damages per yuan of output, followed by transportation with 0.027. Comparing the damages from primary PM versus those from SO_2, we see in table 9.11 that, for the entire economy, primary PM accounts for 56 billion yuan out of the 137 billion. Primary SO_2 and secondary PM (from SO_2 and NO_X) account for 81 billion yuan. For the individual industries, however, the allocation between primary and secondary PM varies quite a bit, as shown in table 9.13.

Recall that we allow the valuation and exposed population to change over time, which means that the marginal damages will also change over time. To account for this, in this policy simulation the externality tax for period t is set proportional to the marginal damage:

$$t_{jt}^{x} = \lambda MD_{jt-1}^{O} \tag{10.12}$$

We use the marginal damage from the previous period $(t - 1)$ so that the tax rate is an exogenous parameter that does not depend on any current period variable. This tax, t_j^x, is applied to the price of output of sector j, and thus the dirtiest sector has the biggest increase in price (for details, see appendix F, equation [F.4]).

Policy of fuel taxes Emissions are not linked one-to-one with outputs. Pollution is a function of fuel use and choice of control strategies, and fuel use per unit output is a function of the choice of capital and production techniques. We next consider a policy of taxing fuels in proportion to the marginal damages produced using the base-case technologies. Because coal is the highest contributor to particulate pollution, it will bear the highest taxes and firms will be encouraged to switch to lower-emission fuels. The higher overall cost of energy will also induce substitution from energy to capital and other inputs.

To implement a simple national tax on each primary fuel, we use the damage per unit fossil fuel calculated in equation (9.22). To summarize, the average marginal damage of fuel f, averaged over all sectors, in the base year, tb, is given by

$$AMD_f = \frac{\sum_j MDF_{fj} FT_{jf,tb}}{\sum_j FT_{jf,tb}}, \tag{10.13}$$

where the sector-specific marginal damage of fuel f is $MDF_{fj} = \sum_x (MDX_{jx} \psi_{jxf} \theta_f)$. The damage per unit pollutant, MDX_{jx}, is given in equation (10.10).

The *AMD* is in yuan per physical unit (per ton of coal or ton of oil) and is given in table 9.15. These marginal damages are quite substantial, up to 94 yuan per ton of coal. To give a better idea of these magnitudes, we also expressed the damages in terms of a yuan of fuel:

$$t_{ft}^{xv} = \xi_f AMD_{ft}, \tag{10.14}$$

where $1/\xi_f$ is the price of a physical unit of fuel. These value ratios are given in the last column of table 9.15, under t_f^{xv}. For coal, the value of damages is a huge 57% of the value of the coal itself (i.e., $t_{coal,1997}^{xv}$ is equal to 0.575 yuan of damage per yuan of coal). In contrast, the damage rate for oil is only 0.025.

If a tax rate equaling the *AMD* is applied, the price of coal will rise by this large amount. As we have discussed above, large changes are not well analyzed with marginal methods and the linearity assumptions that we employ here. When analyzing this policy we set the externality tax to some fraction, λ, of this damage:

$$t_{ft}^{xv} = \lambda \xi_f AMD_{ft}. \tag{10.15}$$

These tax rates are also allowed to change over time reflecting the rising marginal damages.

Externality tax revenue offsets Our approach to analyzing the effects of externality taxes as given in equations 10.12 and 10.14 is to first simulate a base case under the current tax system, without these pollution taxes. In the current system, there are various fees and fines on wastewater and other discharges amounting to 6 billion yuan in 1997. These are not explicitly taken into account in the model but included in taxes on production. Our pollution taxes should be viewed as new taxes on top of the existing charges. We then apply the externality taxes and simulate a counterfactual case. The revenue from an ad valorem externality tax on output is the sum over all sectors:

$$R\_EXT_t = \sum_j t_{jt}^x PI_{jt} QI_{jt}, \tag{10.16}$$

where PI_{jt} denotes the seller's price of the output of j. (This term appears in equations [F.11] and [F.12] in appendix F.)

To maintain revenue neutrality (hence keeping real government spending constant) we cut the value-added tax (VAT), the capital income tax, and the sales tax. (Note equations [F.11]–[F.19] describe the government accounts.) Using superscript C to denote the counterfactual simulation, the new tax rates are a fraction α of the base-case rates:

$$t_t^{v,C} = \alpha_t t_t^v, \quad t_t^{k,C} = \alpha_t t_t^k, \quad t_t^{t,C} = \alpha_t t_t^t \tag{10.17}$$

In the counterfactual simulation, α_t is endogenously chosen such that total government revenue is equal to that found in the base case for each period. The economic and health outcomes in the two cases are then compared (e.g., how much is the GDP higher or lower in each year of the counterfactual case).

10.3 Base-Case Simulation

The central aim of our methodology is to provide estimates of the changes in damages, emissions, and economic performance due to implementing some pollution-control policy. This requires a projection of the future Chinese economy. We now describe the base-case projection of the main variables to give the reader a clear idea of how our approach works, though the base case itself is not a major focus of our research. The projections are made in a relatively simple manner, involving many assumptions regarding population growth, technical progress, changes in preferences, and changes in the world economy. We shall briefly describe these assumptions, which we should note have only second-order effects on the percentage change in costs and benefits that we are trying to estimate.[17]

The social accounting matrix (SAM) for the model is based on the 1997 benchmark input-output table, and we thus set 1997 as the first year of the simulation. That is, we initialize the economy to have the capital stocks that were available at the start of 1997 and have the working age population of 1997 supplying labor. The economic model then calculates, for this period, the output of all commodities; the purchases of intermediate inputs, including energy; the consumption by households and government; exports; and the savings available for investment.[18] This investment augments the capital stock for the next period, and by use of the projected population for the next period we solve the model again. This exercise is repeated for the subsequent thirty years.[19] The level of output (of specific commodities and total GDP) thus calculated depends on our projections of the population, savings behavior, changes in spending patterns as incomes rise, the ability to borrow from abroad, improvements in technology, and so forth.

The main variables for the base case are reported in Table 10.2 for the first, tenth and thirtieth years. The 5.1% annual growth rate of GDP over the next thirty years that results from our assumptions is slightly less optimistic than the 6.7% growth rate projected for China previously by the World Bank (1997), but still implies a very rapid growth in per capita income for such a long period. For the first ten years

Table 10.2
Selected variables from base case simulation, years 1, 10, and 30

Variable	Year 1997	2006	2026	30-Year Growth Rate (%)
Population (million)	1,230	1,318	1,464	0.58
Urban population (million)	368	432	568	1.46
GDP (billion 1997 yuan)	7,630	15,000	33,500	5.06
Energy use (fossil fuels, million tons of standard coal equivalent)	1,340	2,220	3,590	3.34
Coal use (million tons)	1,390	2,090	2,750	2.30
Oil use (million tons)	210	440	960	5.16
CO_2 emissions (million tons)	900	1,460	2,260	3.12
Primary particulate emissions (million tons)	24.23	17.00	15.69	−1.44
Combustion TSP	11.59	8.63	8.54	−1.01
Noncombustion TSP	12.64	8.37	7.16	−1.88
High-height source (electricity)	4.25	3.10	3.57	−0.58
Medium-height sources	17.94	11.66	9.07	−2.25
Low-height sources	2.04	2.24	3.05	1.35
SO_2 emissions (million tons)	19.98	25.76	34.59	1.85
Combustion SO_2 emissions	17.88	23.21	31.01	1.85
NO_X (transportation, million tons)	3.24	6.26	11.60	4.34
Premature deaths (1,000)	97	137	232	2.96
Health damage/GDP (%)	1.9	2.4	3.6	

Source: Calculations of authors.

it is almost identical to the assumed 7.5% growth during 2000–2010 in the Second Generation Model (SGM) as used in Jiang and Hu (2001). While the population is projected to rise at a slow 0.6% annual rate during this period, the urban population is rising much more rapidly. That is, the population most exposed to air pollution is rising at a rapid 1.46% per year.

Our assumptions for energy use improvements are fairly optimistic and, together with changes in the structure of the economy, result in an energy-GDP ratio in 2026 that is about 60% of that of 1997.[20] There is a significant decline in the use of coal per unit output but a rise in oil and electricity use about equal to that of GDP

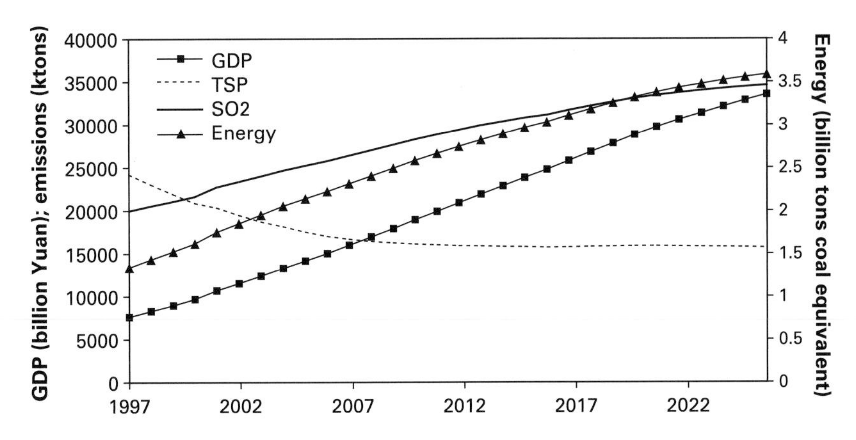

Figure 10.1
Gross domestic product, energy, and emissions projected in base case.

growth and a big rise in the use of gas. These changes are due to our assumptions about changes in transportation demand, in electricity-generation technologies, in space-heating technologies, and improvements in energy efficiency. This shift from coal to oil and gas results in the rate of growth of carbon dioxide (CO_2) emissions even slower than the growth in energy use (3.1% versus 3.3% per year over thirty years).[21] The time paths of GDP, energy use, and emissions are plotted in figure 10.1.

With the industry output and input requirements calculated for each period, we use equations (10.1)–(10.11) to calculate the total emission of pollutants, the dosage implied by the *iF*s, and the health effects of these dosages. Total primary PM_{10} emissions actually fall (at a 1.44% annual rate), despite the increase in energy use. This decline in PM is due to the sharp difference in the assumed emission coefficients for new and old capital (see equation [10.4]), and the shift from coal to oil. Medium-height sources of PM, mostly manufacturing sources, fall dramatically, whereas low-height emissions from transportation, construction, and services actually rise at a 1.35% annual rate. High-height primary emissions from electric power generation fall and then rise due to the opposing trends of lower emissions per unit output and rapidly rising total electricity use.

Projected SO_2 emissions rise much faster than particulates because of a less optimistic estimate of the improvement in the σ_{jx} and ψ_{jxf} emission coefficients; however, there is still a substantial projected improvement, whereas coal use rises at 2.3% per year, SO_2 emissions rise only at 1.85% over this thirty-year period. In figure 10.1 the SO_2 line is less steep than the energy consumption line and a lot less steep than the GDP line. This growth of SO_2 emissions leads to a parallel growth of secondary sulfates of 1.85% per year. The transportation sector is projected to grow rapidly and with it, NO_X emissions. Under our pessimistic assumption of no improvement in NO_X emission factors, emissions grow at 4.3% per year over the next thirty years.

Our base-case estimate of premature mortality in 1997 is 97,000 deaths. This calculation is explained in detail in chapter 9 and is much lower than the World Bank (1997) estimate of 178,000 for the early 1990s, because of our different dispersion and exposure methodology (*iF*), different dose-response assumptions, and changes in urban populations. The growth rates of health effects using our *iF* method and our more optimistic assumptions of energy trends are lower than that projected by Lvovsky and Hughes (1997), for example, after twenty-five years our estimated excess deaths are 2.2 times the first-year level, compared to their 3.7-fold increase. Premature mortality rises at almost 3% per year because of the growth of SO_2 emissions and increased urbanization overcoming the improvements in energy-efficiency and emission factors. The higher death rate multiplied by valuations that rise with per capita incomes means that the health damage:GDP ratio rises quite rapidly.

To reiterate our cautionary note that this description of the base case is not intended as a forecast of economic activity and emissions, but rather is a projection if no big changes in policy are made. We expect both the government and private sectors to have policies and investments that are different from those of today in ways that are not captured. The emphasis of our analysis will be on the changes at the margin due to specific policy changes, not some impractical notion of eliminating all pollution next year or beyond. We turn next to these possible policy interventions.

10.4 Controlling Local Pollution with Corrective Taxes

10.4.1 Effects of Output Taxes Based on Damages

This policy imposes a tax on the gross output of each sector in proportion to the local health damage caused by the marginal unit of output, often referred to as a "green tax." These marginal damages are described above in section 10.2.3. We focus on combustion emissions and then briefly report the case where noncombustion emissions are also included.

We first run a counterfactual simulation in which the externality tax is set according to equation (10.12) above (i.e., a tax on output equal to the marginal damage for combustion $[MD_j^O]$ in table 9.13 with $\lambda = 1$).[22] In light of the estimates of

Table 10.3
Effects of an output tax based on damages (combustion emissions only)

Variable	Effect in First Year (%)	Effect in Twentieth Year (%)
GDP	−0.04	+0.18
Consumption	−0.06	−0.01
Investment	0.06	+0.55
Coal use	−3.95	−3.36
CO_2 emissions	−3.41	−2.41
Primary particulate emissions:	−3.08	−2.30
Combustion TSP	−4.25	−4.00
Noncombustion TSP	−2.01	−1.05
High-height source (electricity)	−9.21	−7.42
Medium-height sources	−1.84	−0.92
Low-height sources	−1.17	−1.21
SO_2 emissions	−4.56	−4.00
NO_X emissions (transportation)	−3.80	−2.93
Premature deaths	−3.21	−2.88
Value of health damages	−3.53	−3.01
Change in other tax rates	−8.5	−6.1
Reduction in damages/GDP	0.07	0.09
Pollution tax/total tax revenue	9.1	6.0

Notes: The entries are percentage changes between the counterfactual tax case and the base case. The last two rows are percentage shares. In the counterfactual case, a tax proportional to the marginal damage per yuan of output is applied.
Source: Calculations of authors.

MD_i^O, this policy is one which imposes an addition of a few percent to the existing sales tax on the "dirtiest" commodities. For electricity this is about 9%, and for transportation, nonmetal mineral products (cement), health-education-other services, and gas supply, this is about 1.5–3%. The "clean" commodities see only a miniscule 0.1% addition.

The economywide effects of using these output taxes are given in table 10.3 for the first and twentieth years, corresponding to 1997 and 2016. The effects on industry prices and output in the first year are given in table 10.4, and the time paths for emissions and damages are plotted in figure 10.2. The initial effect of the taxes is to raise the price of electricity by almost 6% and the prices of the other dirty

Table 10.4
Sector effects of an output tax based on damages (combustion emissions only), percent change in year 1

Sector	Price	Quantity
1 Agriculture	0.00	0.06
2 Coal mining and processing	0.19	−3.85
3 Crude petroleum mining	−0.58	−0.83
4 Natural gas mining	−0.23	−0.63
5 Metal ore mining	0.30	−1.45
6 Nonferrous mineral mining	0.15	−1.15
7 Food products, tobacco	−0.40	0.48
8 Textile goods	−0.15	−0.07
9 Apparel, leather	−0.31	0.23
10 Sawmills and furniture	0.14	−0.38
11 Paper products, printing	0.21	−0.80
12 Petroleum refining and coking	−0.44	−1.20
13 Chemical	0.25	−0.86
14 Nonmetal mineral products	1.63	−2.10
15 Metal smelting and pressing	0.80	−1.68
16 Metal products	0.20	−0.55
17 Machinery and equipment	0.01	−0.37
18 Transport equipment	−0.10	−0.29
19 Electrical machinery	0.01	−0.48
20 Electronic and telecommunications equipment	−0.18	0.01
21 Instruments	−0.14	−0.75
22 Other manufacturing	0.10	−0.83
23 Electricity, steam, hot water	5.76	−6.69
24 Gas production and supply	1.15	−1.42
25 Construction	0.65	−0.28
26 Transport and warehousing	2.24	−3.10
27 Post and telecommunication	−0.11	−0.26
28 Commerce and restaurants	−0.98	0.57
29 Finance and insurance	−1.30	0.81
30 Real estate	0.33	−0.35
31 Social services	0.71	−1.03
32 Health, education, other services	1.54	−1.05
33 Public administration	0.89	0.25

Source: Calculations of authors.

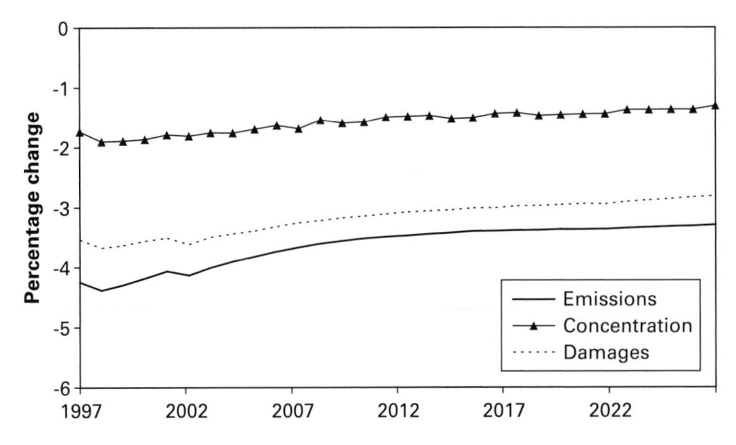

Figure 10.2
Change in particulate matter emissions and concentrations due to a pollution tax on output.

commodities by about 1.5–2.5% compared to the base case.[23] On the other hand, the prices of the cleaner items such as finance, commerce, and some mining industries are reduced by about 0.1–1.3%. Both investment goods such as cement and construction and consumption items such as transportation and services are subject to the higher tax rates. This pattern leads to a tiny change in aggregate GDP but a more substantial change in the composition of demand. GDP falls by less than 0.1%, with total real consumption falling by 0.1% and investment rising by the same offsetting amount. Recall that government spending is held fixed across all scenarios.

In the first year the output of electricity falls by 6.7% and transportation by 3.1%, whereas the other dirty sectors—nonmetal mineral products (cement), social services, and health-education-other services—fall by 1–2%, leading to a reduction in their demand for electricity and coal. Electricity use falls by this large amount because of the direct price effect dampening demand and the indirect effect of the fall in demand for the output of the big users of electricity. For example, the steelmaking industry switches from electricity to other inputs because of the higher costs, and the resulting higher price of steel means that less steel is sold and hence less electricity is used.

Coal mining is not heavily taxed and its price is little changed. Because of these indirect effects, however, coal use falls by a substantial 3.9%. Inputs released from the shrinking sectors go to the cleaner industries, leading to an expansion in finance, food products, and apparel by about 0.2–0.8%. Note that the ability to trace these indirect effects results from using our input-output general equilibrium framework. The alternative method of a partial equilibrium analysis would only capture the direct effects.

The tiny 0.04% reduction in aggregate GDP results from the assumptions of full employment and labor mobility. Workers and other inputs are assumed to find employment in the expanding sectors immediately. A more sophisticated economic model with adjustment costs, where it takes some time to reallocate the inputs that are released, would cause aggregate output to fall by a greater amount during the transition years. In the longer run after the economy adjusts, however, our model will give the more appropriate GDP effects.[24]

These changes lead to a reduction in total primary particulate emissions of 3.1%, with the high-stack-height electricity sector having the largest reduction, 9.2%, whereas low-height emissions fall only 1.2%. The medium-height manufacturing emissions fall by 1.8%. Noncombustion TSP falls by a lot less than combustion TSP, 2.0% versus 4.2%, because most of the process emissions are from the cement industry, which contracts by a lot less than electricity. For SO_2, total emissions fall by a large 4.6% because there is little noncombustion SO_2. The reductions in TSP and SO_2 are not identical to the reduction in total coal use of 3.9% because different industries have different emission factors and mixes of fuels, and their outputs are reduced to different extents. NO_X from transportation is reduced by 3.8%, in light of the sizable green tax on that sector. We assume that the reduction in secondary particles is proportional to the reduction in SO_2 and NO_X; however, one should keep in mind that the reduction may actually be affected by parallel changes in primary particulate emissions and their impact on chemical interactions in the atmosphere.

The effect of these reductions in primary and secondary particulates is to lower the value of health damages by 3.5%. The number of cases of premature mortality falls by 3.2%, whereas CB declines 3.7%. The health effects are not identical because both PM and SO_2 are assumed to contribute to excess deaths, as explained in chapter 9. The value of this reduction in damages in the first year comes to about 0.07% of GDP of that year. This is a very modest reduction compared to the total estimated health damages of 1.9% of GDP (the last row of table 10.2).[25]

Turning to the effects on government finances, the revenue raised from this broad-based tax is substantial, amounting to some 9.1% of total revenues in the first year. With our requirement that government spending be kept fixed at base-case levels, this allows a reduction in value-added and capital income taxes of 8.5%. This tax cut eventually leads to higher retained earnings and hence investment. The higher rate of investment leads to a higher stock of capital and a greater productive capacity for the whole economy. As shown in the column marked "20th Year" in table 10.3, GDP is 0.18% higher by the twentieth year, that is, higher than the base case in the corresponding year. The lower taxes and higher GDP in that year allow annual investment to be 0.55% higher, whereas consumption is reduced by only 0.01%. The lower consumption is due to the higher prices of goods and the assumption that the tax cut goes entirely to enterprises instead of households.

With this slightly higher level of counterfactual aggregate economic activity, more emissions are generated and the reduction in emissions and damages due to the externality tax becomes a bit smaller. By the twentieth year, the reduction in annual damages compared to the base case is 3.0% compared to 3.5% in the first year. As described in section 10.3 above, the base-case assumptions lead to a high productivity growth rate (i.e., inputs in particular fuels per unit output are falling rapidly). In addition, emissions per unit fuel burned are falling. These two trends contribute to a sharp reduction in emissions per unit output along the base case path (see figure 10.1). On the other hand, incomes are rising and the values of health effects (e.g., the value of a case of chronic bronchitis) are adjusted upward, and the urban population is rising at 1.5% per year. The combination of these trends leads to a small reduction in damages per unit output, that is, a lower MD^{O}_{jt}. This leads to a lower tax rate and hence smaller adjustments in the future economy and smaller reductions in health damages, a 3.0% reduction in the twentieth year compared to a 3.5% reduction in the first year.

To recall from table 10.2, the value of total damages is rising rapidly in the base-case projection. The 3.0% reduction in the value of damages in the twentieth year due to this green tax policy corresponds to a large yuan value and, thus the reduction as a share of GDP rises somewhat compared to the first year. That is, even though the huge increase in base-case output over time means a net increase in damages, and the green tax rate is falling, the policy still delivers substantial reductions in pollution damage.

The emissions of the greenhouse gas (GHG) CO_2 are related to the local pollutants. In the first year, CO_2 emissions from fossil fuels fall by 3.4%, a little less than the fall in coal use, because there is some switching toward oil and gas as a result of reallocation of industry activity. We should note that we have not calculated other GHGs such as methane, and the change in total GHGs may be somewhat different.

Why are the reductions in damages so small, in view of the substantial taxes imposed? This may appear counterintuitive and deserves some explanation.

When we put this tax on sales, output falls modestly and the emissions fall correspondingly. We do not obtain a large decline in emissions because this policy provides no incentive to switch fuels or to install scrubbers.[26] (For further comments on pollution effects of taxes on outputs, see Fullerton, Hong, and Metcalf 1999). This is an inefficient policy to reduce emissions of either CO_2 or PM, and we shall consider a more direct policy below.

10.4.2 Sensitivity Analysis of the Role of Noncombustion Emissions

To test the implications of assuming that combustion and process emissions are equally damaging, we repeated the above exercise, setting the pollution tax according to the marginal damages described in table G.4 of appendix G, which are calculated with both emission types. The base case now corresponds to the last two columns of table 9.12. This policy is essentially a significant tax on nonmetal mineral products (cement) and electricity, and a moderate tax on transportation, and metals smelting (iron and steel). The results are reported in table 10.5.

There are some big differences compared to the combustion-only case. The huge tax on cement raises the price of investment goods substantially, and the higher marginal damages (MD^O) means higher overall pollution revenue and tax cuts. The result is a fall in initial real investment of 0.27%, and a rise in consumption of 0.13%. Again the reduction in the output of the dirty sectors leads to a reduction in the use of coal and electricity. Primary PM emissions fall by 6.0% in the first year, while SO_2 emissions fall by 5.6%. These changes reduce total health damages by 4.9% compared to the 3.5% in the combustion-only case. The large initial reduction in investment leads to a smaller capital stock and GDP for many years in this scenario, despite the lower capital income taxes. GDP does not catch up to the corresponding base case levels until after the tenth year in this case, with such a large distortion in the prices of goods; however, the reduction in damages as a share of GDP is larger than the combustion-only case. We can conclude from this experiment that should process emissions be definitively shown to be harmful to human health, the damages may be sizable and one should pay more attention to the cement and metals smelting industries.

10.4.3 Effects of Fuel Taxes Based on Damages

Emissions are a function of output levels, fuel choice, energy efficiency, and control strategies. We considered output taxes above and now turn to a green tax on fossil fuels in proportion to the health damage caused by use of current technologies. A counterfactual simulation is run with taxes set according to the average marginal

Table 10.5
Effects of an output tax based on damages (process and combustion emissions, assumed equally damaging)

Variable	Effect in First Year (%)	Effect in Twentieth Year (%)
GDP	−0.08	0.10
Consumption	+0.13	−0.01
Investment	−0.27	+0.49
Coal use	−5.20	−4.03
CO_2 emissions	−4.50	−2.94
Primary particulate emissions:	−6.05	−3.52
Combustion TSP	−5.77	−4.02
Noncombustion TSP	−6.31	−2.93
High-height source (electricity)	−9.94	−7.86
Medium-height sources	−5.70	−2.68
Low-height sources	−1.12	−1.43
SO_2 emissions	−5.59	−4.55
NO_X emissions (transportation)	−3.95	−3.27
Premature deaths	−4.40	−3.18
Value of health damages	−4.90	−3.30
Change in other tax rates	−12.6	−7.4
Reduction in damages/GDP	0.13	0.12
Pollution tax/total tax revenue	13.1	7.3

Notes: The entries are percentage changes between the counterfactual tax case and the base case. The last two rows are percentage shares. In the counterfactual case, a tax proportional to the marginal damage per yuan of output is applied.
Source: Calculations of authors.

damages of the fuels (AMD_{ft}) described in equation (10.15). As we have discussed in chapter 9 with regard to table 9.15, the damage rate for coal is very high and a full tax equal to the estimated AMD_{ft} will be almost 60% of the market price of coal. Such a high tax is unlikely to be considered by the government.

We therefore begin at a relatively modest level, taxing coal, oil, and gas at a rate equal to 30% of their estimated health damage (i.e. $\lambda = 0.3$ in equation [10.15]). In light of the high marginal damage, this would produce an 18% tax rate on the price of coal, whereas the tax on oil is much smaller. The effects of these taxes on the economy and pollution for the first and twentieth years are shown in tables 10.6

Table 10.6
Effects of a fuel tax based on damages

Variable	Effect in First Year with (%)		Effect in Twentieth Year with (%)	
	Tax Rate = 30% of Damages	Tax Rate = 40% of Damages	Tax Rate = 30% of Damages	Tax Rate = 40% of Damages
GDP	−0.04	−0.06	+0.05	+0.04
Consumption	−0.06	−0.09	−0.02	−0.05
Investment	+0.07	+0.07	+0.20	+0.23
Coal use	−13.2	−16.8	−18.2	−22.9
CO_2 emissions	−10.7	−13.6	−12.5	−15.8
Primary particulate emissions:	−6.0	−7.6	−9.0	−11.4
Combustion TSP	−11.8	−15.0	−16.3	−20.6
Noncombustion TSP	−0.6	−0.9	−0.7	−0.9
High-height source (Electricity)	−12.4	−15.9	−16.3	−20.5
Medium-height sources	−3.9	−4.9	−4.9	−6.2
Low-height sources	−11.0	−14.1	−15.3	−19.3
SO_2 emissions	−10.2	−13.1	−14.0	−17.7
NO_X emissions (transportation)	−1.0	−1.4	−1.6	−2.1
Premature deaths	−10.7	−13.6	−14.3	−18.1
Value of health damages	−10.7	−13.6	−14.0	−17.7
Change in other tax rates	−2.8	−3.5	−2.1	−2.7
Reduction in damages/GDP	0.20	0.26	0.43	0.54
Pollution tax/total tax revenue	2.72	3.46	1.99	2.52

Notes: The entries are percentage changes between the counterfactual tax case and the base case. The last two rows are percentage shares. The columns marked "30%" correspond to having $\lambda = 0.3$ in equation (10.15), while those marked "40%" correspond to $\lambda = 0.4$.
Source: Calculations of authors.

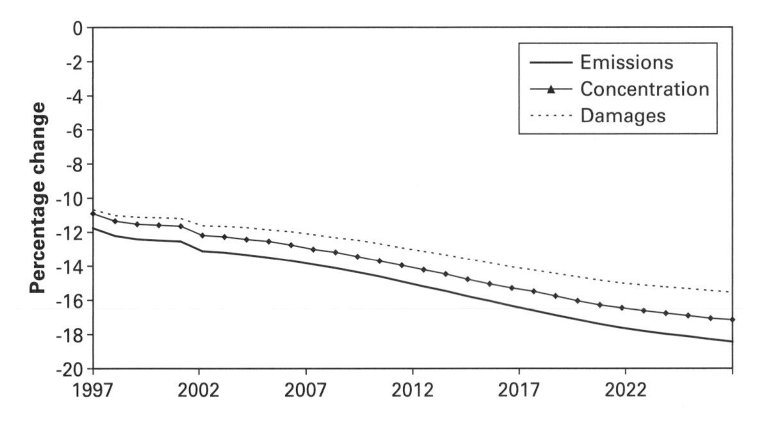

Figure 10.3
Change in particulate matter emissions and concentrations due to a pollution tax on fuels.

and 10.7, and time paths are shown in figure 10.3. These results are quite different from the previous output tax policy simulations.

The heavy tax on coal reduces its use by 13% initially, whereas the small tax on oil reduces the petroleum refining output by 1.3%. The heavy users of these fuels have to raise their output prices to compensate and that causes a reduction in demand for their goods. The electricity sector is the greatest user of coal, and its price rises by 2.4% and output falls by 2.7%. Metals smelting (iron and steel) and nonmetal mineral products (cement) are the next-biggest consumers of coal and their outputs fall by 0.8 and 0.6%, respectively. The reductions in output of these sectors are much smaller than the reductions in their coal inputs because the production functions that we employed allow an easy substitution among inputs. (See appendix F, equation [F.3].) Resources released from these shrinking energy-intensive sectors go to the clean ones such as finance, commerce, food, and agriculture.

The small tax on oil leads only to a small 0.2% reduction in transportation services and a 1.0% reduction in NO_X emissions. This is because the AMD_{oil} is calculated as the economy average marginal damage, which is much lower than the damage from oil consumption in the transportation sector. In contrast, the output tax policy is better targeted for this sector and generates a 3.8% reduction in NO_X emissions. An adjustment of this policy, taxing gasoline and diesel at higher rates compared to crude oil, would be a feasible improvement.

The total additional tax burden is relatively small. Pollution tax revenue comes to only 2.7% of total revenue, and the offsetting cuts for VAT and capital income tax are correspondingly small. The small additional retained earnings of enterprises allow aggregate investment to rise by 0.07% in the first year. This is accompanied by a fall in real consumption of 0.06% as households face mostly higher prices of goods. The changes in composition of aggregate output lead GDP to fall by a small 0.04%.

These changes in industry structure and choice of fuels reduce total primary PM emissions by 6.0%, with high-height emissions from electricity falling the most, 12.4%, and medium-height emissions from manufacturing the least, 3.9%. This reduction of total primary PM comes almost entirely from reductions in combustion PM, the cement industry is modestly affected and thus there is only a 0.6% fall in noncombustion TSP. SO_2 emissions have a similarly large 10.2% reductions.

These large reductions in SO_2 and primary particle emissions lead to a large reduction in total PM, which in turn generates a 10.7% reduction in the value of health damages. The value of reduced damage comes to a large 0.20% of GDP, out of the total damages of 1.9% of GDP in the base case (last row of table 10.2).

The green tax on the three fossil fuels generates only 2.7% of total government revenue in the first year, in contrast to the 9.1% collected in the output tax policy, which taxes all commodities. The output tax case, however, generated only a 3.5% reduction in health damages compared to the 10.7% here. That is, a narrowly based but well-targeted tax that raises only a modest amount of revenue leads to sizable reductions in pollution and related health damages.

This large estimated effect is due in part to our characterization of technology with high elasticities of substitution. That is, our assumed production parameters allow a low-cost switch from one fuel to another and from total energy to capital, labor, and nonenergy intermediate inputs. Although this ease of substitution may be regarded as overly optimistic in the short run, it is reasonable to think that this level of input switching could be achieved over time. As we can see from the results for the twentieth year, the long-run reductions in health damages are even bigger.

The pattern of emission reduction here is quite different from that of the output tax policy. Emissions at all heights are reduced more in this case, in particular low-height (i.e., high-damage) emissions, which are cut 11% compared to 1.2% in the output tax case. Medium-height emissions from manufacturing are cut 3.9% here, compared to 1.8% in the previous case. This is because all sectors can reduce coal use, switch to other fuels and inputs, and reduce combustion emissions. In the output tax case, we see only a shift in demand from dirty to cleaner commodities; those cleaner sectors still use fuels in a manner similar to the base-case usage. That is, although electricity output is reduced substantially in the output tax case, electricity still uses the same mix of coal-oil-gas, because the relative prices of the different fuels are not changed much. The burning of coal is polluting, but coal mining is not particularly polluting, and so coal is not heavily taxed in the output tax policy, leaving fuel prices little changed as shown in table 10.4. The effect is that although electricity output is reduced by only 2.7% in the first year, compared to the 6.7% reduction in the output tax case, the reduction in coal use here leads to a bigger reduction of TSP from electricity.

Another reason for the difference is that the household sector is also hit by the fuel tax, whereas it is not directly affected by the output tax. This sector is an important source of low-height emissions (see table 9.2).

The patterns over time are also very different from the output tax case. Here the reduction in emissions and damages rise over time as shown in figure 10.3, and comparing the "20th Year" and "1st Year" columns in table 10.6. By the twentieth year, even though GDP is 0.05% higher than the base case, the reduction in primary combustion PM emissions is 16%, greater than the 12% reduction in the first year when GDP fell. Similarly, the reduction in SO_2 emissions grows over time. Health damages, driven by these emissions and the growing urban population, fall even more. By the twentieth year, the total health damages are down 14% from the base path.

The main cause of these time trends is the assumed changes in technology. We have rapid technical progress that leads to much higher output using the same inputs, hence lower marginal damage per unit output over time. For the fuel-tax case, the emission per unit fuel is assumed to fall but at a more modest rate than the assumed rate of productivity change. Although emissions per unit fuel are falling, the urban population exposed to pollution is rising rapidly and so the value of the marginal damages per unit fuel is rising after the first few years. This means that the green tax is raised over time and induces more fuel-switching and conservation. The rising green tax rates raise more revenues, but, because the economy is growing even faster, the pollution-tax revenue is falling as a share of all tax revenue.

Considering the emissions of the coincident GHG, CO_2, we see that the reduction in emissions is smaller than the fall in coal use because of the switch to the lesser taxed, but still CO_2-emitting, fuels: oil and gas. In the first year, when coal falls by 13%, CO_2 falls only by 11%. Over time, as the green tax rises, the CO_2 share from oil rises even more. The reduction of CO_2 in the twentieth year is only 12%, whereas coal use falls 18%. Although one may see this as an inefficient instrument to reduce CO_2 compared to an explicit carbon tax, it is actually a very good "second-best" instrument even if one ignores the very important health benefits. In our view we should see these substantial reductions in CO_2 emissions as an

Table 10.7
Sector effects of a fuel tax based on damages, percent change year 1

	Tax = 30% of Damages		Tax = 40% of Damages	
Sector	Price	Quantity	Price	Quantity
1 Agriculture	0.08	0.06	0.11	0.07
2 Coal mining and processing	14.11	−13.25	18.76	−16.89
3 Crude petroleum mining	−0.03	−0.49	−0.05	−0.65
4 Natural gas mining	0.22	−0.13	0.29	−0.19
5 Metal ore mining	0.35	−0.39	0.46	−0.51
6 Nonferrous mineral mining	0.26	−0.84	0.35	−1.09
7 Food products, tobacco	−0.05	0.20	−0.06	0.24
8 Textile goods	0.05	0.08	0.07	0.09
9 Apparel, leather	0.01	0.05	0.02	0.06
10 Sawmills and furniture	0.14	−0.01	0.19	−0.03
11 Paper products, printing	0.17	−0.06	0.23	−0.09
12 Petroleum refining and coking	1.08	−1.28	1.44	−1.7
13 Chemical	0.34	−0.34	0.45	−0.46
14 Nonmetal mineral products	0.77	−0.57	1.01	−0.75
15 Metal smelting and pressing	0.49	−0.78	0.65	−1.02
16 Metal products	0.27	−0.30	0.36	−0.4
17 Machinery and equipment	0.19	−0.20	0.26	−0.27
18 Transport equipment	0.17	−0.04	0.23	−0.07
19 Electrical machinery	0.17	−0.23	0.23	−0.31
20 Electronic and telecommunications equipment	0.08	0.02	0.11	0.01
21 Instruments	0.11	−0.16	0.15	−0.22
22 Other manufacturing	0.17	−0.34	0.23	−0.44
23 Electricity, steam, hot water	2.39	−2.69	3.14	−3.49
24 Gas production and supply	2.54	−2.74	3.33	−3.57
25 Construction	0.29	−0.01	0.39	−0.03
26 Transport and warehousing	0.15	−0.21	0.2	−0.28
27 Post and telecommunication	0.07	−0.08	0.09	−0.11
28 Commerce and restaurants	−0.30	0.31	−0.37	0.39

Table 10.7
(continued)

Sector	Tax = 30% of Damages		Tax = 40% of Damages	
	Price	Quantity	Price	Quantity
29 Finance and insurance	−0.38	0.39	−0.47	0.49
30 Real estate	0.16	−0.06	0.21	−0.08
31 Social services	0.12	−0.09	0.17	−0.13
32 Health, education, other services	0.23	−0.17	0.3	−0.22
33 Public administration	0.16	0.03	0.22	0.03

Note: This is the percent change in output prices (purchaser's price) and quantities in the first year due to a tax on fossil fuels. Two simulations are reported, one where the tax is 30% of the marginal damage of a unit of fuel, and other where it is 40%.
Source: Calculations of authors.

important side benefit, to China and the world, of dealing with the urgent issue of local air quality and public health.

10.4.4 Sensitivity Analysis of Fuel Taxes

Effects of nonlinearities In the fuel tax experiment above we limited the green tax to only 30% of estimated marginal damages. To examine the effects of using a higher tax we set $\lambda = 0.4$ in equation (10.12)—that is, a third more—and redo the simulation. The results would show the extent of nonlinearities in the economic model. These are also reported in tables 10.6 and 10.7, in columns marked "40%." As one can see, the changes are somewhat less than a third more than those estimated for the $\lambda = 0.3$ case. The effect on industry prices and quantities are somewhat less than a proportional change, that is, less than 4/3 times the results of the previous simulation. Combustion TSP emissions in the first year are down 15.0% compared to 11.8% in the low-tax case. Damages are reduced by 13.6% compared to 10.7%. Coal use is 16.8% lower with a 40% tax, compared to 13.2% lower with a 30% tax.[27] The higher pollution tax revenues allow a greater reduction in existing taxes and hence a higher rate of investment. Initial consumption is correspondingly lower in the high-tax case.

The purpose of this exercise is not to argue for a particular tax rate, but to show the use and limits of using economic models like ours. It can capture some nonlinear features of the system, but we should point out that the items that we have not explicitly modeled, such as short-run adjustment costs, could be more than proportionally higher in the high-tax case.

Effects of different dose-response coefficients and valuations In chapter 9 we discussed how the unit damages depend on the parameter values for the DR and health end-point valuations (section 9.5 and table 9.16). When we use the low end of the range of estimates for DR_{hx} and V_h, the damages are about a third of those using the central values. Here we examine how varying these two important parameters affect our analysis of green tax effects.

A new base-case simulation is first run using the low-end values of DR and health valuations given in table 9.10. This has the same GDP and emissions as the original base case reported in table 10.2, but the eleven estimated health effects are between 30% and 100% of those in the central case, and the total damage value is 35%. The health damages from fuels are also 35% of those given in table 9.15, which are derived from central parameter values. We then repeat the fuel tax experiment, where a tax equal to 30% of estimated marginal damages is placed on coal, oil, and gas. Because these damages are only 35% of the central estimates, the taxes now are correspondingly lower. We compare the counterfactual tax results with the new base case, and the differences between the two are given in table 10.8, in the columns marked "Low-End Parameters." For comparison, the earlier results in table 10.6 are reproduced in the "Central Parameters" columns.

As a result of the much lower taxes, coal use falls by only 5.1% in the first year compared to 13% in the central case. Electricity, the biggest user of coal, falls by 1.0% (versus 2.7%). Primary combustion TSP is 4.6% lower (versus 12%). That is, the changes are about 35–40% of the changes in the central case. The green tax reduces the value of health damages by 4.1% from the new base case compared to the 11% reduction using central parameters. The changes in consumption, investment, and aggregate GDP are in the same direction as the central case, and the offsetting tax cuts allow a higher investment, which leads to a higher future GDP. Consumption is reduced in the earlier years but also catches up when the GDP expands faster. Consumption is lower by 0.02% in the first year in this case, compared to a 0.06% reduction in the central case.

To summarize these results, using the central values of these parameters, a fuel tax amounting to less than 3% of revenues can be expected to deliver a reduction in health damages of up to 11% and at the same time reduce consumption quite trivially. Being conservative and using the low-end values of the dose-response and

Table 10.8
Sensitivity analysis of varying dose-response and health valuation parameters: Effects of a fuel tax based on central versus low-end parameters

Variable	Central Parameters (%)		Low-End Parameters (%)	
	First-Year Effects	Twentieth-Year Effects	First-Year Effects	Twentieth-Year Effects
GDP	−0.04	+0.05	−0.02	+0.03
Consumption	−0.06	−0.02	−0.02	+0.01
Investment	+0.07	+0.20	+0.04	+0.09
Coal use	−13.2	−18.2	−5.15	−7.28
CO_2 emissions	−10.7	−12.5	−4.17	−5.00
Primary particulate emissions:	−6.0	−9.0	−2.31	−3.58
Combustion TSP	−11.8	−16.3	−4.57	−6.50
Noncombustion TSP	−0.6	−0.7	−0.24	−0.24
High-height source (Electricity)	−12.4	−16.3	−4.83	−6.48
Medium-height sources	−3.9	−4.9	−1.48	−1.95
Low-height sources	−11.0	−15.3	−4.30	−6.07
SO_2 emissions	−10.2	−14.0	−3.96	−5.58
NO_X emissions (transportation)	−1.0	−1.6	−0.37	−0.55
Premature deaths	−10.7	−14.3	−4.19	−5.67
Value of health damages	−10.7	−14.0	−4.10	−5.57
Change in other tax rates	−2.8	−2.1	1.1	0.8
Reduction in damages/GDP	0.20	0.43	0.03	0.06
Pollution tax/total tax revenue	2.72	1.99	1.06	0.78

Notes: The entries are percentage changes between the counterfactual tax case and the base case. The last two rows are percentage shares. The "central" and "low-end" parameter values for dose-response and health valuations are given in table 9.10.
Source: Calculations of authors.

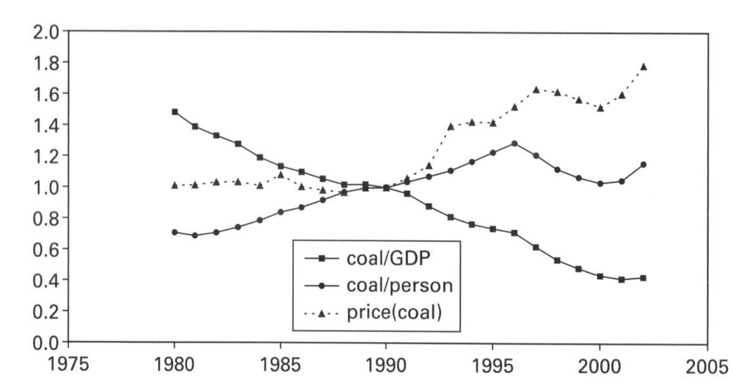

Figure 10.4
Coal use and price of coal.

health valuations, a fuel tax amounting to 1.1% of revenues can be expected to deliver a reduction in health damages of 4.1% and reduce consumption trivially. In other words, the ratio of benefits to revenues in both cases is about 3:1, and the ratio of benefits to long-run GDP or consumption foregone is large.

10.4.5 General Comments on Results and Comparisons to Earlier Estimates

A key factor in the above analysis of green-tax instruments is the responsiveness of industry to price changes. Are our estimates of the long-run reduction in coal use due to higher coal prices reasonable? Two historical trends in energy use and energy prices (out of many examples) are worth noting in this regard. Within the recent Chinese experience, the use of coal per unit GDP has fallen rapidly in parallel with rises in the price of coal, as is well known. In figure 10.4 we plot these trends where the price of coal is the nominal price divided by the GDP deflator.[28] Also plotted is the coal use per person. Between 1990 and 2002 the real coal price rose at an average of 4.8% per year, whereas coal per unit GDP fell at a 7.1% rate. Despite the tremendous increase in incomes, coal use per person only rose 1.2% per year. Although there are many questions about the reliability of the data around 2000—see Sinton 2001, Sinton and Fridley 2003, and discussion in chapter 1—it is reasonable to think that coal use per capita at least flattened from 1996 to 2002, if not fell. The U.S. experience during the oil shocks is also instructive. Between 1973 and 1983, when real oil prices[29] rose by 93%, oil use per capita fell by 22% (Jorgenson, Ho, and Wilcoxen 2004, figure 2).

There has been much debate about these trends, whether the reduction in energy use is due to improvements in energy efficiency at the enterprise level induced by higher prices, or due to changes in industrial structure that are independent of prices (for example, Sinton and Levine 1994; Garbaccio, Ho, and Jorgenson 1999b). Although we cannot resolve these complicated issues here, at the minimum we can say that the historical experience does not contradict that input substitution is feasible in a rather short period of time for the price ranges that we are considering. We certainly believe that enterprises and households are able to change consumption patterns substantially in response to prices. Although we may be overestimating the adaptation during the first few years, the estimated energy conservation over a ten-year horizon in response to the higher prices is reasonable.

Turning now to comparisons with our earlier work—Garbaccio, Ho, and Jorgenson 2000 and Ho, Jorgenson, and Di 2002—the results here are somewhat different. The main difference is that we focus exclusively on using the *iF* method to estimate health damages instead of using the methods of World Bank 1997. As explained in chapter 9, the *iF* method is not calibrated to observed concentrations, which result from both human and natural sources, and thus generates a lower total damage estimate; however, it explicitly accounts for secondary sulfates and, with the projected rise in SO_2 emissions, this is an important distinction. The *iF* method also allows us to take into consideration the effects of NO_X from the transportation sector, a sector that will grow even faster than the rapidly expanding GDP. In our analyses, we also use the more recent estimates of nonindustrial emissions, which are much lower than those reported for 1997. This means we have a lower estimate of the marginal damages from emissions of the service and household sectors.

10.5 Implications for Policy and Conclusions

We have only considered two national policies here. These are chosen because they rely on market instruments that are likely to be more efficient and relatively easy to apply compared to technology regulations or industrial policies. Other major market-based policy mechanisms such as emission permit trading are potentially important complementary policies. Current trading policy experiments, however, cover only SO_2 and for a limited number of industries.

A few lessons stand out from this analysis. First, the benefits of reducing air pollution from our policies likely far exceed the cost of foregone consumption. Although we have not modeled the short-run adjustment costs, these may be mitigated by phasing in the taxes gradually and would not substantially change the high benefit:cost ratio.

Second, the conclusion that long-run net benefits are positive is robust to many of the uncertainties in the parameters underlying this analysis and other similar analyses. It is certainly important to have larger samples and wider coverage to sharpen the estimates, as discussed in the conclusion of chapter 9. In particular, it is important to consider the relative damages by the various sectors. Despite these uncertainties, however, the overall conclusion making an economic case for reducing coal use remains valid.

Third, there may be tradeoffs between effectiveness of an instrument and the ease of implementing it. A broad-based tax may gain greater acceptance, because the pain is shared by many, but it is not very effective. A narrow tax targeted at the main polluters or fuels is more efficient but requires a large adjustment on the part of these polluters. This is not to argue that these are the two best policies from which to choose. In the United States, for example, much of the historical reduction in emissions came from direct end-of-pipe regulations. Such policies mandating the removal of particulates and SO_2 have an important place in the policy toolbox. A proper analysis of these regulations requires knowledge of the capital and operational costs of pollution control technologies, costs that may differ by sector. Such analysis would be a valuable complement to the results reported here.

Even if it is shown that such direct regulation of PM and SO_2 is appropriate, we believe that ultimately, an air pollution–reduction strategy must include some lever to reduce coal use. This applies whether we are considering only local damages or global ones, such as those due to climate change. Green-tax policies are a potent prospective lever. Such a coal-reduction strategy will lead to the closure of some coal producers. The transition cost of making this happen—of relocating miners and of retrofitting or replacing coal-fired boilers and other combustion technologies to use cleaner energy sources, and so forth—should be an urgent topic of research.

Fourth, because of this link to coal use, efforts to reduce local pollution will reduce China's contribution to the GHGs, the agents of global climate change. The large magnitude of the reduction should spur serious consideration by wealthier countries worried about climate change to aid China in these control efforts.

Fifth, the use of pollution-tax instruments may deliver sizable new revenues to the government. This is important, in light of the well-known difficulties of raising government revenue in China. (The size of the Chinese tax revenues relative to the economy is small compared to other countries even though the government has an outsized role.) The use of green-tax revenue to reduce the burden on enterprises would be a natural element of reform of public finances in China. That is because it would replace very distortionary taxes with a less distortionary one.

Sixth, an important result from our analysis is that secondary particles should be explicitly considered in air pollution–damage analysis. When we compare the results using from the *iF* method versus the Lvovsky and Hughes method (World Bank 1997), we find that the latter's inability to attribute observed particulate concentration to secondary sulfates and nitrates causes a large understatement of the benefits of reducing SO_2 and NO_X emissions. This reinforces our conclusion in chapter 9 that more advanced research on air dispersion and secondary chemical transformation should be an important item on the research agenda. An example is a current effort to measure trace gases and model regional air quality by our scientific colleagues.[30]

There are, of course, big gaps in the *iF* implementation here. We have made explicit estimates of *iF*s for the most-polluting industries only. Even for the industries we analyzed, our sample comes from only five cities. We have only been able to differentiate crudely the fine particles from available TSP data, and we have additionally ignored other pollutants such as ozone, nitrogen oxides (for the most part), and mercury. We have also had to make very simple assumptions about long-range effects of air pollution. Closing these gaps is an important next step for our research agenda, building on the start presented here.

There are also gaps in our modeling of production and economic activity. An important factor determining costs is the substitution elasticities. The ability to substitute capital for energy is likely to differ by industry. It is important for analyses of energy and environmental policies, not just our study here, to have good estimates of these elasticities for the various sectors. We have embarked upon such an effort using national-level input-output data. Other studies at the plant level, or at more disaggregated industry levels, would be important complements.

This chapter, like others in this volume, has gone to some length to describe its methods in detail and highlight the sources of uncertainty. We did this in the belief that transparent analysis will be easier to critique and improve upon in the future and will provide more credible research advice to decision makers as they confront the pressing economic and environmental policy choices facing China.

Acknowledgments

The research of this chapter was generously funded by a number of sources. Development of the model was originally supported by the Integrated Assessment Program, Biological and Environmental Research, U.S. Department of Energy (grant DE-FG02-95ER62133). Later research was supported by the U.S. EPA, the Task

Force on Environmental and Natural Resource Pricing and Taxation of the China Council for International Cooperation on Environment and Development, the Harvard Asia Center, and the V. Kann Rasmussen Foundation. We would also like to thank Shantong Li and Fan Zhai of the Development Research Center of the State Council of China, Gordon Hughes and Kseniya Lvovsky of the World Bank, and Richard Garbaccio of the U.S. EPA, for their help and comments. Mun Ho thanks Resources for the Future, Washington, D.C., for graciously hosting him as a visiting scholar.

Notes

1. In addition to those in chapter 1, a partial list of English papers includes Sunman, Monasinghe, and Zhang 2002; Panayotou and Wu 2002; Chandler et al. 1998; Dasgupta, Wang, and Wheeler 1997, containing references to other World Bank analysis; and Sinton et al. 2005, a report by Lawrence Berkeley Laboratory discussing energy strategies.

2. Examples are Aunan et al. 2003, which discusses the impact of a carbon tax, and Ho, Jorgenson, and Di 2002.

3. Jiang and Hu 2001 and Wei and Glomsrød 2002 study the effects of a carbon tax on the Chinese economy. Da Motta 2002 summarizes the results of mitigation policies using various models from the Third Assessment Report of the Intergovernmental Panel on Climate Change (IPCC 2002).

4. See, for example, the U.S. EPA 2004 review of recent international experience with pollution policies based on economic incentives.

5. See, for example, OECD 2000 and EEA 2004.

6. Fisher-Vanden and Ho (forthcoming) give a simple example of such a model for China.

7. Fullerton, Hong, and Metcalf (1999) find that an "imprecise" output tax produces a welfare gain that is less than half that obtained from a direct pollution tax.

8. See Morgenstern et al. 2002 and Biello 2002 on pilot projects in SO_2 trading in Taiyuan, Benxi, and Nantong.

9. For example, to analyze a policy mandating electrostatic precipitators to remove particulate matter, one would need to know the capital costs and operating costs of such equipment for the entire range of factories. This is discussed further in the conclusion.

10. For example, Jorgenson and Wilcoxen (1990) studied the economic effects of regulations in the United States using data on capital and operating costs of equipment that were installed in response to EPA regulations.

11. This simple approach ignores the possibility that cleaner equipment will cost more than dirty equipment. Furthermore, the exogenous energy efficiency improvements described above are set independently of these emission factors. An integrated approach would of course be preferred when such data become available.

12. The applicability of this linear formulation in estimating *total* damages is discussed in chapter 9. For our purposes here of estimating the *change* in damages due to a policy intervention, it is sufficient that this linear approximation is valid just for the range of concentra-

tions realized under the policies. We shall not be considering extreme reductions to, for instance, the low levels seen in wealthier Western countries.

13. A more sophisticated approach would have an explicit utility function that includes health risks and prices them endogenously. The evidence for income elasticities of risk valuations, however, is hard to interpret, as discussed in section 9.3.4. The use of unit income elasticities ignores these complications, and scaling by lagged incomes turns this into a simple exogenous parameter.

14. The terms "Solow" and "dynamic recursive" refer to the features that certain fixed rules determine investment, that there is no forward-looking behavior where expected changes in future policies can affect behavior today. This is simpler than models with intertemporal equilibrium and such features.

15. A variable is said to be "exogenous" if it is predetermined before any solution of the model (e.g., population). A variable is "endogenous" if it is determined within the model simultaneously with other variables (e.g., prices).

16. By inelastic labor supply we mean that the total hours worked is a predetermined number not affected by economic events. The alternative formulation of having hours worked dependent on the wage rate seems to us to be too elaborate to implement sensibly for the current Chinese economy, with its large pool of underemployed workers. A more elaborate model with heterogeneous workers might change the details but not the overall result.

17. The details of the projections of exogenous variables are discussed in more detail in appendix F, section F.7.

18. Specifically, we solve the model to find an equilibrium set of prices. That is, we find a set of prices such that the demand for each commodity equals the supply, the demand for labor equals the exogenously specified work force, the demand for capital equals the stock of capital, and the government accounts meet the specified deficit.

19. If one had input-output tables for a more recent year, one would use it as the initial year; however, the choice of the starting year affects mostly the level of the time path of the economy. It has only a small effect on the rate of growth or composition change. More important, unless there were dramatic changes in economic structure or pollution ratios, the effect of the starting date on the percentage change due to policy changes is small.

20. As explained in appendix F, we assume that the input-output structure gradually changes toward a structure similar to the United States in 1982. It is not identical, however, and we project that the Chinese economy would still use more coal per unit output than the United States in 1982. The official Chinese data show a sharp decline in energy consumption between 1996 and 2000, followed by a huge rise between 2000 and 2003, not just in the energy-output ratio but also in the absolute level of energy use. Sinton and Fridley (2003) give a good discussion of this phenomenon, whether they are real or due to errors in the data. Such short-term fluctuations give no useful indication of a long term trend and we prefer to look at the entire period since 1978 to guide our projections. Our projections may be termed "optimistic," but are in line with other projections, such as those by the Energy Research Institute, National Development and Reform Commission, Beijing (e.g., Jiang and Hu 2001).

21. Our base case is remarkably similar to the "current trends" scenario in Sathaye, Monahan, and Sanstad 1996, which were derived using different methods. They project GDP growth at 4.9% and carbon emissions at 3.1% over a thirty-year period.

22. Other values of λ have been used to examine the possible nonlinearities in the system. The results of varying λ in a setup similar to this one are reported in Ho, Jorgenson, and Di (2002).

23. The rise in the commodity prices is not identical to the new taxes, because both sellers' and buyers' prices are changed as a result of general equilibrium effects noted in section 10.2.2. That is, a new 9% tax on electricity introduces a gap of 9% between sellers' and buyers' prices. The sellers' price, however, is lower compared to the base case because of changes in the rest of the economy, and thus the final buyers' price is less than 9% higher than the base case.

24. GDP may fall even when all labor and capital remain fully employed. Real GDP is defined as an aggregate over "final demand" commodities (i.e., over goods that make up consumption, investment, government, and exports). The aggregation procedure uses both values and prices of these commodities. Given a particular set of prices, two baskets of goods made by the same number of workers will have a different aggregate value.

25. The total damage in the simulation is about 140 billion yuan and corresponds to that reported in chapter 9, table 9.11, using the central parameters. This calculation uses only combustion emissions. If noncombustion emissions are also included, the total damage would be 200 billion yuan.

26. Many other analyses using more complicated economic models would allow for a labor-supply effect, that is, higher goods prices may lead to less work effort and more leisure. This leisure enters the welfare calculation and the welfare effect of the green taxes may thus be greater than our simpler assumption of fixed labor supplies; however, this does not change the essential point that output taxes are not effective pollution taxes.

27. We report three significant digits here to illustrate the scale effect. This should not be taken as the degree of precision in a numerical model with so many imprecisely estimated parameters. The degree of nonlinearity depends on the functional forms and is independent of the errors in the parameters.

28. The price of coal is taken from the Price Indices of Industrial Products in table 9.10 of NBS (1999a, 2003) and coal use is from table 7.2 of the same source. The prices are also given in Sinton et al. 2004 (table 6B1), and a slightly different calculation of coal quantities is in Sinton et al. 2004 (table 4A1).

29. The real oil price here is calculated as the consumer price for petroleum products divided by the U.S. GDP deflator (from EIA 2004, table 3.3).

30. Colleagues in the Harvard China Project, in collaboration with Tsinghua and other universities in China and Hong Kong, are making continuous measurements of trace gases in China and employing them in a high-resolution chemical tracer model to advance understanding of emissions and seasonal atmospheric chemistry on the regional spatial scale.

References

Aunan, Kristin, Terje Berntsen, David O'Connor, Haakon Vennemo, and Fan Zhai. 2003. Agricultural and human health impacts of climate policy in China: A general equilibrium analysis with special reference to Guangdong. Technical Working Papers no. 206, Organization for Economic Co-operation and Development. March. Paris.

Aunan, Kristin, Jinghua Fang, Haakon Vennemo, Kenneth A. Oye, and Hans Martin Seip. 2004. Co-benefits of climate policy: Lessons learned from a study in Shanxi. *Energy Policy* 32:567–581.

Biello, David. 2002. Emission trading with Chinese characteristics. *Environmental Finance* 3 (10):14.

Bovenberg, A. Lans, and Ruud de Mooij. 1994. Environmental levies and distortionary taxation. *American Economic Review* 84 (4):1085–1089.

Bovenberg, A. Lans, and Lawrence H. Goulder. 1996. Optimal environmental taxation in the presence of other taxes: General equilibrium analyses. *American Economic Review* 86 (4): 985–1006.

Chandler, William, Yuan Guo, Jeffrey Logan, Yingyi Shi, and Dadi Zhou. 1998. China's electric power options: An analysis of economic and environmental costs. Report no. PNWD-2433. Advanced International Studies Unit, Pacific Northwest National Lab. Washington, D.C.

China Environment Yearbook. 1998. Beijing: China Environmental Science Press. In Chinese.

Da Motta, Ronaldo Seroa. 2002. Mitigation policies and economic impacts. Paper presented at the International Workshop on Climate Change Mitigation. September 11–13, Beijing.

Dasgupta, Susmita, Hua Wang, and David Wheeler. 1997. Surviving success: Policy reform and the future of industrial pollution in China. World Bank Research Department working paper. March. Washington, D.C.

ECON Centre for Economic Analysis. 2000. An environmental cost model. ECON report no. 16/2000. May. Oslo, Norway.

Energy Information Administratiion (EIA). 2004. *Annual Energy Review*. U.S. Department of Energy, Washington, D.C. Available at http://www.eia.doe.gov/emeu/aer/.

European Environment Agency (EEA). 2004. Exploring the ancillary benefits of the Kyoto Protocol for air pollution in Europe. Working paper, Copenhagen.

Feng, Therese. 1999. Controlling air pollution in China: risk valuation and the definition of environmental policy. Cheltenham: Edward Elgar.

Fisher-Vanden, Karen, and Mun S. Ho. Forthcoming. How do market reforms affect China's responsiveness to environmental policy. *Journal of Development Economics*.

Fullerton, Don, Inkee Hong, and Gilbert Metcalf. 1999. A tax on output of the polluting industry is not a tax on pollution: The importance of hitting the target. NBER working paper no. 7259. National Bureau of Economic Research. Cambridge, Mass.

Garbaccio, Richard F., Mun S. Ho, and Dale W. Jorgenson. 1999a. Controlling carbon emissions in China. *Environment and Development Economics* 4 (4):493–518.

Garbaccio, Richard F., Mun S. Ho, and Dale W. Jorgenson. 1999b. Why has the energy output ratio fallen in China? *Energy Journal* 20 (3):63–91.

Garbaccio, Richard F., Mun S. Ho, and Dale W. Jorgenson. 2000. The health benefits of controlling carbon emissions in China. In *Ancillary Benefits and Costs of Greenhouse Gas Mitigation, Proceedings of an IPCC Co-Sponsored Workshop*. March 27–29. Washington, D.C.: Organization for Economic Co-operation and Development.

Garbaccio, Richard F., Mun S. Ho, and Dale W. Jorgenson. 2001. A dynamic economy-energy-environment model of China, version 2. Unpublished manuscript. Harvard University, Cambridge, Mass.

Goulder, Lawrence, Ian Parry, and Dallas Burtraw. 1997. Revenue raising vs. other approaches to environmental protection: The critical significance of pre-existing tax distortions. *RAND Journal of Economics* 28:708–731.

Hammitt, James K., and John D. Graham. 1999. Willingness to pay for health protection: Inadequate sensitivity to probability? *Journal of Risk and Uncertainty* 18 (1):33–62.

Ho, Mun S., and Dale W. Jorgenson. 2001. Productivity growth in China, 1981–95. Unpublished manuscript. Harvard University, Cambridge, Mass.

Ho, Mun S., Dale W. Jorgenson, and Wenhua Di. 2002. Pollution taxes and public health. In *Economics of the environment in China*, ed. Jeremy J. Warford and Yining Li. for the China Council for International Cooperation on Environment and Development. Boyds, Maryland: Aileen International Press.

Intergovernmental Panel on Climate Change (IPCC). 2002. *Climate change 2001: Mitigation*. Cambridge: Cambridge University Press.

Jiang, Kejun, and Xiulian Hu. 2001. Carbon tax: An integrated analysis for China. Powerpoint presentation of the Center for Energy, Environment and Climate Change, Energy Research Institute, National Development and Reform Commission. Beijing.

Jorgenson, Dale W., and Peter J. Wilcoxen. 1990. Environmental regulation and U.S. economic growth. *RAND Journal of Economics* 21 (2):314–340.

Jorgenson, Dale W., Mun S. Ho, and Peter J. Wilcoxen. 2004. Energy and U.S. economic growth in the next thirty years: The role of information technology and technical change. Report to the National Renewable Energy Laboratory, U.S. Department of Energy.

Levy, Jonathan I., James K. Hammitt, Yukio Yanagisawa, and John D. Spengler. 1999. Development of a new damage function model for power plants: Methodology and applications. *Environmental Science and Technology* 33 (24):4364–4372.

Lvovsky, Kseniya, and Gordon Hughes. 1997. An approach to projecting ambient concentrations of SO_2 and PM-10. Unpublished Annex 3.2 to World Bank, *Clear water, blue skies: China's environment in the new century*. Washington, D.C.: World Bank.

Maddison, David, Kseniya Lvovsky, Gordon Hughes, and David Pearce. 1998. Air pollution and the social cost of fuels. Unpublished manuscript. Washington, D.C.: World Bank.

Metcalf, Gilbert. 2000. Environmental levies and distortionary taxation: Pigou, taxation and pollution. Discussion paper no. 2000-04. Department of Economics, Tufts University. Medford, Mass.

Morgenstern, Richard D., Robert Anderson, Ruth Greenspan Bell, Alan J. Krupnick, and Xuehua Zhang. 2002. Demonstrating emissions trading in Taiyuan, China. *Resources* 148:7–11.

National Bureau of Statistics (NBS). 1999a, 2003. *China statistical yearbook, 1999; 2003*. Beijing: China Statistics Press.

National Bureau of Statistics (NBS). 1999b. *China input output table, 1997*. Beijing: China Statistics Press.

Organization for Economic Co-operation and Development (OECD). 2000. *Ancillary Benefits and Costs of Greenhouse Gas Mitigation, Proceedings of an IPCC Co-Sponsored Workshop*. March 27–29. Washington, D.C.: OECD.

Panayotou, Theodore, and Yajun Wu. 2002. Green taxes. In: *Economics of the environment in China*, ed. Jeremy J. Warford and Yining Li for the China Council for International Cooperation on Environment and Development. Boyds, MD: Aileen International Press.

Pratt, John W. and Richard J. Zeckhauser. 1996. Willingness to pay and the distribution of risk and wealth. *Journal of Political Economy* 104 (4):747–763.

Sathaye, Jayant, Patty Monahan, and Alan Sanstad. 1996. Costs of reducing carbon emissions from the energy sector: A comparison of China, India and Brazil. *Ambio* 25 (4):262–265.

Sinton, Jonathan E. 2001. Accuracy and reliability of China's energy statistics. *China Economic Review* 12 (4):347–354.

Sinton, Jonathan E., and David G. Fridley. 2003. *Comments on recent energy statistics from China*. Report no. LBNL-53856. Lawrence Berkeley National Laboratory. Berkeley, Calif.: LBNL.

Sinton, Jonathan E., David Fridley, Jieming Lin, Joanna Lewis, Nan Zhou, and Yanxia Chen, eds. 2004. *China energy databook 6.0*. Lawrence Berkeley National Laboratory, China Energy Group, LBNL-55349. Berkeley, Calif.: LBNL.

Sinton, Jonathan E., and Mark D. Levine. 1994. Changing energy intensity in Chinese industry. *Energy Policy* 22 (3):239–255.

Sinton, Jonathan E., Rachel E, Stern, Nataniel T. Eden, and Mark D. Levine. 2005. *Evaluation of China's energy strategy options*. Report no. LBNL 56609. Berkeley, Calif.: Lawrence Berkeley National Laboratory.

Sunman, H., M. Monasinghe, and S. Q. Zhang. 2002. Economics, environment, and industry. In *Economics of the environment in China*, ed. Jeremy J. Warford and Yining Li for the China Council for International Cooperation on Environment and Development. Boyds, Maryland: Aileen International Press.

U.S. Environmental Protection Administration (U.S. EPA). 2004. International experiences with economic incentives for protecting the environment. Report no. EPA-236-R-04-001. Washington, D.C.

Venners, Scott, Binyan Wang, Zhonggui Peng, Yu Xu, Lihua Wang, and Xiping Xu. 2003. Particulate matter, sulfur dioxide and daily mortality in Chongqing, China, *Environmental Health Perspectives* 111 (4):562–567.

Wang, Xiaodong, and Kirk R. Smith. 1999a. Secondary benefits of greenhouse gas control: Health impacts in China. *Environmental Science and Technology* 33 (18):3056–3061.

Wang, Xiaodong, and Kirk R. Smith. 1999b. Near-term health benefits of greenhouse gas reductions: A proposed assessment method and application in two energy sectors of China. World Health Organization report no. WHO/SDE/PHE/99.01. March. Geneva, Switzerland.

Wei, Taoyuan, and Solveig Glomsrød. 2002. The impact of carbon tax on the Chinese economy and reductions of greenhouse gases. *World Economics and International Politics* 8:47–49.

World Bank. 1994. *China: Issues and options in greenhouse gas emissions control.* Washington, D.C.: World Bank.

World Bank. 1995. *Macroeconomic stability in a decentralized economy*. Washington, D.C.: World Bank.

World Bank. 1996. *The Chinese economy: Fighting inflation, deepening reforms*. Washington, D.C.: World Bank.

World Bank. 1997. *Clear water, blue skies: China's environment in the new century*. Washington, D.C.: World Bank.

World Bank. 2002. *Health, nutrition, and population statistics*. World Bank Human Development Network. Available at http://devdata.worldbank.org/hnpstats/.

Zhai, Fan, and Shantong Li. 2000. The implications of accession to WTO on China's economy. Paper presented at Third Annual Conference on Global Economic Analysis, June 27–30, Melbourne, Australia.

Contributors

John S. Evans is Senior Lecturer on Environmental Science and Director of Environmental Science and Risk Management, Department of Environmental Health, Harvard School of Public Health, Boston, from which he received his Sc.D. in environmental health sciences.

Susan L. Greco is Senior Analyst, Risk Analysis, Abt Associates, Bethesda, Maryland. She has an Sc.D. in environmental health sciences from the Harvard School of Public Health, Boston.

James K. Hammitt is Professor of Economics and Decision Sciences, Harvard School of Public Health, and Director, Harvard Center for Risk Analysis, Boston. He has a Ph.D. in Public Policy from Harvard University.

Jiming Hao is a member of the Chinese Academy of Engineering and Dean of the Institute of Environmental Science and Engineering, Tsinghua University, Beijing. He received a Ph.D. in environmental engineering from the University of Cincinnati.

Mun S. Ho is Visiting Fellow at the Institute for Quantitative Social Science, Harvard University, Cambridge, and Visiting Scholar at Resources for the Future, Washington, D.C. He earned a Ph.D. in economics from Harvard University.

Dale W. Jorgenson is Samuel W. Morris University Professor, Harvard University, Cambridge. He received his Ph.D. in economics from Harvard University.

Jonathan I. Levy is Mark and Catherine Winkler Associate Professor of Environmental Health and Risk Assessment, Departments of Environmental Health and Health Policy and Management, Harvard School of Public Health, Boston. He holds an Sc.D. in environmental health sciences from the Harvard School of Public Health.

Ji Li is Associate Professor, Harbin Institute of Technology Shenzhen Graduate School, Shenzhen, China. He received a Ph.D. in environmental engineering from Tsinghua University in Beijing.

Bingjiang Liu is Division Chief of Air Pollution and Noise Control, State Environmental Protection Administration, Beijing. He earned a Ph.D. in environmental engineering from Tsinghua University.

Yongqi Lu is Research Scientist, Department of Civil and Environmental Engineering, University of Illinois at Urbana Champaign. He has a Ph.D. in environmental engineering from Tsinghua University.

Chris P. Nielsen is the Harvard Kernan Brothers Fellow and Executive Director of the China Project, University Center for the Environment and Division of Engineering and Applied

Sciences, Harvard University, Cambridge. He has an S.M. in technology and policy from the Massachusetts Institute of Technology.

Shuxiao Wang is Assistant Professor, Department of Environmental Science and Engineering, Tsinghua University, Beijing, from which she received her Ph.D. in environmental engineering.

Ying Zhou is a Research Fellow, Department of Environmental Health, Harvard School of Public Health, Boston, from which she received an Sc.D. in environmental health sciences.

Index

Italicized page numbers indicate boxes, figures, or tables. The integrated assessment described in chapters 2–3 and 9–10, incorporating results of other chapters, is labeled "*Clearing the Air* integrated study" in the index.